SPECIAL PUBLICATION
THE MUSEUM OF
SOUTHWESTERN BIOLOGY

NUMBER 3, pp. 1–290 24 MARCH 1997

Life Among the Muses: Papers in Honor of James S. Findley

Edited by:

Terry L. Yates, William L. Gannon,

and Don E. Wilson

James Smith Findley, 1993, on his back porch in Corrales, New Mexico.

Table of Contents

Preface

Terry L. Yates

In Greek mythology, the Muses were nine goddesses who presided over the arts and sciences and provided a source of genius and inspiration for those who worked among them. Dr. James S. Findley often reflected on how conducting research and teaching activities in the Museum of Southwestern Biology was like living among the Muses. The ability to take current information beyond the obvious and to synthesize raw data into a vision for the future has been a hallmark of Jim's career, one which he shared freely with his many students. His belief in the value of research collections and his explorations of biological diversity have never faltered and persist today in his own active research program and in those of his academic offspring, which some have called The Findley Net. We hope the readers of this volume also will be touched by the Muses and, in turn, will better appreciate the impact that Professor Findley has had on the biological sciences in general and the discipline of mammalogy in particular.

The original idea for this volume was inspired by the retirement of Dr. Findley from UNM in 1992 and resulted from a meeting that Don Wilson and I had in Washington, DC. We decided to honor Jim with a symposium of papers to be given by former students. That event took place in October, 1992, at the Department of Biology's Sevilleta Field Station in the Middle Rio Grande Valley in New Mexico. Not all of Jim's graduate students were able to attend, but the number and diversity of papers presented was impressive. Don and I then invited Bill Gannon as a co-editor, and considered whether this publication should be focused on a specific area. We finally decided that to do so would be to lose much of the vision that Dr. Findley has for the impact that the field of mammalogy has on the rest of biology and his belief that graduate students should be free to develop their own ideas. That diversity is reflected in this volume and represents a testament to the breadth and depth of Professor Findley's influence.

The volume is arranged into 17 chapters, all authored by at least one former student of Dr. Findley. A total of 27 authors contributed to the volume. The chapters are grouped loosely by discipline, but the order of placement of disciplines within the book is more or less random. Although the range of subjects included here is impressive, it by no means reflects the range of activities represented by the Findley legacy. For example, not a single paper on reef fishes or the field camps required to study fish is listed here. The volume begins with an overview of the academic offspring of Professor Findley, followed by an annotated bibliography. Next are papers broadly defined as beloning to the field of systematics, including works in biodiversity, biogeography, and taxonomy. Then there are several papers in conservation biology. These papers reflect another of Jim's life-long interests, as his current involvement with the Nature Conservancy, the Corrales Bosque Reserve, and the State Legislature will attest.

The remainder of the volume includes papers on natural history, functional morphology, ecology, techniques, and behavior. Taxonomically, most chapters focus on mammals, but reptiles and mollusks are also represented. Despite the diversity of topics covered by this *Festschrift*, they are all unified by the common influence of Professor Findley.

We are particularly indebted to Associate Provost for Research Nasir Ahmed and Dean of the College of Arts and Sciences Michael Fischer for their support of this volume and the MSB Publication series in general. We would also like to acknowledge the efforts of Anne Rice, who handled all of the layout and typesetting for the volume and incorporated various rounds of corrections. We also wish to thank Fonceal Gutierrez for handling all correspondence and billing related to reprint and book orders.

All manuscripts in this volume received independent, anonymous reviews. We would like to thank the following colleagues for their assistance in this regard: J. Scott Altenbach, Sydney Anderson, Troy Best, Michael Bogan, James Brown, Jerry Choate, James S. Findley, Lee Fitzgerald, Richard Forbes, Trish Freeman, Jennifer Frey, Kenneth Geluso, Hugh Genoways, David Hafner, Edward Heske, John Hubbard, Clyde Jones, Gordon Kirkland, Timothy Lawlor, Spencer Lucas, Manuel Molles, Robert Parmenter, Michael Patrick, Travis Perry, Luis Ruedas, Greg Schmitt, Rick Smartt, Felisa Smith, William Stanley, and Patty Wilber. We also would like to thank Paul Polechla, Ellen Roots, Anne Brown, and Michael Bogan for a variety of technical assistance. We also thank the UNM Department of Biology for allowing the use of the field station for the original symposium, the Sevilleta Long–Term Ecological Research Program for logistical support, and Ted Stans and the Sevilleta National Wildlife Refuge for their continued collaboration.

I am particularly happy to see the publication of this volume honoring Jim Findley. Although I was not a graduate student of his, I have learned many things from him during the 18 years that have passed since he hired me to succeed him as Curator of Mammals at the MSB. I inherited a collection and a program with a firm foundation, rich in tradition. The science of mammalogy continues to benefit from the institution he began in New Mexico more than four decades ago. I, along with all of those who played a role in this volume, wish to express our appreciation and gratitude to Jim for all he has given to us and to the field of mammalogy.

The Academic Offspring of James S. Findley

KENNETH N. GELUSO AND DON E. WILSON

Abstract

Professor James S. Findley supervised 26 doctoral students and 34 master's degree students during his 37-year tenure at the University of New Mexico. Those students exemplify a wide range of interests, and their current positions represent a diversity that serves as a fitting tribute to Professor Findley's own wide-ranging interests and talents. In addition to a brief update on each of the students, a sampling of their published contributions also speaks to this diversity.

Figure 1. James S. Findley measuring skulls of *Sorex vagrans* at the University of Kansas Museum of Natural History in 1953.

INTRODUCTION

As a biologist and naturalist, James Smith Findley has had a long and distinguished career, which continues unabated today. Although his international reputation is based primarily on his research contributions, Jim also is the academic patriarch of 50 postgraduate students. This scholastic spawn covers a broad range of interests and subsequent careers, providing testimony to the broad interests and flexibility of Professor Findley. During his career, Jim served

Life Among the Muses: Papers in Honor of James S. Findley. Edited by T.L. Yates, W.L. Gannon & D.E. Wilson.

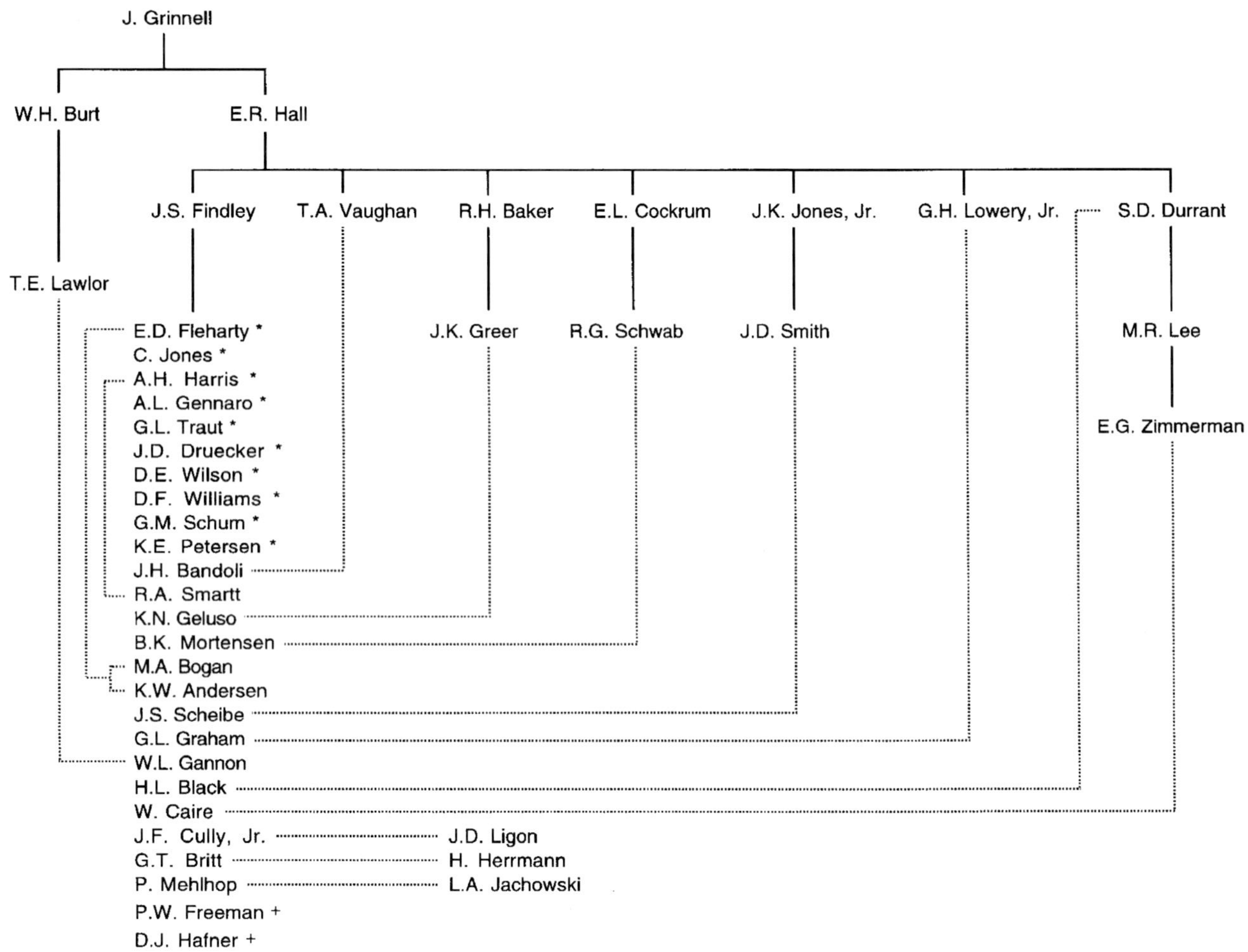

Figure 2. James S. Findley, his doctoral students, and their M.S. advisors. Solid lines connect students with their doctoral advisors and dotted lines with their M.S. advisors. Asterisks indicate students who received both graduate degrees under JSF. Plus signs denote students who received doctorates without obtaining a master's degree.

as the major advisor of 26 doctoral students and 34 students who received their M.S. degrees. Ten students completed both degrees under his guidance.

Jim taught at the University of South Dakota for a year before completing his doctorate. In 1955, he received his Ph.D. at the University of Kansas (Fig. 1) and then joined the faculty at the University of New Mexico. Jim's first graduate student was James Sands, who received his M.S. in 1956. Eugene Fleharty was the first to complete a Ph.D. in 1963.

The doctoral students of Jim Findley form an intricate web of relationships considering where they obtained their master's degrees (Fig. 2). Although incomplete in some respects (e.g., we made no attempt to trace Findley's academic grandchildren), the partial genealogy shown in Figure 2 demonstrates the influence of Findley's own academic forefathers, E. Raymond Hall of the University of Kansas and Joseph Grinnell of the University of California at Berkeley. The overall influence of that line recently was explicated by one of Jim's academic siblings, the late J.K. Jones, Jr. (1991, Genealogy of twentieth-century systematic mammalogists in North America: The descendants of Joseph Grinnell. Pp. 48-56, *in* Latin American mammalogy: History, biodiversity, and conservation [M.A. Mares and D.J. Schmidly, eds.]. University of Oklahoma Press, Norman, 468 pp.).

Below is a chronological list of James S. Findley's doctoral and master's degree students. To compile these profiles, we solicited information from Jim's students about their careers. For those whom we were unable to contact, we obtained pertinent information from records at the University of New Mexico and from JSF. Although a complete bibliography listing the published scientific output of these students would be impressive, space constraints forced us to limit each to a maximum of five publications. Nevertheless, the breadth of coverage of these publications is substantial.

DOCTORAL STUDENTS OF JAMES S. FINDLEY

1. **Eugene D. Fleharty**—B.A., Hastings College, 1956; M.S., University of New Mexico, 1958; Ph.D., University of New Mexico, 1963. For his master's research, Gene studied the distribution and variation in chipmunks of the *Eutamias quadrivittatus* group in New Mexico. His dissertation dealt with the comparative biology of three species of New Mexican garter snakes. Before completing his doctorate, Gene taught at Nebraska Wesleyan University for two years and then became associated with Fort Hays State University in 1962. At the latter institution, Gene has been active in studies of the distribution, ecology, and population dynamics of small mammals in western Kansas. He served as Chair of the Department of Biological Sciences for eleven years (1979-90) and was the recipient of the President's Scholar Award in 1991. Gene is presently a Professor of Zoology at Fort Hays State University and has had over 50 graduate students complete their master's research under his guidance.

Jones, J.K., Jr., E.D. Fleharty, and P.B. Dunnigan. 1967. The distributional status of bats in Kansas. Museum of Natural History, University of Kansas, Miscellaneous Publication, 46:1-33.

Fleharty, E.D., J.R. Choate, and M.A. Mares. 1972. Population fluctuations in *Sigmodon hispidus*: Factors influencing a "crash." Bulletin of Southern California Academy of Science, 71:132-138.

Baar, S.L., and E.D. Fleharty. 1976. A model of the daily energy budget and energy flow through a population of the white-footed mouse. Acta Theriologica, 21:179-193.

Fleharty, E.D., and G.K. Hulett. 1977. Can man survive? An inquiry into the impact of western man upon the environment. Independent Study, Division of Continuing Education, University of Kansas, Lawrence, 534 pp.

Fleharty, E.D., and K.W. Navo. 1983. Irrigated cornfields as habitat for small mammals in the sandsage prairie region of western Kansas. Journal of Mammalogy, 64:376-379.

2. **Clyde Jones**—B.A., Hastings College, 1957; M.S., University of New Mexico, 1960; Ph.D., University of New Mexico, 1964. For his master's thesis, Clyde studied variation and distribution of *Microtus mexicanus* in New Mexico and adjacent areas. His dissertation involved studies of distribution, ecology, and activity periods of bats of the Mogollon Mountains area of New Mexico and adjacent Arizona. After completing his doctorate, Clyde taught at Tulane University for five years, during which time he conducted field studies of primates, bats, and other mammals in Rio Muni, West Africa. In 1970, he joined the U.S. Fish and Wildlife Service at the Biological Survey Unit housed at the National Museum of Natural History. He served as Chief of the Mammal Section from 1970-73, Director of the National Fish and Wildlife Laboratory from 1973-79, and Director of the Denver Wildlife Research Center from 1979-82. Since 1982, he has been associated with Texas Tech University where he served as Director of The Museum from 1982-85 and Chair of the Department of Museum Science from 1982-87. Clyde is currently a Professor of Biological Sciences and Curator in the Natural Science Research Laboratory. Clyde conducts biodiversity surveys and studies systematics, ecology, and biogeography of Recent mammals; he also is interested in conservation and management of mammals in their natural habitats as well as the care and management of specimens in collections.

Findley, J.S., A.H. Harris, D.E. Wilson, and C. Jones. 1975. Mammals of New Mexico. University of New Mexico Press, Albuquerque, 360 pp.

Jones, J.K., Jr., D.M. Armstrong, R.S. Hoffmann, and C. Jones. 1983. Mammals of the northern Great Plains. University of Nebraska Press, Lincoln, 379 pp.

Jones, J.K., Jr., R.W. Manning, C. Jones, and R.R. Hollander. 1988. Mammals of the northern Texas panhandle. Occasional Papers, The Museum, Texas Tech University, 126:1-54.

Cato, P.S., and C. Jones (eds.). 1991. Natural history museums: Directions for growth. Texas Tech University Press, Lubbock, 252 pp.

Jones, J.K., Jr., and C. Jones. 1992. Revised checklist of Recent land mammals of Texas, with annotations. Texas Journal of Science, 44:53-74.

3. **Arthur H. Harris**—B.S., University of New Mexico, 1958; M.S, University of New Mexico, 1959; Ph.D., University of New Mexico, 1965. Art's master's thesis established much of the framework for the *Mammals of New Mexico*, published by the University of New Mexico Press in 1975. His dissertation was primarily a biogeographic/systematic study on the origin of

mammalian faunas in the modern grasslands of the San Juan Basin. He spent two years teaching at Fort Hays Kansas State College while finishing his dissertation and then moved to Texas Western College (now the University of Texas at El Paso) in 1965. At this institution, Art is currently a Professor of Biological Sciences. He also serves as Curator of Higher Vertebrates and Curator of Paleobiology within the Laboratory for Environmental Biology, an arm of the El Paso Centennial Museum. Most of Art's research in recent years has dealt with vertebrate fossil faunas of the Southwest, dealing primarily with ecological and biogeographical topics with some occasional entanglements with systematics.

Harris, A.H. 1985. Late Pleistocene vertebrate paleoecology of the West. University of Texas Press, Austin, 293 pp.

Russell, B.D., and A.H. Harris. 1986. A new leporine (Lagomorpha: Leporidae) from Wisconsinan deposits of the Chihuahuan Desert. Journal of Mammalogy, 67:632-639.

Harris, A.H. 1989. The New Mexican late Wisconsin—East versus West. National Geographic Research, 5:205-217.

Harris, A.H. 1990. Fossil evidence bearing on southwestern mammalian biogeography. Journal of Mammalogy, 71:219-229.

Harris, A.H. 1993. Wisconsinan pre-pleniglacial biotic change in southeastern New Mexico. Quaternary Research, 40:127-133.

4. Antonio L. Gennaro—B.S., New Mexico State University, 1957; M.S., University of New Mexico, 1961; Ph.D., University of New Mexico, 1966. Tony studied geographic variation of *Sigmodon hispidus* in New Mexico for his thesis and investigated factors influencing the northern limits of four desert rodents in the Rio Grande Valley for his dissertation. In 1966, he joined the faculty of Eastern New Mexico University and is currently a Professor of Biology and Director of the Natural History Museum. Tony served as Chair of the Department of Biology from 1983-90 and was instrumental in establishing the B.S. degree in Wildlife and Fishery Science, the Natural History Museum & Exhibit, and the university's Grassland Preserve. He also has hosted more than 100 wildlife episodes for educational television sponsored by the university. At present, Tony writes a weekly script for a radio show entitled "Nature's Way," which he hosts on KENW-FM and KMTH-FM in Portales. Besides teaching full time and conducting studies on Mississippi kites, Tony recently published a book based on his radio shows.

Gennaro, A.L. 1968. Northern geographic limits of four desert rodents of the genera *Peromycus*, *Perognathus*, *Dipodomys*, and *Onychomys* in the Rio Grande Valley. American Midland Naturalist, 80:477-493.

Aday, B.J., Jr., and A.L. Gennaro. 1973. Mammals of the New Mexican Llano Estacado, excluding bats. Studies in Natural Sciences, Eastern New Mexico University, Portales, 1:1-33.

Gennaro, A.L. 1974. Growth, size, and age at sexual maturity of *Holbrookia maculata maculata* in eastern New Mexico. Herpetologica, 30:85-90.

Gennaro, A.L. 1988. Breeding biology of an urban population of Mississippi kites in New Mexico. Pp. 188-190, *in* Proceedings of the Southwest raptor management symposium and workshop (R.L. Glinski, B.G. Pendleton, M.B. Ross, M.N. LeFranc, B.A. Millsap, and S.W. Hoffman, eds.). National Wildlife Federation, Washington, DC, 395 pp.

Botelo, E., A.L. Gennaro, and P.C. Arrowood. 1993. Parental care, nestling behaviors, and nesting interactions in a Mississippi kite (*Ictinia mississippiansis*) nest. Journal of Raptor Research, 27:16-20.

5. Gerald L. Traut—B.S., University of Wisconsin at Milwaukee, 1960; M.S., University of New Mexico, 1963; Ph.D., University of New Mexico, 1968. For his master's thesis, Jerry studied the comparative morphology of two species of *Onychomys* in New Mexico, and he studied geographic variation in *Pipistrellus hesperus* for his dissertation. In 1968, he joined the faculty at UNM as Assistant Professor, with the help of a massive student campaign to have UNM make an exception to the unwritten policy of not hiring their own recent graduates. In 1972, Jerry accepted a strictly teaching position at Phoenix College where he still teaches general biology for non-majors. During his tenure at Phoenix College, Jerry and colleagues published two laboratory manuals and a collection of "Readings" for their introductory courses. Since the mid-1970s, Jerry has spent most of his summer breaks exploring the national parks of the western states and has offered several five-week field courses for his students in these parklands.

Findley, J.S., and G.L. Traut. 1970. Geographic variation in *Pipistrellus hesperus*. Journal of Mammalogy, 51:741-765.

6. Don E. Wilson—B.S., University of Arizona, 1965; M.S., University of New Mexico, 1967; Ph.D., University of New Mexico, 1970. For his master's thesis, Don studied ecological relationships among *Peromyscus* in the Sandia Mountains of New Mexico. His doctoral research dealt with *Myotis nigricans*, which he studied in Panama at the Smithsonian Tropical Research Institute. Don began a long-term as-

sociation with the Organization for Tropical Studies in 1968 and has returned to Costa Rica to teach in their tropical biology program on a regular basis. In addition, he studied seed predation while in Costa Rica on a postdoctoral fellowship. This research was done with Dan Janzen and sponsored by the National Science Foundation through the University of Chicago. In 1971, Don was hired as a research zoologist by the U.S. Fish and Wildlife Service and stationed with the Biological Survey Unit at the National Museum of Natural History. He served as Chief of the Mammal Section from 1973-78, Chief of the Museum Section from 1978-83, and Chief of the Biological Survey from 1983-90. Don was named Director of Biodiversity Programs at the National Museum of Natural History in September of 1990.

Wilson, D.E. 1979. Reproductive patterns. Pp. 317-378, *in* Biology of bats of the New World family Phyllostomatidae. Part III (R.J. Baker, J.K. Jones, Jr., and D.C. Carter, eds.). The Museum, Texas Tech University, Special Publication, 16:1-441.

Timm, R.M., D.E. Wilson, B.L. Clawson, R.K. LaVal, and C.M. Vaughan. 1989. Mammals of the Braulio Carrillo–La Selva Complex, Costa Rica. North American Fauna, 75:1-162.

Handley, C.O., Jr., D.E. Wilson, and A.L. Gardner. 1991. Demography and natural history of the common fruit bat, *Artibeus jamaicensis*, on Barro Colorado Island, Panama. Smithsonian Contributions to Zoology, 511:1-173.

Wilson, D.E. 1991. Mammals of Tres Marías Islands. Bulletin of the American Museum of Natural History, 206:214-250.

Wilson, D.E., and D.M. Reeder (eds.). 1993. Mammal species of the world, a taxonomic and geographic reference. Smithsonian Institution Press, Washington, DC, 1206 pp.

7. Daniel F. Williams—B.A., Central Washington State College, 1966; M.S., University of New Mexico, 1968; Ph.D., University of New Mexico, 1971. Dan studied geographic variation of the Apache pocket mouse in New Mexico for his master's thesis. His dissertation concerned the systematics and evolution of the *Perognathus fasciatus* group of pocket mice.

Dan has held a teaching position at California State University, Stanislaus since 1971; he is currently a Professor of Zoology at this institution. Dan spent a year as a research fellow of the Carnegie Museum of Natural History and has held a faculty research post with the Pacific Southwest Research Station, U.S. Forest Service. Dan's research focus has slowly shifted from systematics and evolution to population ecology and conservation biology. Currently, he is directing the San Joaquin Valley Endangered Species Recovery Planning Program, a cooperative research program of the U.S. Fish and Wildlife Service and the U.S. Bureau of Reclamation.

Williams, D.F., and M.A. Mares. 1978. A new genus and species of phyllotine rodent (Mammalia: Muridae) from northwestern Argentina. Annals of the Carnegie Museum, 47:193-221.

Williams, D.F. 1978. Karyological affinities of the species groups of silky pocket mice (Rodentia, Heteromyidae). Journal of Mammalogy, 59:599-612.

Williams, D.F. 1991. Habitats of shrews (genus *Sorex*) in forest communities of the western Sierra Nevada, California. Pp. 1-14, *in* The biology of the Soricidae (J.S. Findley and T.L. Yates, eds.). Museum of Southwestern Biology, University of New Mexico, Special Publication, 1:1-91.

Williams, D.F., S. Byrne, and T.A. Rado (eds.). 1992. Endangered and sensitive species of the San Joaquin Valley, California: Their biology, management, and conservation. California Energy Commission, Sacramento, 388 pp.

Williams, D.F., H.H. Genoways, and J.K. Braun. 1993. Taxonomy. Pp. 38-196, *in* Biology of the Heteromyidae (H.H. Genoways and J.H. Brown, eds.). American Society of Mammalogists, Special Publication, 10:1-719.

8. Hal L. Black—B.S., University of Utah, 1966; M.S., University of Utah, 1968; Ph.D., University of New Mexico, 1972. One of the last master's degree students of Stephen D. Durrant, Hal studied the relative abundance of small rodents on native and introduced rangelands in Utah. His doctoral dissertation dealt with food habits of a diverse community of insectivorous bats in the Southwest. After graduation, Hal went to the University of Zambia for two years and studied various African mammals. He returned to a position at Brigham Young University in 1975 and began his research on the use of dogs to protect livestock from coyotes. Regularly returning to the Smithsonian Tropical Research Institute in Panama, much of Hal's current research deals with the biology and ecology of the large tropical ant, *Paraponera clavata*. Other research in progress deals with the ecology of non-native Hawaiian ants and the behavioral ecology of black bears. Hal is a Professor of Zoology and teaches mammalogy, vertebrate zoology, and nature appreciation.

Black, H.L. 1974. A north temperate bat community: Structure and prey populations. Journal of Mammalogy, 55:138-157.

Black, H.L. 1981. Navajo sheep- and goat-guarding dogs: A New World solution to the coyote problem. Rangelands, 3:235-237.

Findley, J.S., and H.L. Black. 1983. Morphological and dietary structuring of a Zambian insectivorous bat community. Ecology, 64:625-630.

Warren, S.D., H.L. Black, D.A. Eastmond, and W.H. Whaley. 1988. Structural function of buttresses of *Tachigalia versicolor*. Ecology, 69:532-536.

Thurber, D.K., M.C. Belk, H.L. Black, C.D. Jorgensen, S.P. Hubbell, and R.B. Foster. 1993. Dispersion and mortality of colonies of the tropical ant *Paraponera clavata*. Biotropica, 25:215-221.

9. Gwendolyn T. Britt—B.S., Eastern New Mexico University, 1955; M.S., University of Colorado, 1958; Ph.D., University of New Mexico, 1972. Gwen, who now goes by her middle name Tracy, worked under the direction of Heinz Herrmann for her master's degree. She studied nitrogen accumulation in chick embryos that were cultivated *in vitro*. After receiving her M.S., she worked in the cancer research laboratory at Mt. Sinai Hospital in Chicago for a while and then returned to the University of Colorado to obtain a teaching certificate. Before returning to graduate school, Tracy taught in secondary schools in Colorado, Texas, and Florida. For her dissertation, Tracy conducted population studies of *Perognathus flavus* and *Dipodomys ordii* on a grassland mesa outside of Albuquerque. During and following her doctoral research, she had two extended bouts with the debilitating disease, polymyositis. These erased her goal of a college career. She returned to teaching at an all-Indian secondary school and has since retired due to a laryngectomy. Tracy now resides in Albuquerque.

Britt, L.G., and H. Herrmann. 1959. Protein accumulation in early chick embryos grown under different conditions of explantation. Journal of Embryology and Experimental Morphology, 7:66-72.

10. Jay D. Druecker—B.A., Nebraska Wesleyan University, 1962; M.S., University of New Mexico, 1966; Ph.D., University of New Mexico, 1972. For his master's research, Jay studied the distribution and ecology of bats in southwestern New Mexico. His dissertation dealt with the reproductive biology of *Lasiurus cinereus*, *Myotis volans*, and *Lasionycteris noctivagans*. In 1968, Jay joined the faculty at Chadron State College in Nebraska, and currently he is a Professor in the Department of Biology. In the early 1970s, Jay took a leave of absence from Chadron to finish up his doctorate at UNM, and in the late 1970s, he spent 13 months as a postdoctoral student in the Department of Anatomy at the University of Nebraska Medical Center. While at Chadron, Jay has been advising students interested in entering the health professions and has been actively recruiting students into pre-professional programs. Jay was a major contributor in the development and implementation of the Rural Health Opportunities Program. This cooperative program with the University of Nebraska Medical Center is designed to recruit and train rural Nebraskans in seven health professions with the understanding that they will return to rural Nebraska to establish a practice. Jay also established a similar program with the Health Sciences Center at Creighton University. As part of these activities, he also established a Health Professions Office at his college. In his first few years at Chadron State, Jay was involved in research dealing with mammals and looks forward to returning to such activities in the near future.

Williams, D.F., J.D. Druecker, and H.L. Black. 1970. The karyotype of *Euderma maculatum* and comments on the evolution of the plecotine bats. Journal of Mammalogy, 51:602-606.

Case, R., J. Stubbendieck, and J.D. Druecker. 1975. Pocket gophers: A grassland nuisance. Farm, Ranch, and Home Quarterly, 22:16-18.

Czaplewski, N.J., J.P. Farney, J.K. Jones, Jr., and J.D. Druecker. 1979. Synopsis of bats of Nebraska. Occasional Papers, The Museum, Texas Tech University, 61:1-24.

Desy, E.A., and J.D. Druecker. 1979. The estrous cycle of the plains pocket gopher, *Geomys bursarius*, in the laboratory. Journal of Mammalogy, 60:235-236.

Freeman, P.W., J.D. Druecker, and S. Tvrz. 1993. *Sorex merriami* in Nebraska. The Prairie Naturalist, 25:291-294.

11. Kenneth N. Geluso—B.A., University of Vermont, 1967; M.S., University of Oklahoma, 1970; Ph.D., University of New Mexico, 1972. For his master's research, Ken studied the distribution of five species of *Peromyscus* in the Black Mesa region of Oklahoma, under the direction of J. Keever Greer. His dissertation involved an ecophysiological study dealing with renal form and function in insectivorous bats. In 1973, Ken began studies on the effects of pesticides on *Tadarida brasiliensis* at Carlsbad Caverns National Park. This research was done with Scott Altenbach and Don Wilson and was sponsored by the World Wildlife Fund, U.S. Fish and Wildlife Service, and National Park Service. Ken still regularly visits the park and is currently working on the *Mammals of Carlsbad Caverns National Park*. Hired in 1977, Ken is presently an Associate Professor of Biology at the University of Nebraska–Omaha.

Geluso, K.N. 1971. Habitat distribution of *Peromyscus* in the Black Mesa region of Oklahoma. Journal of Mammalogy, 52:605-607.

Geluso, K.N., J.S. Altenbach, and D.E. Wilson. 1976. Bat mortality: Pesticide poisoning and migratory stress. Science, 194:184-186.

Geluso, K.N. 1980. Renal form and function in bats: An ecophysiological appraisal. Pp. 403-414, *in* Proceedings of the fifth international bat research conference (D.E. Wilson and A.L. Gardner, eds.). Texas Tech Press, Lubbock, 434 pp.

Geluso, K.N., J.S. Altenbach, and R.C. Kerbo. 1987. Bats of Carlsbad Caverns National Park. Carlsbad Caverns Natural History Association, Carlsbad, New Mexico, 33 pp.

Manning, R.W., and K.N. Geluso. 1989. Habitat utilization of mammals in a man-made forest in the Sandhill Region of Nebraska. Occasional Papers, The Museum, Texas Tech University, 131:1-34.

12. Kenneth W. Andersen—B.S., Fort Hays Kansas State College, 1963; M.S., Fort Hays Kansas State College, 1965; Ph.D., University of New Mexico, 1973. Under the guidance of Eugene Fleharty, Ken studied the distribution of mammals in north-central Kansas for his master's thesis. His dissertation involved a comparative study of renal efficiencies of *Peromyscus* from the Sandia Mountains of New Mexico. After completion of his doctoral work, Ken worked as chief mammalogist for a major environmental consulting firm for three years and then in 1976, was hired by Gannon University in Erie, Pennsylvania. He is presently the Chair of the Department of Biology at this institution. Ken's current research activities include monitoring the activity of bats in Pennsylvania using ultrasonic detectors and investigating the ecology of bats on Andros and San Salvador islands in the Bahamas. During the past several years, Ken and his students have made yearly pilgrimages to the Bahamian Field Station to conduct various studies in ecology.

Andersen, K.W., and E.D. Fleharty. 1964. Additional fox records for Kansas. Transactions of the Kansas Academy of Science, 67:193-194.

Andersen, K.W., and E.D. Fleharty 1967. Mammalian distribution within biotic communities of northeastern Jewell County, Kansas. Fort Hays Studies, Science Series, 6:1-46.

Andersen, K.W., and J.K. Jones, Jr. 1971. Mammals of northwestern South Dakota. University of Kansas Publications, Museum of Natural History, 19:361-393.

Andersen, K.W. 1990. Bats of San Salvador Island. Bahamian Field Station, Occasional Paper, 1:1-4.

Zeakes, S.J., K.W. Andersen, and D.T. Gerace. 1990. New locality record for a species of the genus *Periglischrus* (Acarina: Spincternicidae: Mesostigmata) on the buffy flower bat (*Erophylla sezekorni*) from the Bahamas. Bat Research News, 31:23-25.

13. Michael A. Bogan—B.S., Baker University, 1964; M.S., Fort Hays State University, 1966; Ph.D., University of New Mexico, 1973. John Nickel at Baker recommended that Mike give Fort Hays a try for a graduate degree in some aspect of field biology. At Hays, Mike fell under the tutelage of Eugene Fleharty and assessed differences in small mammals on grazed and ungrazed grasslands for his master's research. At UNM, Mike studied geographic variation in *Myotis californicus*, starting an affection for *Myotis* that continues today. After receiving his doctorate, Mike went to work for the Fish and Wildlife Survey at the National Museum of Natural History. Here, he worked for the Service nine years and then for eleven years while stationed in Fort Collins. Now Mike is employed by the National Biological Service at the University of New Mexico. While occupying a variety of positions throughout his career, Mike has always served as a curator of mammals and is currently Curator of the Biological Service Collections housed in the Museum of Southwestern Biology. Mike's research interests continue to center around traditional Findleyesque questions of systematics, distribution, and zoogeography.

Bogan, M.A. 1975. Geographic variation in *Myotis californicus* in the southwestern United States and Mexico. U.S. Fish and Wildlife Service, Wildlife Research Report, 3:1-31.
Bogan, M.A. 1978. A new species of *Myotis* from the Islas Tres Marias, Nayarit, Mexico, with comments on variation in *M. nigricans*. Journal of Mammalogy, 59:519-530.
Bogan M.A., R.B. Finley, Jr., and S.J. Petersburg. 1988 [1989]. The importance of biological surveys in managing public lands in the western United States. Pp. 254-261, *in* Management of amphibians, reptiles, and small mammals in North America. United States Department of Agriculture Forest Service, General Technical Report RM-166, 458 pp.
Seal, U.S., E.T. Thorne, M.A. Bogan, and S.H. Anderson (eds.). 1989. Conservation biology and the black-footed ferret. Yale University Press, New Haven, Connecticut, 328 pp.
Lara, M., M.A. Bogan, and R. Cerqueira. 1992. Sex and age components of variation in *Proechimys cuvieri* (Rodentia: Echimyidae) from northern Brazil. Proceedings of the Biological Society of Washington, 105:882-893.

14. G. Michael Schum—B.S., University of New Mexico, 1971; M.S., University of New Mexico, 1972; Ph.D., University of New Mexico, 1975. For his master's thesis, Mike studied vocalizations and echolocating sounds of *Nyctinomops macrotis* at a nursery colony near Los Lunas. His doctoral research involved a multivariate morphometric study of phenetic packing in assemblages of North American bat faunas. From 1974-78, Mike worked at the Lovelace Inhalation Toxicology Research Institute in Albuquerque as a biomathematician developing models to predict deposition of environmental pollutants in mammalian lungs. After a brief postdoctoral appointment in 1979 in the Radiobiology Laboratory at the University of California–Davis, Mike joined the Department of Biology at San Diego State University where he served as an Assistant Professor of Biology from 1979-85. After working as an independent computer consultant for a while, Mike was hired by the California Envi-

ronmental Protection Agency and is currently an environmental toxicologist in the Ecological and Human Risk Section of the Department of Toxic Substances Control. In this capacity, Mike acts as a computer specialist involved in modeling the movement of toxic chemicals through the environment. Since 1988, Mike also has been actively involved in a research project sponsored by NIH and dealing with effects of age and body size on particle deposition in the lung. This project is being conducted at the Air Pollution Health Effects Laboratory at the University of California–Irvine.

Schum, G.M., and H.C. Yeh. 1980. Theoretical evaluation of aerosol deposition in anatomical models of mammalian lung airways. Bulletin of Mathematical Biology, 42:1-15.

Schum, G.M. 1982. Mammalia. Pp. 525-529, *in* Aquatic biota of Mexico, Central America, and the West Indies—An annotated taxonomic bibliography (S.L. Hurlbert, ed.). San Diego State University Press, San Diego, 529 pp.

Schum, G.M. 1984. Phenetic structure and species richness in North and Central American bat faunas. Ecology, 65:1315-1324.

Schum, G.M., R.F. Phalen, and M.J. Oldham. 1991. Effect of dead space on particle deposition. Journal of Aerosol Medicine, 4:297-311.

Schum, G.M., R.F. Phalen, and M.J. Oldham. 1994. Risk assessment of inhaled particles: Integrating time-activity patterns with respiratory tract particle deposition models. Annals of Occupational Hygiene, 38 (supplement 1):973-978.

15. Richard A. Smartt—B.A., University of Texas at El Paso, 1970; M.S., University of Texas at El Paso, 1972; Ph.D., University of New Mexico, 1976. Under the direction of Arthur Harris, Rick's thesis involved the reconstruction of past environments based on Pleistocene microtines. For his dissertation, he compared the ecological structure of a rodent assemblage with the morphological makeup of the rodent community. Rick continues to work on rodent morphology but with an emphasis on Pleistocene paleo-ecology. After teaching at the University of Texas at El Paso from 1977-83, Rick accepted a position at the New Mexico Museum of Natural History where he served as Chair of the Science Division and Curator of Invertebrate Zoology; currently, he is Director of the Museum. Rick's present research involves gastropod communities in Central America.

Smartt, R.A. 1978. A comparison of ecological and morphological overlap in a *Peromyscus* community. Ecology, 59:216-220.

Smartt, R.A. and C. Lemen. 1980. Intrapopulational morphological variation as a predictor of feeding behavior in deer mice. American Naturalist, 116:291-293.

Smartt, R.A., and D.J. Hafner. 1989. Preliminary report on a Late Pleistocene (gastropod and rodent) microfauna from Whut Cave, Cibola County, New Mexico. Pp. 281-285, *in* Southeastern Colorado plateau (O.J. Anderson, S.G. Lucas, D.W. Love, and S.M. Cather, eds.). New Mexico Geological Society Guidebook, 40th Field Conference, 345 pp.

Smartt, R.A., S.G. Lucas, and D.J. Hafner. 1990. A bighorn sheep, *Ovis canadensis*, from the Late Pleistocene of the Mesa del Oro, Cibola County, New Mexico. Texas Journal of Science, 42:107-109.

Smartt, R.A., and R.M. Sullivan. 1990. Distribution and ecology of *Pecosorbis kansasensis* in eastern New Mexico. Journal of Arid Environments, 19:181-187.

16. Patricia W. Freeman—A.B., Randolph–Macon Woman's College, 1969; Ph.D., University of New Mexico, 1977. After receiving her A.B. degree, Trish worked in the Department of Mammalogy at the American Museum of Natural History as Scientific Assistant to Richard G. Van Gelder, Sydney Anderson, Karl F. Koopman, Guy G. Musser, and Hobart M. Van Deusen. She returned to graduate school to study with Jim Findley in 1972. Trish was particularly interested in functional morphology and morphometrics, and she used multivariate techniques to examine molossid bats for her dissertation. Trish was hired in 1977 by the Field Museum of Natural History where she was Assistant Curator and Head of the Division of Mammals. In 1981, Trish became Curator of Zoology at the University of Nebraska State Museum in Lincoln where she is now both Curator and Associate Professor. Trish works on feeding mechanisms in bats and has begun experimentally testing teeth to examine the relationship between form and function at the tooth-food interface. Recently she has been involved with faunal surveys of Nebraska, particularly mammals and mollusks. Past research

has included collaborative work with spouse Cliff Lemen in community ecology, macroevolution, and development of a method to track mammals with fluorescent pigments.

Freeman, P.W. 1981. A multivariate study of the family Molossidae (Mammalia, Chiroptera): Morphology, ecology, evolution. Fieldiana: Zoology, New Series, 7:1-173.

Lemen, C.A., and P.W. Freeman. 1984. The genus: A macroevolutionary problem. Evolution, 38:1219-1237.

Lemen, C.A., and P.W. Freeman. 1985. Tracking mammals with fluorescent pigments: A new technique. Journal of Mammalogy, 66:134-136.

Freeman, P.W. 1988. Frugivorous and animalivorous bats (Microchiroptera): Dental and cranial adaptations. Biological Journal of the Linnean Society (London), 33:249-272.

Freeman, P.W. 1992. Canine teeth of bats (Microchiroptera): Size, shape, and role in crack propagation. Biological Journal of the Linnean Society (London), 45:97-115.

17. B. Kim Mortensen—B.A., University of California at Riverside, 1969; M.S., University of California at Davis, 1971; Ph.D., University of New Mexico, 1977. Kim studied activity patterns of desert rodents in California for his master's thesis, under the direction of Robert G. Schwab. His doctoral topic involved an analy-

sis of morphology as a predictor of feeding and foraging strategies of leaf-nosed bats. Kim did much of his doctoral research while stationed at the Smithsonian Institution, National Museum of Natural History. Here, he also worked for Charles O. Handley and helped to computerize data on the mammals of Venezuela and assisted in field studies of primates in Guyana and cave-dwelling bats in Hispaniola. In 1978, Kim became associated with the University of Arizona where he conducted a variety of studies for the National Park Service and National Science Foundation. For the NPS, he conducted vegetative inventories for Organ Pipe Cactus National Monument, Grand Canyon National Park, and Saguaro National Monument. For NSF, he worked on technological and impact assessments for the commercialization of guayule in Mexico and development of jojoba on Indian reservations in the United States. Since 1983, Kim has worked in environmental epidemiology and toxicology. He began in the M.P.H. Program in Epidemiology at Emory University and later worked at the Centers for Disease Control in Atlanta. Eventually he assumed his current position as Chief, Bureau of Epidemiology and Toxicology, Ohio State Health Department.

Mortensen, B.K., G.W. Walter, W.A Garrett, J.E. Anderson, and V. Burse. 1987. Exposure study of persons possibly exposed to polychlorinated biphenyls in Paoli, Pennsylvania. U.S. Department of Health and Human Services, Agency for Toxic Substances and Disease Registry, November 1987, 48 pp.

Hopkins, R.S., J.V. Writer, B.K. Mortensen, and R.W. Indian. 1990. Childhood injury mortality in Ohio, 1979 to 1986: Setting priorities for prevention. American Journal of Diseases of Children, 144:79-82.

Hopkins, R.S., L.E. Tyler, and B.K. Mortensen. 1990. Effects of maternal cigarette smoking on birth weight and preterm birth–Ohio, 1989. Morbidity and Mortality Weekly Report, 39:662-665.

Bausterian, K.M., R.A. Etzel, M.M. Agocs, E.M. Socie, M.A. Rouse, and B.K. Mortensen. 1991. Indoor air mercury concentrations following application of interior latex paint. Archives of Environmental Contamination and Toxicology, 21:62-64.

18. William Caire—B.S., Texas Tech University, 1969; M.S., University of North Texas, 1972; Ph.D., University of New Mexico, 1978. Under the direction of Earl G. Zimmerman, Bill's master's research involved a case of circular overlap in chromosomal and morphological traits in *Peromyscus* in Texas and Oklahoma. His dissertation dealt with the distribution and zoogeography of mammals of Sonora, Mexico. Before completing his doctorate, Bill was hired by the University of Central Oklahoma and today, he is a Professor of Biology and Assistant Dean of the College of Mathematics and Science at this institution. He and his students have been involved in a wide variety of research

projects including mammalian distributions, bat fly behavior, cactus proteins, distribution and construction of woodrat houses, radio telemetry studies of home ranges, hemoglobin electrophoresis, and ion concentration changes in the urine of hibernating *Myotis velifer*.

Caire, W., and E. Zimmerman. 1975. Circular overlap in *Peromyscus maniculatus* in Texas and Oklahoma as evidenced by chromosomal and morphological variations. Systematic Zoology, 24:89-95.

Caire, W., L. Hornuff, and N. Sohrabi. 1985. Stimuli used by *Trichobius major* (Diptera: Streblidae) to locate its bat host, *Myotis velifer*. Southwestern Naturalist, 30:405-412.

Caire, W., J. Tyler, B.P. Glass, and M. Mares. 1989. The mammals of Oklahoma. University of Oklahoma Press, Norman, 544 pp.

Caire, W., and K. Haydari. 1990. Absence of electrophoretically distinct fetal hemoglobin in the bats *Lasiurus borealis* and *Tadarida brasiliensis*. Journal of Mammalogy, 71:695-697.

Thies, M., and W. Caire. 1991. Nearest neighbor analysis of the spatial distribution of houses of *Neotoma micropus* in southwestern Oklahoma. Southwestern Naturalist, 36:233-237.

19. David J. Hafner—B.A., University of California, Berkeley, 1974; Ph.D., University of New Mexico, 1981. Dave's dissertation detailed the evolution and historical zoogeography of *Ammospermophilus* relative to the development of North American deserts. His research has focused on late Pleistocene historical zoogeography of western North America, using genetic analyses of disjunct mammalian populations to determine relationships. In 1983, Dave was hired as Curator of Zoology at the New Mexico Museum of Natural History and was immersed in the planning, construction, and development of the new museum, which opened in 1986. While serving as Chair of the Science Department, Dave conducted NSF-supported research on North American *Ochotona* from New Mexico to Alaska, studied pocket gophers in New Mexico and Central America, and led public trips to Costa Rica and the Galápagos Islands. As Chief Curator of the Museum, Dave currently conducts studies on the mammals of

the Baja California Peninsula and adjacent islands.

Hafner, D.J., J.C. Hafner, and M.S. Hafner. 1979. Systematic status of kangaroo mice, genus *Microdipodops*: Morphometric, chromosomal, and protein analyses. Journal of Mammalogy, 60:1-10.

Hafner, D.J., and K.N. Geluso. 1983. Systematic relationships and historical zoogeography of the desert pocket gopher, *Geomys arenarius*. Journal of Mammalogy, 64:405-413.

Hafner, D.J. 1984. Evolutionary relationships of the Nearctic Sciuridae. Pp. 3-23, *in* Biology of ground-dwelling squirrels: Annual cycles, behavioral ecology, and sociality (J.O. Murie and G.R. Michener, eds.). University of Nebraska Press, Lincoln, 459 pp.

Hafner, D.J. 1992. Speciation and persistence of a contact zone in Mojave Desert ground squirrels, subgenus *Xerospermophilus*. Journal of Mammalogy, 73:770-778.

Hafner, D.J. 1993. North American pika (*Ochotona princeps*) as a late Quaternary biogeographic indicator species. Quaternary Research, 39:373-380.

20. Patricia Mehlhop—B.S., University of Maryland, 1967; M.S., University of Maryland, 1969; Ph.D., University of New Mexico, 1981. Pat's research for her master's degree dealt with

ecological parasitology and was done under the supervision of Leo A. Jachowski. After receiving this degree, she worked in the mammal collection at the United States National Museum and helped Clyde Jones and Don Wilson prepare maps for the *Mammals of New Mexico*. Pat then accepted a position at the Smithsonian Environmental Research Center and conducted ecological studies with James F. Lynch on rodents, birds, and insects in the floodplain forests of Chesapeake Bay. During this time, she also accompanied Clyde Jones on survey trips to southwestern New Mexico and decided to turn her attention to graduate studies in desert ecology. For her dissertation, Pat studied foraging ecology in seed-harvesting ants, with Jim Findley and Norman J. Scott, Jr. as co-chairs of her committee. She then returned to the Smithsonian Institution as a postdoctoral fellow and continued studies in ant ecology. For the past decade, Pat has worked for The Nature Conservancy, studying rare and declining species and establishing state programs to track the status of these species and their habitats. In 1990, she established such a program in New Mexico in conjunction with The Nature Conservancy and the University of New Mexico. Presently, Pat heads this conservation program known as the New Mexico Natural Heritage Program.

Mehlhop, P., and N.J. Scott, Jr. 1983. Temporal patterns of seed use and availability in a guild of desert ants. Ecological Entomology, 8:69-85.

Bogan, M.A., and P. Mehlhop. 1983. Systematic relationships of gray wolves (*Canis lupus*) in southwestern North America. Occasional Papers, Museum of Southwestern Biology, 1:1-21.

Mehlhop, P., and J.F. Lynch. 1986. Bird/habitat relationships along a successional gradient in the Maryland coastal plain. American Midland Naturalist, 116:225-239.

Moss, S.P., and P. Mehlhop–Cifelli. 1990. Status of the Texas kangaroo rat, *Dipodomys elator* (Heteromyidae) in Oklahoma. Southwestern Naturalist, 35:356-358.

Mehlhop, P., and C.C. Vaughn. 1994. Threats to and sustainability of ecosystems for freshwater molluscs. Pp. 68-77, *in* Sustainable ecological systems: Implementing an ecological approach to land management. United States Department of Agriculture Forest Service, General Technical Report, RM-247, 363 pp.

21. John S. Scheibe—B.A., California State University at Fullerton, 1974; M.A., California State University at Fullerton, 1977; Ph.D., University of New Mexico, 1982. For his master's thesis, John studied the ecology of a small mammal community in the Tehachapi Mountains of California, under the direction of James Dale Smith. But at the University of New Mexico, he satisfied his growing interest in reptiles by studying lizards under the guidance of Jim Findley. For his doctoral work, he attempted to identify biotic and abiotic parameters associated with patterns of species diversity in lizard com-

munities. John joined the faculty at Southeast Missouri State University in Cape Girardeau in 1983. He is presently a Professor of Biology at this institution and teaches experimental design, mammalogy, herpetology, and ecological modeling. John's current research projects include effects of chaos and history on expected patterns of structure in lizard communities, niche dynamics in *Peromyscus leucopus* and *Glaucomys volans,* and the evolution of gliding locomotion in sciurids, anomalurids, petaurids, and burramyids.

Scheibe, J.S. 1984. The effects of weather, sex and season on the nocturnal activity of *Peromyscus truei* (Rodentia). Southwestern Naturalist, 29:1-5.

Scheibe, J.S. 1984. Sexual differences in the home ranges of *Peromyscus truei* and *Dipodomys panamintinus* (Rodentia). Southwestern Naturalist, 29:7-13.

Scheibe, J.S. 1987. Climate, competition, and the structure of temperate zone lizard communities. Ecology, 68:1424-1436.

Kullberg, R.G., and J.S. Scheibe. 1989. The effects of succession on niche breadth and overlap in a hot spring algal community. American Midland Naturalist, 121:21-31.

Scheibe, J.S., D. Figgs, and J. Heiland. 1990. Morphological attributes of gliding rodents: A preliminary analysis. Transactions, Missouri Academy of Science, 24:49-55.

22. James H. Bandoli—B.S., University of San Diego, 1973; M.S., Northern Arizona University, 1977; Ph.D., University of New Mexico, 1983. As an undergraduate, Jim became associated with the International Biological Program and spent nearly a year studying mammals and reptiles at the Desert Biome Research Station in Nevada. As a graduate student, his master's research dealt with seasonal activity of *Thomomys bottae* and was completed under the guidance of Terry A. Vaughan. Jim continued to work with pocket gophers at UNM where his doctoral research addressed geographical variation and maintenance of secondary sexual size dimorphism. He joined the faculty at the University of Southern Indiana in 1982 and is currently an Associate Professor of Biology. Here, Jim teaches courses in ecology, evolution, com-

parative anatomy, animal behavior, and vertebrate biology. For the last few years his research interests have centered around habitat requirements and mating strategies of spottail darters, a state endangered species of percid fish.

Bandoli, J.H. 1981. Factors influencing seasonal burrowing activity in the pocket gopher, *Thomomys bottae*. Journal of Mammalogy, 62:293-303.

Bandoli, J.H. 1987. Activity and plural occupancy of burrows in Botta's pocket gopher, *Thomomys bottae*. American Midland Naturalist, 118:10-14.

Bandoli, J.H., T. Lanigan, and T. Sheckles. 1991. Reproduction in the spottail darter in Indiana: Use of artificial nest sites. Proceedings of the Indiana Academy of Science, 100:65-75.

23. Karen E. Petersen—B.S., University of Nebraska at Lincoln, 1975; M.S., University of New Mexico, 1977; Ph.D., University of New Mexico, 1983. Karen's thesis involved a morphological, behavioral, and ecological study of *Sigmodon hispidus* and *S. fulviventer* in the Rio Grande Valley of New Mexico. For her doctoral research, she tested for patterns in community morphologies in selected guilds of mammals, by comparing morphological distances from sampled communities with those from randomly generated communities. After completing her

doctorate, Karen worked with Rex Cates at UNM and studied the effects of nutrient and water stresses on Douglas fir. Subsequently, Karen was a Visiting Assistant Professor at the University of South Florida from 1986-88 and at Austin Peay State University in Tennessee from 1988-89. Karen then joined the faculty at the University of Central Arkansas as an Assistant Professor during the 1989-90 academic year. In 1990, she gave up this position to get married and moved to Seattle, Washington. Here, Karen began working at the University of Washington, and she is currently a lecturer in the Department of Zoology, teaching comparative vertebrate anatomy and human physiology.

Hafner, D.J., K.E. Petersen, and T.L. Yates. 1981. Evolutionary relationships of jumping mice (genus *Zapus*) of the southwestern United States. Journal of Mammalogy, 62:501-512.

Hafner, D.J., and K.E. Petersen. 1985. Song dialects and gene flow in the white-crowned sparrow, *Zonotrichia leucophrys*. Evolution, 39:687-694.

Sullivan, R.M., and K.E. Petersen. 1988. Systematics of southwestern populations of least chipmunks (*Tamias minimus*) reexamined: A synthetic approach. Occasional Papers, Museum of Southwestern Biology, 5:1-27.

Henderson, C.B., K.E. Petersen, and R.A. Redak. 1988. Spatial and temporal patterns in the seed bank and vegetation of a desert grassland community. Journal of Ecology, 76:717-728.

24. Jack F. Cully, Jr.—B.S., University of New Mexico, 1967; M.S., University of New Mexico, 1972; Ph.D., University of New Mexico, 1984. For his master's research, Buck compared the mobbing behavior of gray-breasted jays (*Aphelocoma ultramarina*), a highly social species, with the congeneric scrub jay (*A. coerulescens*), a territorial species. This work was done under the supervision of J. David Ligon. Buck's doctoral subject dealt with habitat relationships among rodents in the San Juan Basin of northwestern New Mexico. Following graduation, Buck continued his association with UNM as an Assistant Professor of Research and began studies of social behavior and population ecology of Gunnison's prairie dog in northern New Mexico. This project was funded by New Mexico Share With Wildlife and eventually expanded to include investigations of the ecology of plague in prairie dog ecosystems. In 1989, Buck moved to the University of Notre Dame with a fellowship from the National Institutes of Health and continued to work on the ecology of vector borne diseases. In 1991, Buck went to work for the U.S. Fish and Wildlife Service in the Endangered Species Program; three years later, he moved to Kansas State University where he cur-

rently is employed by the National Biological Service.

Cully, J.F., Jr., and J.D. Ligon. 1976. Comparative mobbing behavior of scrub and Mexican jays. Auk, 93:116-125.

Cully, A.C., and J.F. Cully, Jr. 1989. Spatial and temporal variability in perennial and annual vegetation at Chaco Canyon New Mexico. Great Basin Naturalist, 49:113-122.

Cully, J.F., Jr. 1991. Response of raptors to reduction of a Gunnison's prairie dog population by plague. American Midland Naturalist, 125:140-149.

Cully, J.F., Jr., J.P. Greco, and D. Kissel. 1991. Defensive behavior of eastern chipmunks against *Aedes triseriatus* (Diptera: Culicidae). Journal of Medical Entomology, 28:410-416.

Cully, J.F., Jr., and W.A. Austin. 1993. Final rule to list the Mexican spotted owl as a threatened species. Federal Register, 58:14248-14271.

25. Gary L. Graham—B.S., Louisiana State University, 1975; M.S., Louisiana State University, 1979; Ph.D., University of New Mexico, 1986. Under the direction of the late George H. Lowery, Jr., Gary was introduced to the tropics during his master's research, which dealt with the reproductive seasonality of Peruvian bats. He continued working with this group of bats for his doctoral research but this time investigated their use of diurnal roosts. During his graduate years, Gary also was interested in conservation and conducted research in this area for The Nature Conservancy and Conservation International. He coordinated the tax checkoff for the wildlife program in New Mexico for two years before moving to Texas to become Associate Science Director for Bat Conservation International. While at BCI, Gary coordinated a successful effort to protect eight species of flying foxes under the Convention on International Trade in Endangered Species. He also organized a conference on Pacific Island flying foxes to assist land-use managers in bat conservation. In 1991, Gary worked for Defenders of Wildlife to coordinate the Texas Watchable Wildlife Program and write the *Texas Wildlife Viewing Guide*. Gary is currently Chief of Endangered Resources for the Texas Parks and Wildlife Department where he has been able to work on neotropical, migratory songbirds and ecological challenges posed by the North American Free Trade Agreement.

Graham, G.L. 1988. Interspecific associations among Peruvian bats at their diurnal roosts and roost sites. Journal of Mammalogy, 69:711-720.

Graham, G.L. 1990. Bats vs. birds: Elevational variations among Peruvian volant vertebrate faunas. Journal of Biogeography, 17:657-668.

Graham, G.L. 1992. Texas wildlife viewing guide. Falcon Press, 160 pp.

Graham, G.L. 1992. Conservation and subsistence harvesting of Pacific Island flying foxes. Pp. 46-49, *in* Pacific Island flying foxes: Proceedings of an international conservation conference (D.E. Wilson and G.L. Graham, eds.). U.S. Fish and Wildlife Service, Biological Report, 90:1-176.

Graham, G.L. 1994. Bats of the world. Golden Press, New York, 160 pp.

26. William L. Gannon—B.A., Humboldt State University, 1980; M.S., Humboldt State University, 1986; Ph.D., University of New Mexico, 1997. Bill first became interested in vocalizations of small mammals while attending Humboldt State University and eventually studied geographic variation of alarm calls in

three species of *Tamias* for his master's thesis. This research was completed under the supervision of Timothy E. Lawlor. After defending his thesis in 1986, Bill headed for the University of New Mexico where Terry L. Yates hired him as Collections Manager in the Division of Mammals at the Museum of Southwestern Biology. Bill became interested in pursuing his doctorate after several discussions with his current advisor, Jim Findley. For his dissertation, Bill studied character displacement in two species of *Myotis*, using evidence from ecomorphology and sonic analysis. When Jim retired, James H. Brown became the co-chair of Bill's doctoral committee. Today, Bill is busy both with his own research and with projects involving the Museum of Southwestern Biology.

Gannon, W.L., and T.E. Lawlor. 1989. Variation in the chip vocalization of three species of Townsend chipmunks (genus *Eutamias*). Journal of Mammalogy, 70:740-753.

Gannon, W.L., and W.T. Stanley. 1991. Chip vocalization of Palmer's chipmunk (*Tamias palmeri*). Southwestern Naturalist, 36:315-317.

Frey, J.K., T.L. Yates, D.W. Duszynski, W.L. Gannon, and S.L. Gardner. 1992. Designation and curatorial management of type host specimens (symbiotypes) for new parasite species. Journal of Parasitology, 78:930-932.

Gannon, W.L., R.B. Forbes, and D.E. Kain. 1993. *Tamias ochrogenys*. Mammalian Species, 445:1-4.

Glass, P.J., and Gannon, W.L. 1995. Description of *M. uropataginalis* (new muscle) with additional comments from a microscopy study of the uropatagium of the fringed myotis (*Myotis thysanodes*). Canadian Journal of Zoology, 72:1752-1754.

MASTER'S DEGREE STUDENTS OF JAMES S. FINDLEY

1. **James L. Sands**—B.S., University of New Mexico, 1953; M.S., University of New Mexico, 1956. For his master's research, Jim studied the relationship of two subspecies of pocket gophers in central New Mexico. Directly after graduation, he was hired by the New Mexico Department of Game and Fish. During his tenure with the department, Jim held various positions in game management, including Area Game Manager from 1963-67 and Game Bird Project Biologist from 1968-82. During the latter period, he studied migratory game birds with emphasis on band-tailed pigeons and waterfowl. Jim served as the New Mexico representative to the Central Flyway Waterfowl Technical Committee from 1972-82. Jim retired from the Department of Game and Fish in 1982 and now resides in Pennsylvania. Since retirement, Jim served

a term on the Carrolltown Municipal Authority from 1985-91.

Sands, J.L., and J.S. Findley. 1959. The relationships of two subspecies of pocket gopher in central New Mexico. Journal of Mammalogy, 40:331-337.

Sands, J.L. 1967. Bighorn sheep. Pp. 69-82, *in* New Mexico wildlife management. New Mexico Department of Game and Fish, Santa Fe. 250 pp.

Sands, J.L. 1968. Status of the lesser prairie chicken. Audubon Field Notes, 22:454-456.

Sands, J.L. 1970. First record of the black brant (*Branta nigricans*) for New Mexico. Condor, 72:110.

2. Eugene D. Fleharty (see doctoral student #1)

3. Arthur H. Harris (see doctoral student #3)

4. John S. Mohlhenrich—B.S., University of New Mexico, 1957; M.S., University of New Mexico, 1959. For his master's research, John determined the distributional limits and ecological preferences of *Sigmodon hispidus* and *S. fulviventer* in New Mexico. We were unable to obtain current information about John.

Mohlhenrich, J.S. 1961. Distribution and ecology of the hispid and least cotton rats in New Mexico. Journal of Mammalogy, 42:13-24.

5. Clyde Jones (see doctoral student #2)

6. Richard B. Forbes—A.B., University of South Dakota, 1958; M.S., University of New Mexico, 1961; Ph.D., University of Minnesota, 1964. For his master's thesis, Dick studied morphological variation, molt, foods, and reproductive activity of silky pocket mice in New Mexico. His doctoral research, under the direction of James C. Underhill, involved a comparative study of morphological variation and ecology of chipmunks in the Itasca region of Minnesota. Following graduation, Dick joined the faculty at Portland State University in Oregon where he is presently Professor of Biology. Dick has studied development and behavior of Townsend's chipmunks, collaborated in physi-

ological studies of Cascades frogs and Chilean tinamous, and guided graduate students in ecological studies of vertebrates from frogs to mammals. He teaches courses in general biology, ecology, and vertebrate zoology. Although Dick continues to have a broad interest in vertebrate biology, his recent work has focused on terrestrial vertebrates in the metropolitan area of Portland, Oregon.

Hopkins, D.D., and R.B. Forbes. 1980. Dietary patterns of the Virginia opossum in an urban environment. Murrelet, 61:20-30.

Barss, J.M., and R.B. Forbes. 1985. A spotted bat (*Euderma maculatum*) from northcentral Oregon. Murrelet, 65:60.

Withers, P.C., R.B. Forbes, and M.S. Hedrick. 1987. Metabolic, water and thermal relations of the Chilean tinamou (*Nothoprocta perdicaria*). Condor, 89:424-426.

Wollmuth, L.P., L.I. Crawshaw, R.B. Forbes, and D.A. Grahn. 1987. Temperature selection in *Rana cascadae* tadpoles. Physiological Zoology, 60:472-480.

Kirk, J.J., and R.B. Forbes. 1991 *Batrachoseps wrightii* (Oregon slender salamander). Herpetological Review, 22:22-23.

7. Antonio L. Gennaro (see doctoral student #4)

8. Eugene E. Schroeder—B.A., Saint Olaf College, 1959; M.S., University of New Mexico,

1961. For his thesis, Gene determined the ecological distribution of mammals on Mount Taylor in Valencia County, New Mexico. We were unable to obtain current information about Gene.

9. Frances V. Sheppard—A.B., State University of New York, University at Albany, 1957; M.S., University of New Mexico, 1962. For her thesis, Frances studied the natural history of bats occurring in Bernalillo County, New Mexico. We were unable to obtain current information about Frances.

10. Gerald L. Traut (see doctoral student #5)

11. Thomas H. Follingstad—B.S., University of New Mexico, 1959; M.S., University of New Mexico, 1964; M.D., University of Colorado, 1966. For his master's thesis, Tom studied geographic variation of *Thomomys bottae* in the Rio Grande Valley of New Mexico and Conejos County, Colorado. While working on his master's degree, Tom decided to pursue a medical career and received an M.D. at the University of Colorado. After an internship at St. Paul-Ramsey Hospital in Minnesota, he completed his residency at the University of Minnesota in 1970. Tom's speciality is internal medicine, and currently he is associated with the Lovelace Health System in Albuquerque.

12. David M. Niles—B.S., University of New Mexico, 1961; M.S., University of New Mexico, 1964; Ph.D., University of Kansas, 1972. For his thesis, Dave studied geographic variation of horned larks (*Eremophila alpestris*) in New Mexico and continued to study this species for his dissertation. His doctoral work was done under the supervision of Richard F. Johnston. For some time, Dave was Curator of Birds at the Delaware Museum of Natural History. We were unable to obtain current information about Dave.

Niles, D.M. 1963. A Ross goose in New Mexico. Condor, 65:166.

Niles, D.M. 1966. Additional distributional records from New Mexico. Condor, 68:213-215.

Niles, D.M. 1972. Determining age and sex of purple martins. Bird-banding, 43:137-138.

Niles, D.M. 1973. Adaptive variation in body size and skeletal proportions of horned larks of the southwestern United States. Evolution, 27:405-426.

Niles, D.M. 1985. The survival of nests of American robins in northern Delaware. Delmarva Ornithologist, 18:10-19.

13. Gerald G. Wheeler—B.S., Colorado State University, 1961; M.S., University of New Mexico, 1964. For his master's research, Jerry investigated various aspects of saltatory locomotion in *Dipodomys ordii* and *D. merriami* and examined the relationship between speed of locomotion and limb morphology. We were unable to obtain current information about Jerry.

14. Robert M. Hunt, Jr.—B.A., College of Wooster, 1963; M.S., University of New Mexico, 1965; Ph.D., Columbia University, 1971. For his master's research, Bob studied geographic variation in *Peromyscus maniculatus* in New Mexico. Jim Findley's enthusiasm for mammals led Bob to become interested in the paleontology of the group, and at the American Museum of Natural History, he completed his doctorate working on the evolution of mammalian carnivores. This work was done under the direction of Malcolm McKenna and Richard Tedford. From 1970-73, Bob taught human anatomy at Columbia University and in 1973, became Curator of Vertebrate Paleontology at the University of Nebraska in Lincoln. His research interests at Nebraska have focused on the evolution of Cenozoic Carnivora and on the reconstruction of Oligocene and Miocene environments of the Great Plains. Bob has served as Chair of the Government Liaison Committee of the Society of Vertebrate Paleontology and currently as Secretary-Treasurer. As Professor of Geology, he teaches classes in vertebrate paleontology and human origins in the Department of Geology at UNL. Bob's current concerns center on the effort by the Society of Vertebrate Paleontology to protect and conserve fossil vertebrates on Federal lands in the United States that are threatened by commercial fossil businesses.

Hunt, R.M., Jr., X.X. Xue, and J. Kaufman. 1983. Miocene burrows of extinct beardogs: Indication of early denning behavior of large mammalian carnivores. Science, 221:364-366.

Hunt, R.M., Jr. 1990. Taphonomy and sedimentology of Arikaree (lower Miocene) fluvial, eolian, and lacustrine paleoenvironments, Nebraska and Wyoming: A paleobiota entombed in fine-grained volcaniclastic rocks. Geological Society of America Special Paper, 244:69-111.

Hunt, R.M., Jr. 1991. Evolution of the aeluroid Carnivora: Viverrid affinities of the Miocene carnivoran *Hespestides*. American Museum Novitates, 3023:1-34.

Hunt, R.M., Jr. 1992. Death at a 19-million year old waterhole: The bonebed at Agate Fossil Beds National Monument, western Nebraska. Museum Notes, University of Nebraska State Museum, 83:1-6.

Hunt, R.M., Jr., and R. Tedford. 1993. Phylogenetic relationships within the aeluroid Carnivora and implications of their temporal and geographic distribution. Pp. 53-73, *in* Mammal phylogeny (placentals) (F. Szalay, M. Novacek, and M. McKenna, eds.). Springer–Verlag, New York, 321 pp.

15. William Stalheim—B.A., Augustana College, 1960; M.S., University of New Mexico, 1965. For his thesis, Bill studied the natural history of *Spermophilus variegatus* in Embudo Canyon of the Sandia Mountains, New Mexico. We were unable to obtain current information about Bill.

16. Modesto del Castillo—B.S., University of New Mexico, 1964; M.S., University of New Mexico, 1966. For his master's thesis, Modesto studied developmental aspects of the bacula of *Lasiurus cinereus* and *Antrozous pallidus*. After graduation in 1966, he was hired by Elizabethtown Community College of the University of Kentucky. Modesto's interests have changed from vertebrate zoology to microbiology and then to the development of educational computer software. In 1972, he was the recipient of the Great Teacher Award sponsored by the University of Kentucky Alumni Association. He also received an excellence award from the IBM Corporation and League for Innovation in

1988. This grant supported the development of computer software that described mechanisms of viral replication cycles in animals. Modesto has been Professor of Biology since 1978 and presently teaches courses in both biology and microbiology.

17. Jay D. Druecker (see doctoral student #10)

18. John L. Tatschl—B.S., University of New Mexico, 1964; M.S., University of New Mexico, 1966; Ph.D., Kansas State University, 1973. For his master's thesis, John studied the winter and breeding birds of all the major plant communities in the Sandia Mountains east of Albuquerque. John's dissertation, under the direction of John Zimmerman, involved niche separation and competition between three species of flycatching birds (*Tyrannus*) found throughout Kansas. After receiving his doctorate and exploring several career options, John began working for the U.S. Army Corps of Engineers where he spent six years as an animal ecologist in the Planning Division and seven as a wetland specialist in the Operation's Division. After two years with the Federal Aviation Administration as the Environmental Specialist for the Central Region (Iowa, Kansas, Missouri, and Nebraska), John decided that it was time to step out of the environmental arena. John now resides in Kansas and has his own business called Multiple Funding in which he buys accounts receivables from businesses.

Tatschl, J.L. 1967. Breeding birds of the Sandia Mountains and their ecological distributions. Condor, 69:479-490.

Tatschl, J.L. 1968. Unusual nesting site for house sparrows. Auk, 85:514.

Tatschl, J.L. 1974. Flycatchers—where and when they fly. Bird Populations Institute, 2:2-3.

Zimmerman, J.L., and J.L. Tatschl. 1975. Floodplain birds of Weston Bend, Missouri River. Wilson Bulletin, 87:196-206.

19. Keith B. Grisham—B.S., University of New Mexico, 1960; M.S., University of New Mexico, 1967; Ph.D., Tulane University, 1972. Keith studied geographic variation in *Dipodomys ordii* in the upper Rio Grande Valley for his master's thesis. His dissertation dealt with the embryology of hair follicles as it related to seasonal changes in the pelage of *Microtus montanus nanus*. Specifically, he used photoperiod and temperature as variables to study hair replacement in laboratory-reared voles from Wyoming. This research was accomplished under the guidance of Norman C. Negus. Keith recently retired from the Department of Science at Del Norte High School in Albuquerque where he was both Chair and a teacher of biology. In addition to his consuming interest in science education of secondary school students, he also has pursued his interest in the design and construction of solar adobe homes and landscape architecture.

20. Don E. Wilson (see doctoral student #6)

21. Marilyn K. Dunham—B.S., State University of New York, College at Oneonta, 1966; M.S., University of New Mexico, 1968. For her thesis, Marilyn determined the diet of *Dipodomys ordii* and *D. merriami* in New Mexico, including areas in the state where the species are sympatric. We were unable to obtain current information about Marilyn.

22. Neil S. Weber—A.B., Rutgers University, 1966; M.S., University of New Mexico, 1968. Neil studied seasonal changes in fat content of *Eptesicus fuscus* for his master's thesis and immediately after graduation, began his twenty-five year career with the State of New Mexico.

Since 1992, Neil has been Chief of the Department of Energy Environmental Oversight and Monitoring Program with the New Mexico Environment Department. He also has held the following positions with this department and its predecessor agencies: Program Manager of the Insect and Rodent Control Program from 1968-76; District 2 Environmental Manager from 1976-84; Deputy Director of the New Mexico Environmental Improvement Division from 1984-87; and Chief of the Solid Waste Bureau from 1988-92.

Weber, N.S., and J.S. Findley. 1970. Warm-season changes in fat content of *Eptesicus fuscus*. Journal of Mammalogy, 51:160-162.

Weber, N.S. 1970. A key to the rodents, insectivores, and lagomorphs of New Mexico. New Mexico Environmental Improvement Agency Publication, Santa Fe, 24 pp.

von Reyn, C.F., N.S. Weber, A.M. Barnes, T. Quan, and W.J. Dean. 1976. Bubonic plague from direct exposure to a naturally infected coyote. American Journal of Tropical Medicine and Hygiene, 25:626-629.

Weber, N.S. 1977. Plague in New Mexico. New Mexico Environmental Improvement Agency Publication, Santa Fe, 33 pp.

Mann, J.M., N.S. Weber, W.J. Martone, J.M. Boyce, A.F. Kaufman, and A.M. Barnes. 1979.

Endemic human plague in New Mexico: Risk factors associated with infection. Journal of Infectious Diseases, 140:397-401.

23. Daniel F. Williams (see doctoral student #7)

24. G. Michael Schum (see doctoral student # 14)

25. Sandra Husar Ligon—B.A., Knox College, 1971; M.S., University of New Mexico, 1973. For her master's thesis, Sandy looked at the diets of two species of closely related insectivorous bats, *Myotis evotis* and *M. auriculus*, in areas of allopatry and one known location of sympatry. She found evidence of behavioral character displacement when the two species were found together. In 1973, she was hired by Clyde Jones, of the U.S. National Fish and Wildlife Laboratory, to determine the status of the four remaining species of sirenians. She spent one year based at the Smithsonian Institution and then returned to the University of New Mexico to finish her research. In 1975, she married J. David Ligon. They immediately set out for Kenya to begin what became a 10-year study of the green woodhoopoe. In 1980, Sandy was hired by Jim Findley, then department chairman, to be the freshman laboratory coordinator for the department. Sandy still holds this position and enjoys teaching and revising introductory courses for both majors and non-majors.

Husar, S.L. 1975. A review of the literature of the dugong (*Dugong dugon*). U.S. Fish and Wildlife Research Reports, Washington, DC, 4:1-30.

Husar, S.L. 1976. Behavioral character displacement: Evidence of food partitioning in insectivorous bats. Journal of Mammalogy, 57:331-338.

Ligon, S.H. 1976. A survey of dugongs (*Dugong dugon*) in Queensland. Journal of Mammalogy, 57:580-582.

Ligon, J.D., and S.H. Ligon. 1982. The cooperative breeding behavior of the green woodhoopoe. Scientific American, 247:126-134.

Ligon, J.D., S.H. Ligon, and H.A. Ford. 1991. An experimental study of the bases of male philopatry in the cooperatively breeding superb fairy-wren, *Malurus cyaneus*. Ethology, 87:134-148.

26. Terrence M. May—B.A., California State University, Stanislaus, 1972; M.S., University of New Mexico, 1977. As an undergraduate student, Terry was influenced by former Findley student Dan Williams and proceeded to study mammals at UNM. Based upon principal component analysis of morphology, Terry studied community structure of ten southwestern rodent faunas for his master's thesis. Upon graduation, he taught biology at El Paso Community College for several years and instituted an Anatomy and Physiology program at this college. For the next couple of years, Terry worked as a biomechanical engineer, developing both sport and medical products. For the last twelve years, Terry has been involved in sales of scientific instrumentation. Currently, he is stationed in Colorado and is the Western Regional Manger for Hitachi Instruments, Inc.

27. Amy S. Mills—B.A., Miami University, 1972; M.S., University of New Mexico, 1977; M.D., University of New Mexico, 1982. While at UNM, Amy decided to obtain her Master of Science under Plan II, which requires more course work but no thesis. We were unable to obtain current information about Amy.

28. Karen E. Petersen (see doctoral student #23)

29. David W. Reduker—B.S., University of Massachusetts at Amherst, 1977; M.S., University of New Mexico, 1979; Ph.D., University of New Mexico, 1984. For his master's research, Dave compared the masticatory apparatus in two species of *Myotis*, relating jaw mechanics to foraging behavior. Dave continued at UNM for his Ph.D., but his research interests switched to parasitology. His dissertation dealt with evolutionary relationships among eimerian parasites infecting cricetid rodents. This research was completed under the direction of Donald W. Duszynski and Terry L. Yates. After graduation, Dave spent two years as a postdoctoral fellow in the Veterinary Research Laboratory at Montana State University (1984-86) and another two years as a fellow in the Department of Veterinary Microbiology and Pathology at Washington State University (1986-88). He then worked as a Research Associate for the USDA-Agricultural Research Service for a year before being hired by the Department of Pathology at Colorado State University in 1989. At this institution, Dave was an Assistant Professor until his untimely death on October 25, 1990.

Reduker, D.W. 1983. Functional analysis of the masticatory apparatus in two species of *Myotis*. Journal of Mammalogy, 64:277-286.

Reduker, D.W., and D.W. Duszynski. 1985. *Eimeria ladronensis* n.sp. and *E. albigulae* (Apicomplexa; Eimeriidae) from the woodrat, *Neotoma albigula* (Rodentia; Cricetidae). Journal of Protozoology, 32:548-550.

Reduker, D.W., L. Hertel, and D.W. Duszynski. 1985. *Eimeria* species (Apicomplexa: Eimeriidae) infecting *Peromyscus* rodents in the southwestern United States and northern Mexico with description of a new species. Journal of Parasitology, 71:604-613.

Reduker, D.W., D.W. Duszynski, and T.L. Yates. 1987. Evolutionary relationships among *Eimeria* spp. (Apicomplexa) infecting cricetid rodents. Canadian Journal of Zoology, 65:722-735.

Reduker, D.W., D.J. Jasmer, W.L. Goff, L.E. Perryman, W.C. Davis, and T.C. McGuire. 1989. A recombinant surface protein of *Babesia bovis* elicits bovine antibodies that react with live merozoites. Molecular and Biochemical Parasitology, 35:239-248.

30. Charles C. Reith—B.A., University of California at Santa Barbara, 1977; M.S., University of New Mexico, 1980; Ph.D., University of New Mexico, 1984. For his master's research, Charlie studied the use of space by *Myotis yumanensis* as they foraged in various habitats accessible to their diurnal roost. His doctoral research was accomplished under the supervision of Loren D. Potter and dealt with the revegetation and stabilization of land disturbed by surface coal mining in the San Juan Basin of New Mexico. After completing his doctorate, Charlie worked for a number of environmental consulting firms, performing environmental baseline studies and managing hazardous waste cleanups in the United States and overseas. He is currently vice president of a Gulf Coast oil company and oversees programs for environmental protection, health and safety, fire prevention, and emergency readiness.

Reith, C.C. 1980. Shifts in times of activity by *Lasionycteris noctivagans*. Journal of Mammalogy, 61:104-108.

Reith, C.C. 1982. Insectivorous bats fly in shadows to avoid moonlight. Journal of Mammalogy, 63:685-688.

Reith, C.C., and L.D. Potter (eds.). 1986. Principles and methods of reclamation science with case studies from the arid southwest. University of New Mexico Press, Albuquerque, 224 pp.

Reith, C.C., and B.M. Thomson (eds.). 1990. Deserts as dumps? The disposal of hazardous materials in arid ecosystems. University of New Mexico Press, Albuquerque, 330 pp.

Caldwell, J.A., and C.C. Reith. 1992. Principles and practice of waste encapsulation. Lewis Press, Boca Raton, Florida, 390 pp.

31. Kerry S. Kilburn—B.S., California State University, Stanislaus, 1982; M.S., University of New Mexico, 1984; Ph.D., University of New Mexico, 1988. Kerry became interested in mammalogy as an undergraduate student working with Dan Williams. For her master's research,

Kerry used morphological differences among three species of *Sorex* to generate predictions about their ecological relationships. This work stimulated her interest in functional morphology. Her dissertation, directed by J. Scott Altenbach, involved comparing forelimb morphology between two sympatric species of *Dipodomys* and using that information to test hypotheses about habitat partitioning. Since 1988, Kerry has worked as Assistant Professor of Biology at West Virginia State College. Her primary responsibility has been undergraduate instruction; in addition to reviving the zoology curriculum, she is active in efforts to improve science instruction in courses for non-majors. Kerry also works to stimulate interest in natural history by taking her "travelling snake and bat show" to up to a dozen elementary schools and community groups each year.

Williams, D.F., and K.S. Kilburn. 1984. Sensitive, threatened, and endangered mammals of riparian and other wetland communities in California. Pp. 950-956, *in* California riparian systems ecology, conservation, and productive management (R.E. Warner and K.M. Hendrix, eds.). University of California Press, Berkeley, 1034 pp.

Williams, D.F., and K.S. Kilburn. 1991. *Dipodomys ingens*. Mammalian Species, 377:1-7.

Kilburn, K.S., and T. Hutto. 1992. Process-oriented laboratory activities for introductory biology. D.C. Heath and Company, Lexington, Massachusetts, 39 pp.

Williams, D.F., and K.S. Kilburn. 1992. The conservation status of the endemic mammals of the San Joaquin faunal region, California. Pp. 329-345, *in* Endangered and sensitive species of the San Joaquin Valley, California: Their biology, management, and conservation (D.F. Williams, S. Byrne, and T.A. Rado, eds.). California Energy Commission, Sacramento. 388 pp.

32. José I. Sanchez-Vasquez—B.A., Universidad Nacional Autónoma de México, 1972; M.S., University of New Mexico, 1984. While attending UNM, Ignacio decided to obtain his Master of Science by the non-thesis route. Presently, Ignacio is employed as a biologist at the Centro de Investigaciones Ecologicas del Sureste in San Cristóbal de las Casas, Chiapas.

33. Randy D. Jennings—B.A., University of South Florida, 1977; M.S., University of New Mexico, 1985; Ph.D., University of New Mexico, 1991. For his master's thesis, Randy studied relationships among populations of the desert tortoise, *Gopherus agassizii*, throughout its range using protein electrophoresis and morphology (Thomas H. Fritts and James S. Findley served as co-advisors). Randy's dissertation focused on the influence of the environment on larval morphology of the Chiricahua leopard frog, *Rana chiricahuensis,* and assessed the potential for environmentally induced traits to be adaptive (Norman J. Scott, Jr. and Howard L. Snell were co-advisors with Jim serving as a committee member). Much of Randy's interest remains focused on the natural history, ecomorphology, and population structure of anurans in the deserts of the Southwest. Outside this region, he has been associated with studies of the herpetofauna of the Lago de Yojoa region in Honduras and studies of ecological and morphological correlates of locomotory performance and wariness in the Galapagos lava lizard, *Tropidurus albemarlensis*. In 1991, Randy was hired as Curator of Herpetology at the Barrick Museum of Natural History at the University of Nevada in Las Vegas. In 1993,

Randy became an Assistant Professor of Biology at Western New Mexico University in Silver City.

Jennings, R.D. 1982. Pelagic sightings of Risso's dolphin, *Grampus griseus*, in the Gulf of Mexico and Atlantic Ocean adjacent to Florida. Journal of Mammalogy, 63:522-523.

Jennings, R.D. 1985. Seasonal abundance of hammerhead sharks off Cape Canaveral, Florida, USA. Copeia, 1985:223-225.

Scott, N.J., Jr., and R.D. Jennings. 1985. The tadpoles of five species of New Mexican leopard frogs. Occasional Papers, Museum of Southwestern Biology, 3:1-21.

Snell, H.L., R.D. Jennings, and H.M. Snell. 1988. Intrapopulation variation in predator-avoidance performance: Relationships between vegetation, sex, wariness, and sprint speed in the Galapagos lava lizards. Evolutionary Ecology, 2:353-369.

Jennings, R.D., and N.J. Scott, Jr. 1993. Ecologically correlated morphological variation in tadpoles of the leopard frog, *Rana chiricahuensis*. Journal of Herpetology, 27:285-293.

34. Arnold T. Gerstell—B.A., University of California at Santa Cruz, 1984; M.S., University of New Mexico, 1988. For his master's research, Tay studied the coexistence of two sympatric raptors during a decline of their prey. We were unable to obtain current information about Tay.

Kenneth N. Geluso, Department of Biology, University of Nebraska, Omaha, Nebraska 68182, and Don E. Wilson, Smithsonian Institution, National Museum of Natural History, Washington, DC 20560

Annotated Bibliography of James Smith Findley

William L. Gannon and Don E. Wilson

Abstract

In his first 50 years as a mammalogist, Jim Findley published 99 papers in 24 journals, nine books or book chapters, and assorted popular articles, including a description of mammals for *Encyclopedia Britannica*. From simple observations of natural history ("The Interesting Fate of a Flying Squirrel") to insightful, paradigm-making papers ("Phenetic Packing as a Measure of Faunal Diversity"), Jim's work covers many taxa and many biological disciplines (mammalogy, ornithology, geology, paleontology, herpetology, biogeography, etc.). His first paper was in *The Auk*, but he proceeded to publish half of his papers in the *Journal of Mammalogy*. Jim Findley epitomizes his own favorite saying, "If you want it done right, ask a mammalogist!" Jim did do it right and is an example of one man's way to combine insightful observations with critical thinking. Although we have added comments regarding the circumstance of particular citations, his publications continue to speak for themselves. In fact, since his retirement in 1992, Jim has contributed six papers and published a book, *Bats: A Community Perspective*, for Cambridge Press.

INTRODUCTION

Born in Cleveland, Ohio, Jim published his first two papers as a teenager based on observations made while he was still in high school Fig. 1. His second paper, published in late 1945, was the first of 49 contributions he made to the *Journal of Mammalogy*. Many naturalists have made notes about natural history or animal behavior during their careers, but Jim's "A Record of Moose Speed" has been a continual reminder that even the simplest observations are notable.

Jim's education was a varied one. He finished high school (1946) at Kobe Central School in Japan. Jim studied at the Rocky Mountain Biological Laboratory in Colorado (1947), and graduated *cum laude* from Western Reserve University in 1950 with his BA degree. In 1951, he moved to the University of Kansas (KU) in Lawrence and began his doctorate program. That same year Jim's "Habitat Preferences of Four Species of *Microtus* in Jackson Hole, Wyoming" was one of the earliest examinations of niche partitioning in mammals. In 1953, Findley began his first of many collaborations with colleagues. His first co-authored paper was with a boyhood friend, Norm Negus, on mammals from Gunnison County, Colorado. Dr. Negus currently is a professor at the University of Utah. Rollin "Hoot" Baker was a professor at the University of Kansas while Jim was a graduate student. Together they wrote on the vagrant shrew, a topic related to Jim's doctoral dissertation. Following this paper was Jim's first contribution to paleontology on Pleistocene soricids.

Jim published 17 papers on a number of different topics during the time he was at KU. He published his first of five papers on South Dakota mammals, wrote of the interaction of birds and mammals in, "Utilization of a Bird Nest by the Meadow Vole," and penned the first papers of a series of contributions on shrews (he has published more than 14 on shrews with one more in press), and on the bat genus *Myotis* (the first of 10 papers). During this time he published an early paper on the influence of competition on mammalian systems. Several notable collaborations also began at this time. He co-authored with Dick Van Gelder, who was a postdoctoral research assistant to E.R. Hall at KU, and went on to chair the Mammalogy Department at the American Museum of Natural History in New York. In addition, Findley collaborated with Syd Anderson, another of his fellow graduate stu-

Life Among the Muses: Papers in Honor of James S. Findley. Edited by T.L. Yates, W.L. Gannon & D.E. Wilson.

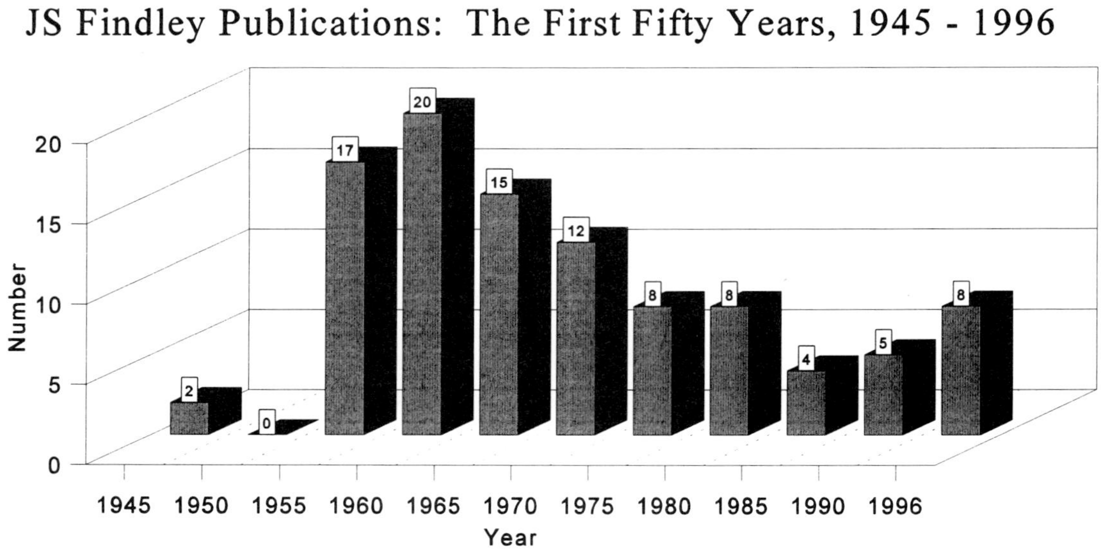

Fig. 1. Number of Findley publications by year.

dents at KU. Syd, who recently retired from the American Museum, is currently a research associate of the MSB and has the "Mammals of Bolivia," a 20-year scientific and literary effort, in press.

Findley was hired by the Department of Biology at the University of New Mexico in 1955, a department he stayed with and helped to develop over the next 37 years until his retirement in 1992. In 1956, he published three papers on South Dakota mammals that resulted from work done while teaching at the University of South Dakota (1954–55). These papers included his first contribution to the *Journal of Wildlife Management* and a complete regional survey ("Mammals of Clay County" [South Dakota]). He also published on a spade-foot toad with another of his University of Kansas fellow graduate students, Pete Chrapliwy. Finally, that same year, Jim published his first paper in the *American Midland Naturalist* and the first of 34 papers on New Mexico natural history ("Merriam Shrew in New Mexico").

During the next few years (1956–59), Jim worked to establish himself at the University of New Mexico (UNM). He published papers with several UNM students in those early years (dwarf shrews with John Poorbaugh, and harvest mice with Barry Pullen). He also completed some earlier work done with friend Norm Negus compiled during summer vacations and work in Wyoming. He published a paper with his first masters-level graduate student, Jim Sands (see *Academic Offspring*, this volume for biographies on Findley students). The next year (1960), Jim described geographic variation in the yellow-nosed cotton rat with his second doctoral student and longtime friend, Clyde Jones. Dr. Jones currently is a Professor of Biology at Texas Tech University.

In the mid-1960s (1962–68), Findley coauthored 12 papers mostly with his students (nine papers with Clyde Jones, one with Art Harris—Findley's third doctoral student—and one with Dan Williams—his seventh doctoral student). During this time he also wrote the first (of 10) publication on geographic variation in mammals, one (of eight) on new mammal records to the New Mexico fauna, the first (of seven) contributions to the *Southwestern Naturalist*, a revised survey of the mammal collections of North America, the first (of three) paper on fossil mammals from cave deposits, the first documentation of migratory behavior of a North American bat (*Lasiurus*), one (of three) publication on coat color in mammals, and the first (of 10) publication on the systematics of southwestern U.S. mammals. In true Jim Findley style, he continued to write not only on mammals but also on toads (one of three papers about amphibians), snakes, and invertebrates. Several reviews and important reference works also were published (one was a chapter on Insectivores and Dermopterans in *Recent Mammals of the World*).

Beginning in 1969 and continuing through much of the 1970s, Jim's interest in bats expanded to include research on ecomorphology (seven papers), physiology, phenetics, and community structure. During this time, Findley co-authored his first (of 10) paper with Don Wilson (Findley's sixth doctoral student) and published four papers as a result of National Science Foundation support to examine the genus *Myotis*. He published his first paper (of four) in the *American Naturalist*, and completed several publications as a result of teaching a field course for the Organization of Tropical Studies in Costa Rica. Several hallmarks stand out from this period. A seminal paper on ecomorphology, "Phenetic Packing as a Measure of Faunal Diversity," was published in 1973. In 1975, "The Mammals of New Mexico," co-authored with former students Harris, Wilson, and Jones, was published as a culmination of their joint studies of New Mexico mammals. "The Structure of Bat Communities," published in 1976, was a major contribution on community structure in bats. These three works constituted foundations for studies on mammals for the next two decades.

In the late 1970s and early 1980s, Jim published with Mike Bogan (his thirteenth doctoral student), Harry Setzer (a fellow KU graduate and long-time curator at the National Museum of Natural History), John Whittaker (a colleague from Indiana State University), and Hal Black (his eighth doctoral student). One wonderful example of Findley's eclectic interests was a paper published in *Science* (1977) on "Comparisons of Frogs, Humans, and Chimpanzees." He pointed out that estimates of evolutionary rates among taxa ought to include a diversity of phenotypic traits as well as genotypic traits. A discussion ensued in that journal in which other authors countered and insisted that there were inconsistencies in the results of numerical taxonomy and proposed that protein sequences might be more reliable in determining mechanisms and timing of evolutionary divergences.

In the mid-1980s, Jim published (with Don Wilson) a cogent summary of his interests on the interplay of ecology and morphology in Tom Kunz's book, *The Ecology of Bats*. The next year (1983), he published (also with Wilson) a continuation of his interest in bat community structure and zoogeography in "Are Bats Rare in Tropical Africa?" This interest in community structure continued with a study of coral reef fishes co-authored with his wife (married 1949), Tommie Findley; an interest he continues today. From Africa to tropical islands and back to New Mexico, where Jim published an updated but abbreviated volume of *Mammals of New Mexico* that included much more natural history (*The Natural History of New Mexican Mammals*, 1987). He also published another paper on rodent community structure in southwestern North America. In 1989, Jim and fellow KU graduate student J. Knox Jones, Jr., wrote an obituary about their former major adviser E.R. Hall. They described Hall as "a farmer, trapper, and naturalist at heart, and a prodigiously successful scientist by profession." As an academic farmer, Hall produced a good crop of young mammalogists properly trained and prepared to go forth as leaders in their discipline. Jim is one of those leaders.

Jim's retirement came in June 1992. The next year, he published a book through Cambridge University Press that was a synthesis of his views on community structure of bats. This book summarizes Jim's style: it is precise, detailed, and highly organized, but overall brief—only 167 pages. Also he published (with colleagues Luis Herrera and Ted Fleming from the University of Miami) an interesting paper on the differential foraging behavior of Pallid bats (*Antrozous pallidus*) as determined from the carbon content in their tissues, specifically toes that were carefully clipped from museum specimens. In 1996, he has at least six papers that either have been published or are submitted for publication. Several papers co-authored with Gordon Kirkland (at Shippensburg University) were on shrew community structure and fossil remains. Others were an update of Jim's views on southwestern North American biogeography, a book review, and a colleague's obituary. Perhaps this obituary for fellow KU doctoral graduate student, J. Knox Jones, Jr. is the most poignant of any paper Jim has written (co-authored by fellow University of Kansas graduate students).

Jim is still actively writing, and he has begun a revision of *Mammals of New Mexico* (1975) with colleagues Terry Yates, Jennifer Frey, and William Gannon (his last doctoral student). Richard Forbes (his sixth master's student) is contributing color plates of all New Mexican mammals for this revision. From a beautiful home in Corrales, New Mexico, Jim takes advantage of the solitude and the opportunity for reflection to ponder the complexities of science. His golden years also include the activities of four children and seven grandchildren, and the successful art career of his wife, Tommie. He is probably busier now than when he had his hands full of students and Administration at UNM. Even so, he continues to make impressive and important contributions to the fields of ecology, morphology, and natural history. The fields of Corrales and the fields of science provide him the balance and perspective of a full life. The Muses are at work for Jim just as they were fifty years ago.

BIBLIOGRAPHY

1945

1. Western Grebe in Michigan. The Auk, 62:312.
2. The interesting fate of a flying squirrel. Journal of Mammalogy, 26:437.

1951

3. A record of moose speed. Journal of Mammalogy, 32:116.
4. A tame red fox cub. Journal of Mammalogy, 32:117.
5. Habitat preferences of four species of *Microtus* in Jackson Hole, Wyoming. Journal of Mammalogy, 32:118-120.

1953

6. (and N.C. Negus) Notes on the mammals of the Gothic region, Gunnison County, Colorado. Journal of Mammalogy, 34:235-239.
7. (and R.H. Baker) *Sorex vagrans* first reported from South Dakota. Journal of Mammalogy, 34:382.
8. Pleistocene Soricidae from San Josecito Cave, Nuevo Leon, Mexico. University of Kansas Publications, Museum of Natural History, 5:635-639.

1954

9. (and R.J. Russell) Swimming ability of the Grasshopper mouse. Journal of Mammalogy, 35:118-119.
10. (and J.K. Jones, Jr.) Geographic distribution of the Short-tailed shrew, *Blarina brevicauda*, in the Great Plains. Transactions of the Kansas Academy of Science, 57:208-211.
11. (and J.K. Jones, Jr. and T.A. Vaughan) Records of mammals from southwestern Iowa. Transactions of the Kansas Academy of Science, 57:212-213.
12. Tree roosting of the Eastern Pipistrelle. Journal of Mammalogy, 35:433.
13. Competition as a possible limiting factor in the distribution of *Microtus*. Ecology, 35:418-420.
14. Reproduction in two species of *Myotis* in Jackson Hole, Wyoming. Journal of Mammalogy, 35:434.
15. (and R.H. Baker) Mammals from southeastern Alaska. University of Kansas Publications, Museum of Natural History, 7:475-477.

1955

16. Taxonomy and distribution of some American shrews. University of Kansas Publications, Museum of Natural History, 7:613-618.
17. Speciation of the Wandering shrew. University of Kansas Publications, Museum of Natural History, 9:1-68 (doctoral dissertation).

18. (and R.G. Van Gelder) Short-tailed shrew in South Dakota. Journal of Mammalogy, 36:452.

19. Birds of Washington State (review). Wilson Bulletin, 67:230-231.

1956

20. (and S. Anderson) Zoogeography of the montane mammals of Colorado. Journal of Mammalogy, 37:80-82.

21. Utilization of a bird's nest by the Meadow vole. Journal of Mammalogy, 37:120.

22. Distribution of some South Dakotan mammals. Chicago Academy Sciences, Natural History Miscellanea, 155:1-2.

23. Comments on the winter food of red foxes in eastern South Dakota. The Journal of Wildlife Management, 20:216.

24. Mammals of Clay County South Dakota. University of South Dakota Publications in Biology, 1:1-45.

25. (and P.S. Chrapliwy) Records of the Central Plains Spadefoot toad, *Scaphiopus bombifrons* Cope, in South Dakota and Wyoming. Herpetologica, 12:124.

26. Possible Himalayan color pattern in the Masked shrew. Journal of Mammalogy, 37:275-276.

27. Merriam shrew in New Mexico. Journal of Mammalogy, 37:277.

28. (and J.K. Jones, Jr.) Molt of the Short-tailed shrew, *Blarina brevicauda*. The American Midland Naturalist, 56:246-249.

29. (and R.H. Baker) Dwarf shrew first reported in South Dakota. Journal of Mammalogy, 37:543.

1957

30. (and J. Poorbaugh) Another Dwarf shrew from New Mexico. Journal of Mammalogy, 38:513.

31. The Hog-nosed bat in New Mexico. Journal of Mammalogy, 38:513-514.

1958

32. (and B.E. Pullen) The Fulvous Harvest mouse in New Mexico. Journal of Mammalogy, 39:306-307.

1959

33. The Bailey Pocket mouse in New Mexico. Journal of Mammalogy, 40:145.

34. (and N.C. Negus) Mammals of Jackson Hole, Wyoming. Journal of Mammalogy, 40:371-381.

35. A new station for the Sacramento Mountain salamander in New Mexico. The Southwestern Naturalist, 4:155.

36. (and J.L. Sands) The relationships of two subspecies of pocket gopher in central New Mexico. Journal of Mammalogy, 40:331-337.

1960

37. Identity of the Long-eared *Myotis* of the Southwest and Mexico. Journal of Mammalogy, 41:16-20.

38. The Fox squirrel in New Mexico. Journal of Mammalogy, 41:403.

39. (and C.J. Jones) Geographic variation in the Yellow-nosed Cotton rat. Journal of Mammalogy, 41:462-469.

1961

40. Geographic variation in New Mexican chickarees. Journal of Mammalogy, 42:313-322.

41. (and C.J. Jones) New United States record of the Mexican Big-eared bat. Journal of Mammalogy, 42:97.

1962

42. (with A.H. Harris) Status of *Myotis lucifugus phasma* and comments on variation in *Myotis yumanensis*. Journal of Mammalogy, 43:192-199.

43. (and C.J. Jones) Distribution and variation of voles of the genus *Microtus* in New Mexico and adjacent areas. Journal of Mammalogy, 43:154-166.

1963

44. (and C.J. Jones) Geographic variation in the Least Cotton rat in New Mexico. Journal of Mammalogy, 44:307-315.

45. (and C. Jones) The Long-nosed bat in New Mexico. The Southwestern Naturalist, 8:174-175.

46. (and C. Jones) Second record of the Lyre snake *Trimorphodon vilkinsoni* in New Mexico. The Southwestern Naturalist, 8:175-177.

47. (and S. Anderson and J.K. Doutt) Collections of mammals in North America. Journal of Mammalogy, 44:471-500.

1964

48. Verification of the occurrence of *Bufo microscaphus* Cope in New Mexico. The Southwestern Naturalist, 9:107.

49. (and A.H. Harris) Pleistocene-Recent fauna of the Isleta Caves, Bernalillo County, New Mexico. American Journal of Science. 262:114-120.

50. (and C. Jones) Seasonal distribution of the Hoary bat. Journal of Mammalogy, 45:461-470.

1965

51. (and C. Jones) Northernmost records of some neotropical bat genera. Journal of Mammalogy, 46:330-331.

52. (and C. Jones) Comments on spotted bats. Journal of Mammalogy, 46:679-680.

53. An inventory of the mammals. (Review) Ecology, 46:759-760.

54. Shrews from Hermit Cave, Guadalupe Mountains, New Mexico. Journal of Mammalogy, 46:206-210.

1966

55. Northernmost limit of the River crab, *Pseudothelphusa* (Decapoda), in western Mexico. The Southwestern Naturalist, 11:413.

1967

56. A black population of the Goldman Pocket mouse. The Southwestern Naturalist, 12:191-192.

57. Insectivores and dermopterans. Pp. 87-109, *in* Recent Mammals of the World; A Synopsis of Families (S. Anderson and J.K. Jones, Jr., eds.). Ronald Press, New York, 450 pp.

58. (and C. Jones) Taxonomic relationships of bats of the species *Myotis fortidens*, *M. lucifugus*, and *M. occultus*. Journal of Mammalogy, 48:429-444.

59. (and C.J. Jones) New United States record of the Mexican Big-eared bat. Journal of Mammalogy, 42:97.

1968

60. (and D.F. Williams) The Plains Pocket mouse (*Perognathus flavescens*) in New Mexico. Journal of Mammalogy, 49:771.

1969

61. Brain size in bats. Journal of Mammalogy, 50:340-344.

62. Biogeography of Southwestern boreal and desert mammals. Pp. 113-130, *in* Contributions in Mammalogy; A Volume Honoring Professor E. Raymond Hall (J.K. Jones, Jr., ed.). University of Kansas Museum of Natural History, Miscellaneous Publication No. 51, University of Kansas Press, Lawrence, Kansas. 428 pp.

1970

63. (and N.S. Weber) Warm-season changes in fat content of *Eptesicus fuscus*. Journal of Mammalogy, 51:160-162.

64. (and D.E. Wilson) Reproductive cycle of a neotropical insectivorous bat, *Myotis nigricans*. Nature, 225:1155.

65. Phenetic relationships in the genus *Myotis*. Bijdragen tot de Dierkunde (Proceedings of the Second International Bat Research Conference), 40:26-29, Amsterdam.

66. (and G.L. Traut) Geographic variation in *Pipistrellus hesperus*. Journal of Mammalogy, 51:741-765.

1971

67. (and D.E. Wilson) Spermatogenesis in neotropical species of *Myotis*. Journal of Mammalogy, 52:420-426.

1972

68. (and D.E. Wilson) Randomness in bat homing. The American Naturalist, 106:418-424.

69. Phenetic relationships among bats of the genus *Myotis*. Systematic Zoology, 21:31-52.

70. (and D.E. Wilson and E.H. Studier) Morphologic properties of bat wings. Journal of Mammalogy, 53:429-444.

1973

71. Phenetic packing as a measure of faunal diversity. The American Naturalist, 107:580-584.

1974

72. Insectivora. Pp. 622-628, *in* Encyclopedia Britannica. 15th edition.

73. (and D.E. Wilson) Observations on the neotropical Disk-winged bat, *Thyroptera tricolor* Spix. Journal of Mammalogy, 55:562-571.

1975

74. (and A.H. Harris, D.E. Wilson, and C. Jones) Mammals of New Mexico. University of New Mexico Press, Albuquerque, New Mexico, 360 pp.

1976

75. Mammals of the World (review). Journal of Mammalogy, 57:420-421.

76. The structure of bat communities. The American Naturalist, 110:129-139.

1977

77. (and W. Caire) The status of mammals in the northern region of the Chihuahuan Desert. Pp. 127-139, *in* Transactions of the Symposium on the biological resources of the Chihuahuan Desert region United States and Mexico (R.H. Wauer and D.H. Riskind, eds.). National Park Service Transactions and Proceedings Series, Number Three, 658 pp.

78. (and D.E. Wilson) *Thyroptera tricolor*. Mammalian Species, 71:1-3.

1978

79. (and M.A. Bogan, H.W. Setzer, and D.E. Wilson) Phenetics of *Myotis blythi* in Morocco. Pp. 217-230, *in* Proceedings of the Fourth International Bat Research Conference, Kenya National Academy Advancement of Arts and Science, 354 pp.

1979

80. Comparisons of frogs, humans, and chimpanzees. Science, 204(4391):434-435.

81. (and D.F. Williams) Sexual size dimorphism in vespertilionid bats. The American Midland Naturalist, 102:113-126.

1980

82. (and J.O. Whitaker, Jr.) Foods eaten by some bats from Costa Rica and Panama. Journal of Mammalogy, 61:540-544.

1982

83. (and D.E. Wilson) Ecological significance of chiropteran morphology. Pp. 243-260, *in* Ecology of Bats (T.H. Kunz, ed.). Plenum Publishing Corporation. 533 pp.

1983

84. (and H. L. Black) Morphological and dietary structuring of a Zambian insectivorous bat community. Ecology, 64:625-630.

85. (and D.E. Wilson) Are bats rare in tropical Africa? Biotropica 15:299-303.

1985

86. (and M.T. Findley) A search for pattern in butterfly fish communities. The American Naturalist, 126:800-816.

1987

87. The natural history of New Mexican mammals. University of New Mexico Press, Albuquerque, New Mexico, 164 pp.

88. The Corrales Bosque Preserve. Village of Corrales. Cottonwood Press, 9 pp (not peer reviewed).

1989

89. Morphological patterns in rodent communities of southwestern North America. Pp. 253-263, *in* Patterns in the Structure of Mammalian Communities (D.W. Morris, Z. Abramsky, B.J. Fox, and M.R. Willig, eds.). Special Publication No. 28, The Museum, Texas Tech University Press, Lubbock, Texas. 266 pp.

90. (and M.T. Findley) Circumtropical patterns in butterflyfish communities. Environmental Biology of Fishes, 25:33-46.

91. (and J. Knox Jones, Jr.). Eugene Raymond Hall: 1902–1986 (Obituary). Journal of Mammalogy, 70:455-458.

1993

92. Bats: A community perspective. Cambridge University Press, Cambridge, 167 pp.

93. (and L.G. Herrera M., T.H. Fleming). Geographic variation in carbon composition of the Pallid bat, *Antrozous pallidus*, and its dietary implications. Journal of Mammalogy, 74:601-606.

1995

94. (and T.H. Kunz, J. Gaisler, K.F. Koopman, P.A. Racey, G. Marimuthu, M.L. Augee, and G.R. Horst). International bat research conferences: Historical glimpses. Bat Research News, 36:132-136.

1996

95. Seventy-five years of mammalogy (1919–1994) (Review). Journal of Mammalogy, 77:282-286.

96. (and C. Jones, H.H. Genoways, E.C. Birney, and R.J. Baker). J. Knox Jones, Jr.: 1929–1992. (Obituary). Journal of Mammalogy, 77:578-593.

97. Mammalian biogeography in the American Southwest. Pp. 297-308, *in Festschrift* for J. Knox Jones, Jr. (H.H. Genoways and R.J. Baker, eds.). Texas Tech University Press, Lubbock. 266 pp.

98. (and G.L. Kirkland, Jr.). 1996. First Holocene record for Preble's shrew (*Sorex preblei*) in New Mexico. The Southwestern Naturalist, 41 (3):320-322.

99. (and G.L. Kirkland, Jr.). In Preparation. Transcontinental comparisons of forest small mammal communities: North New Mexican mountains and southern Pennsylvania compared. Pp. in (P. Meserve and P. Marquet, eds.), Large spatial and temporal scales in mammalian ecology: perspectives from the Americas. Proceedings from the Seventh International Theriological Congress, Acapulco, MX.

William L. Gannon, Museum of Southwestern Biology, University of New Mexico, Albuquerque, New Mexico, 87131, and Don E. Wilson, Smithsonian Institution, National Museum of Natural History, Washington, DC 20560

Biogeography of Baja California Peninsular Desert Mammals

David J. Hafner and Brett R. Riddle

Abstract

The distinctiveness of the desert-adapted mammalian fauna of the Baja California peninsula, mirrored in the other biota of the region, indicates a long evolutionary history of the Peninsular Desert separate from that of the Sonoran regional desert. The independent evolutionary trajectory of some elements of the Peninsular biota may have begun as early as the middle Miocene separation of the peninsula from the Sonoran–Sinaloan mainland. During the late Miocene and early Pliocene formation of regional semideserts, the emergent peninsula was isolated from adjacent semideserts by marine transgressions. More recently, Peninsular Desert taxa were isolated in one or more desert refugia during late Pleistocene full-pluvial intervals. A biogeographic model is developed that reflects alternating periods of isolation and secondary contact, dispersal, and range shifts for Peninsular Desert mammals. Combined evidence from fossil records and phylogeographic patterns of mtDNA or allozyme complements of four taxa of arid-adapted rodents (*Ammospermophilus, Chaetodipus, Neotoma,* and *Peromyscus* [*Haplomylomys*]) indicate both early (Miocene) and more recent (Pliocene-Pleistocene) periods of isolation of Peninsular Desert mammals.

INTRODUCTION

Studies of North American deserts traditionally have followed the initial description of regional desert floras by Shreve (1942), who defined regions based on a combination of plant species composition, climate, and topography. Axelrod (1979, 1983), studying the evolution of regional deserts, defined their climatic parameters in more detail but retained the mostly phenetic classification of Shreve. Four major regional deserts have been recognized (Fig. 1): two warm deserts (Sonoran and Chihuahuan), the cool Mojave Desert, and the cold Great Basin Desert. In addition, two smaller, isolated regional deserts are occasionally included: the San Joaquin (southern Central Valley of California; Axelrod, 1967) and the Hidalgan (Hidalgo, Guerrero, and Puebla, Mexico; Axelrod, 1979; Shreve, 1940).

Axelrod (1979, 1983) emphasized that components of the North American regional deserts may be quite ancient, even though desert ecosystems are extremely young. According to his model, desert floras (and faunas, presumably) developed in local dry sites during the Tertiary drying trend, drawing arid-adapted species from surrounding communities. Semideserts attained their maximum area during the early Pliocene (3-5 million years ago, following time scale of Woodburne, 1987) and were reduced in area during the moist late Pliocene and during Pleistocene pluvial intervals. Full regional deserts formed during interglacials, and reached their maximum extent following the Wisconsinan glacial during the warm interval variously referred to as the altithermal, xerothermic, hypsithermal, or climatic optimum. Following Axelrod's (1966) scenario, sister taxa in alternate regional deserts might be of Pliocene age, having developed *in situ* along with regional semideserts. Alternatively, regional sister-taxa may have been isolated during more recent Pleistocene pluvial periods in desert refugia. Finally, sister-taxa may be only geographic races of post-Wisconsinan immigrants into adjacent regional deserts.

Life Among the Muses: Papers in Honor of James S. Findley. Edited by T.L. Yates, W.L. Gannon & D.E. Wilson.

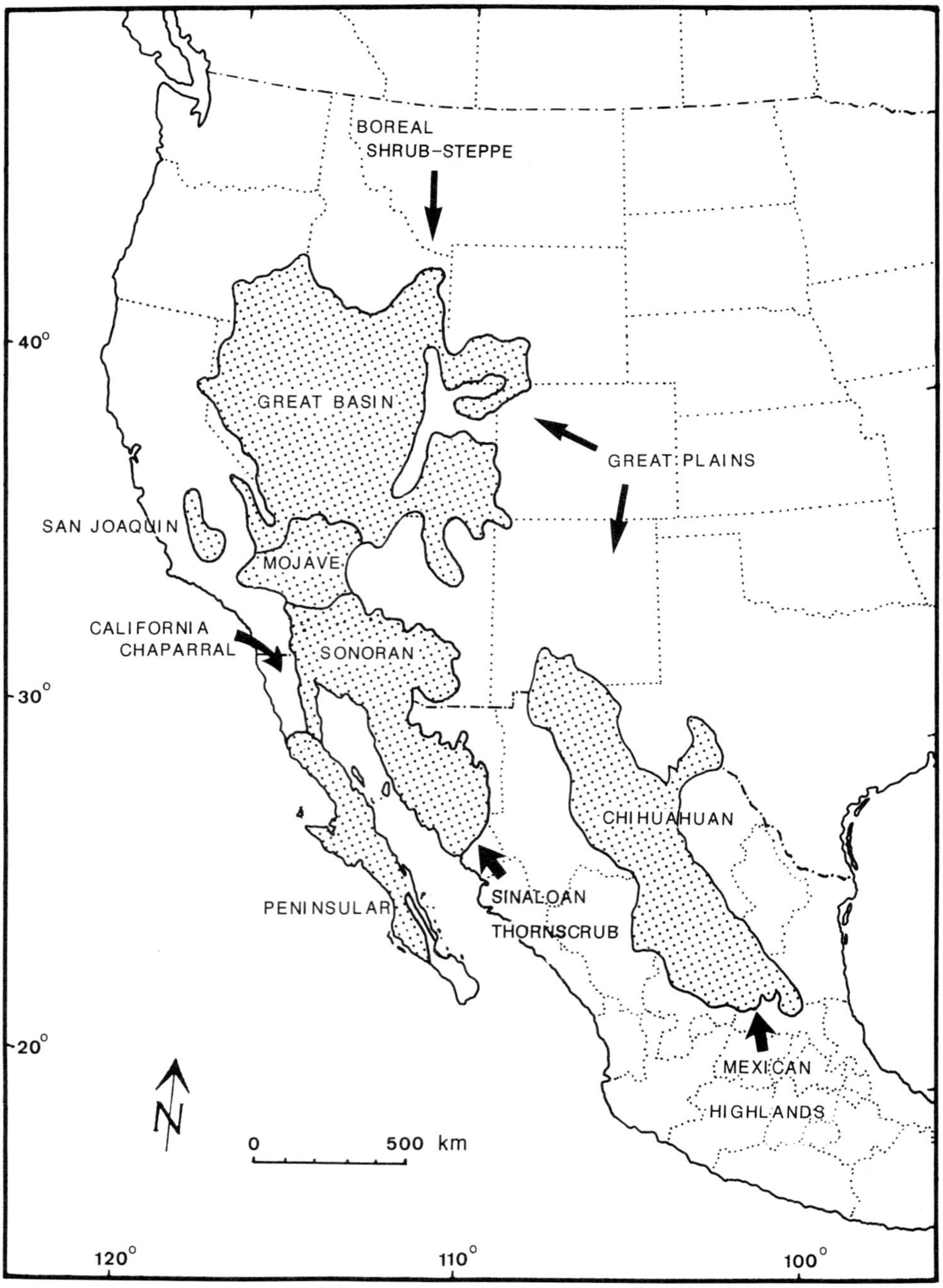

Fig. 1. North American regional deserts (stippled) and neighboring physiographic regions discussed in text. The Peninsular Desert previously has been considered as a subset of the Sonoran Desert.

Despite the potentially great antiquity of vertebrate lineages of the regional deserts, most researchers have assumed that divergence of regional desert lineages was coincident with glacial-interglacial cycles of the late Pleistocene: e.g., Findley (1969), Orr (1960), and Schmidly et al. (1994) for mammals; Hubbard (1974) for birds; and Savage (1960) for reptiles. Murphy (1983a,b) and Grismer (1994) expanded models for the evolution of the reptilian fauna of the Baja California peninsula to include all of the Neogene. They emphasized the impact of Miocene and Pliocene vicariant events and deemphasized Pleistocene effects on the reptiles of Baja California. Only Riddle (1995) has considered the role of Miocene and Pliocene events in lineage divergence of arid-adapted mammals. He concluded that "correspondence between molecular, distributional, paleoecological, and geological evidence suggests that divergence and current distributions of most extant lineages can be attributed to late Tertiary–early Quaternary development of the western North American cordillera" rather than late Pleistocene glacial cycles.

In order to understand the biogeographic history of any widespread desert taxon, it is critical to associate phyletic branching patterns with regional deserts, which should be equivalent to areas of endemism (*sensu* Humphries and Parenti, 1986). To the extent that speciation occurred during isolation of regional deserts and has been detected by systematists, the regional deserts should also be delineated by coincident species boundaries. Following this phyletic approach to delineation of regional deserts, Hafner (1981) proposed recognition of an additional regional desert in North America based on arid-adapted, non-volant small mammals and desert reptiles: the Peninsular Desert of the Baja California peninsula south of about 30°N latitude (Fig. 1). Hafner (1992) argued that the Peninsular Desert, which traditionally has been considered as a subset of the Sonoran Desert, experienced an evolutionary history more independent from the Sonoran Desert than those of the Mojave, San Joaquin, and Great Basin Deserts.

We present additional evidence for recognition of the Peninsular regional desert, identify probable autochthonous and allochthonous mammals of the region and their sources, present a general model for the evolution of the Peninsular Desert biota, and summarize genetic evidence in support of an ancient history of some North American regional desert mammals. Recognition of regional deserts should apply to a wide array of taxa, such that broad regional floras and faunas are defined. Although we base our arguments primarily on mammalian distributions, we contend that these patterns are consistent with patterns observed for desert reptiles, plants, and invertebrates for which distributional data are available.

METHODS

We restricted our faunal analysis to small mammals of low vagility, excluding both large species and bats. Species with low vagility are more likely to be sensitive to geographic barriers to dispersal, and to disperse more slowly into alternate areas, thus preserving the record of past geographic restriction. We further restricted our analysis of regional deserts to species that we considered to be highly arid-adapted: they occur throughout at least one of the regional deserts and are generally restricted to the lower life zones of that desert. Mammal distributions were taken from Hall (1981), modified as necessary following taxonomic changes included in Wilson and Reeder (1992). Six regional deserts were considered: Chihuahuan, Sonoran, Mojave, Great Basin, San Joaquin, and Peninsular. Only the Hidalgan Desert (Axelrod, 1979; Shreve, 1940) was omitted from the analysis, as it is considered by Axelrod (1979) to be a relict late Tertiary semidesert, substantially different in character from the other regional deserts.

Mammalian species were assigned to one of three general habitat groupings: habitat generalists (occurring in a wide variety of habitats including desert regions); xeric (restricted to deserts); and peripheral invaders (more widespread in a habitat other than desert). Peripheral invaders were further assigned to a source regional habitat based on the most common occurrence of the species.

In order to eliminate the effects of species richness in a phenetic comparison of faunal similarity among regional deserts, we employed

Simpson's (1960) faunal similarity index, C/N_l. This index represents the proportion of the total species in the more depauperate region (N_l) that is shared in common (C) with a neighboring, more speciose region. Thus, a depauperate fauna that is a subset of a neighboring, richer regional fauna would share 100% of its species (C/N_l = 1.00).

Small, non-volant mammals of the Baja California peninsular mainland were assigned to five general habitats based on comparison of their geographic range with phytogeographic regions (Brown and Lowe, 1980; Küchler, 1970; Roberts, 1989): habitat generalists, montane, chaparral, coastal, and desert. For habitat generalists, interpretation of biogeographic history based on geographic range alone is not possible. Montane, chaparral, and coastal species were assumed to represent relicts of cooler, more mesic pluvial intervals. Drought-resistant chaparral habitat is considered to be mesic only in comparison to desert regions; chaparral habitat receives significantly higher winter precipitation (Steinhauser, 1979). Desert taxa were assumed to be recent arrivals to the Baja California peninsula if they did not occur on either the peninsular mainland or landbridge islands significantly below 30°N latitude. These recent arrivals were of two sorts: Sonoran Desert taxa virtually restricted to the San Felipe region of northeastern Baja California and those of indeterminate origin. To further test our hypotheses, comparisons and contrasts were made between mammalian and other biota (birds, invertebrates, and plants).

RESULTS

North American Regional Desert Mammals

Overall, mammalian species richness of regional deserts (Table 1) varies from the geographically small and relatively depauperate San Joaquin (15 species) to the Chihuahuan Desert (with nearly twice the number of species). When only xeric species are considered, species richness is highest in the centrally located Sonoran and Mojave regions (together termed Mohavia by Axelrod, 1950, 1979), and is more even across all deserts. The Mojave Desert possesses only a single endemic, *Spermophilus mohavensis*, which may have speciated in a relatively cool, mesic pluvial refugium in the northwestern Mojave (Hafner, 1992). Two pocket mice (*Perognathus parvus xanthonotus* and *P. alticola*) occur at higher elevations nearby (Hall, 1981), and may be products of isolation in this same pluvial refugium. Other than the single endemic, the Mojave Desert is a transitional mixture of Great Basin and Sonoran species. Three deserts include a significant proportion of peripheral invaders from neighboring physiographic provinces: Peninsular (California chaparral), Great Basin (boreal shrub-steppe and Great Plains), and Chihuahuan (Great Plains and Mexican highlands). Most genera are shared among all regional deserts (62.5%) or between at least five of the six regions (81.3%). Considering typical desert genera, the Peninsular Desert is notably rich in spiny pocket mice (*Chaetodipus*), while having no silky pocket mice (*Perognathus*) or grasshopper mice (*Onychomys*). The latter two genera occur only in the San Felipe region of the peninsula, and do not extend into the Peninsular Desert proper (i.e., south of 30°N latitude).

Overall similarity among regional desert mammalian faunas exclusive of insular forms (C/N_l; Table 2) range from 0.231 (Great Basin versus Chihuahuan) to ≥ 0.700 (Mojave to Great Basin and Sonoran). A phenogram constructed from these similarity values using the unweighted pair-groups method with arithmetic averaging (Fig. 2a; cophenetic correlation coefficient = 0.80) groups the three continuous (but climatically distinct) central deserts (Sonoran, Mojave, and Great Basin), followed by the San Joaquin. The Peninsular Desert is second in distinctiveness only to the Chihuahuan Desert. The topology of this phenogram is identical to that based on desert reptiles (from Hafner, 1981; C/N_l values repeated in Table 2, phenogram repeated in Fig. 2b; cophenetic correlation coefficient = 0.75).

Peninsular Desert Mammalian Fauna

Of 54 species reported from the Baja California peninsula (Table 3), 14 (25.9%) are endemic to near-shore islands and 40 (74.1%) are found on the peninsular mainland. Two of the mainland species (5.0%) are habitat generalists: *Thomomys bottae* and *Peromyscus mani-*

Table 1. Non-volant small mammals of North American regional deserts (excluding insular endemics).

	Regional Deserts					
	San Joaquin	Mojave	Peninsular	Great Basin	Sonoran	Chihuahuan
1. Total species	15	20	18	26	25	29
2. Endemic species	5	1	4	5	3	5
3. Generalist species	4	3	2	4	4	4
4. Peripheral invaders	–	–	3	5	2	11
California chaparral	–	–	3	–	–	–
Boreal shrubsteppe	–	–	–	3	–	–
Great Plains	–	–	–	2	1	7
Sinaloan thornscrub	–	–	–	–	1	–
Mexican highlands	–	–	–	–	–	4
5. Xeric species ([1] – [3+4])	11	17	13	17	19	14
6. Taxonomic Distribution:						
Insectivora	1	1	1	1	1	1
Lagomorpha	2	2	3	3	3	2
Rodentia	12	17	14	22	21	26
Sciuridae	2	3	2	4	2	3
Geomyidae	1	1	1	1	1	1
Heteromyidae	5	7	7	10	9	10
Perognathus	2	1	–	2	3	1
Chaetodipus	–	2	5	1	3	4
Microdipodops	–	–	–	2	–	–
Dipodomys	3	4	2	5	3	5
Muridae	4	6	4	7	9	12
Reithrodontomys	1	1	–	1	2	2
Peromyscus	1	3	3	2	4	3
Onychomys	1	1	–	2	1	1
Sigmodon	–	–	–	–	–	3
Neotoma	1	1	1	1	2	3
Lemmiscus	–	–	–	1	–	–

Table 2. **Similarity of faunas of small, non-volant mammals (above diagonal) and desert-adapted reptiles (below diagonal) of North American regional deserts, based on Simpson's (1960) Index, C/N_1.**

	Regional Deserts					
	San Joaquin	Mojave	Peninsular	Great Basin	Sonoran	Chihuahuan
San Joaquin	–	0.600	0.400	0.533	0.533	0.333
Mojave	0.889	–	0.556	0.750	0.700	0.400
Peninsular	0.778	0.621	–	0.444	0.444	0.333
Great Basin	0.667	1.000	0.833	–	0.440	0.231
Sonoran	0.889	0.931	0.576	1.000	–	0.520
Chihuahuan	0.778	0.556	0.407	0.883	0.519	–

culatus. We consider 18 (45.0%) of the mainland species to be probable pluvial relicts: species that depend on relatively mesic habitat (montane coniferous forest, chaparral, or coastal habitats) that must have spread down the peninsula during pluvial periods (Cody et al., 1983) and were isolated during subsequent warming. Six species categorized as pluvial relicts have relictual populations in the central peninsula, Cape Region, or on central or southern nearshore islands. Nine other mainland species (22.5%) are considered to be post-pluvial or recent arrivals: desert species that have entered only the northern portion of the peninsula. The single island population of a recently arriving species, *Peromyscus crinitus*, is found on Isla Willard, near 30°N latitude in the northern Sea of Cortéz. Four of the pluvial relictual species, all from chaparral habitat, and a single recent arrival are widespread throughout the Peninsular Desert and are here considered as components of the Peninsular Desert fauna. Eleven (27.5%) of the mainland species form the core species of the Peninsular Desert fauna. Of these species, only *Sylvilagus auduboni* lacks a related population on the near-shore islands (perhaps a result of post-isolation extinction). *Chaetodipus dalquesti*, known only from the Cape Region of Baja California, is included here under *C. arenarius*, following Wilson and Reeder (1992); taxonomy of heteromyid rodents follows Williams et al. (1993).

Species considered to be recent arrivals to the Baja California peninsula (Figs. 3a and 3b) reach their highest density at the head of the Sea of Cortéz and (by definition) generally have not penetrated far south of 30°N latitude, the northern limit of the Peninsular Desert. Species confined to montane, coastal, or chaparral habitats (Figs. 4a, 4b, and 4c) reach their highest density in the northwestern portion of the peninsula, considered as part of the California Biotic Province (Munz and Keck, 1949, 1950; Wiggins, 1960). The more mesic montane and coastal species are absent from most of the peninsula, represented by only two species (*Peromyscus truei* and *Sorex ornatus*) in the Sierra Laguna of the southern Cape Region.

Desert mammals that are widespread throughout North American regional deserts are rather evenly distributed throughout the peninsula. Those species that occur on both shores of the Sea of Cortéz (Fig. 5a) or that are virtually endemic to Baja California (Fig. 5b) tend to be distributed linearly along the peninsula, as are the major phytogeographic regions (Fig. 6; Roberts, 1989; Wiggins, 1980).

Table 3. Non-volant small mammals of the Baja California Peninsula and adjacent islands.

	% Total	% Mainland		
1. Insular endemics (14 species)	25.9			*Sylvilagus mansuetus*
				Lepus insularis
				Ammospermophilus insularis
				Peromyscus caniceps
				Peromyscus dickeyi
				Peromyscus guardia
				Peromyscus interparietalis
				Peromyscus pseudocrinitus
				Peromyscus sejugis
				Peromyscus slevini
				Neotoma anthonyi
				Neotoma bryanti
			†	*Neotoma bunkeri*
			†	*Neotoma martinensis*
2. Peninsular mainland (40)	74.1			
A. Habitat generalists (2)		5.0	[3,4]	*Thomomys bottae*
			[3,4]	*Peromyscus maniculatus*
B. Pluvial relicts (18)		45.0		
Montane				*Scapanus latimanus*
				Tamias merriami
			[1]	*Tamias obscurus*
				Sciurus griseus
				Tamiasciurus mearnsi
				Peromyscus boylii
			[2]	*Peromyscus truei*
California chaparral			[3,4]	*Sylvilagus bachmani*
				Spermophilus beecheyi
			[1,4]	*Spermophilus atricapillus*
			[3,4]	*Chaetodipus fallax*
				Chaetodipus californicus
			[4]	*Dipodomys simulans*
				Dipodomys gravipes
				Peromyscus californicus
				Neotoma fuscipes
Coastal				*Microtus californicus*
			[2]	*Sorex ornatus* (including *S. juncensis*)
C. Recent arrivals (9)		22.5		
San Felipe region				*Spermophilus tereticaudus*
			[4]	*Chaetodipus formosus*
				Chaetodipus penicillatus
				Dipodomys deserti
			[3]	*Peromyscus crinitus*
				Neotoma albigula
Indeterminate origin				*Perognathus longimembris*
				Onychomys torridus
				Reithrodontomys megalotis

Table 3. Continued

D. Peninsular Desert (11)	27.5	
Widespread in regional deserts		[3,4] *Notiosorex crawfordi*
		[4] *Sylvilagus auduboni*
		[3,4] *Lepus californicus*
Circum-Gulf distribution		[3,4] *Chaetodipus baileyi*
		[3,4] *Dipodomys merriami*
		[3,4] *Peromyscus eremicus*
		[3,4] *Neotoma lepida*-group
		[3,4] *Ammospermophilus* spp.
Endemic to Baja California		[3,4] *Chaetodipus spinatus*
		[3,4] *Chaetodipus arenarius*
		[3,4] *Peromyscus eva*

[1] relictual isolate in central Peninsula
[2] relictual isolate in Sierra Laguna, Cape Region
[3] insular population(s) on Baja California island(s)
[4] widespread throughout Peninsular Desert
† extinct

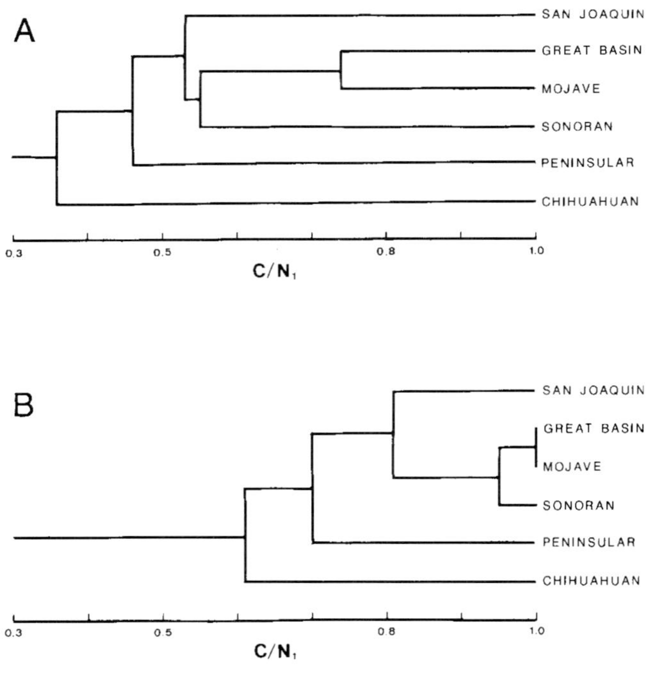

Fig. 2. UPGMA phenograms depicting similarity of (a) desert-adapted, non-volant small mammals and (b) desert reptiles of North American regional deserts, based on Simpson's (1960) index, C/N_l.

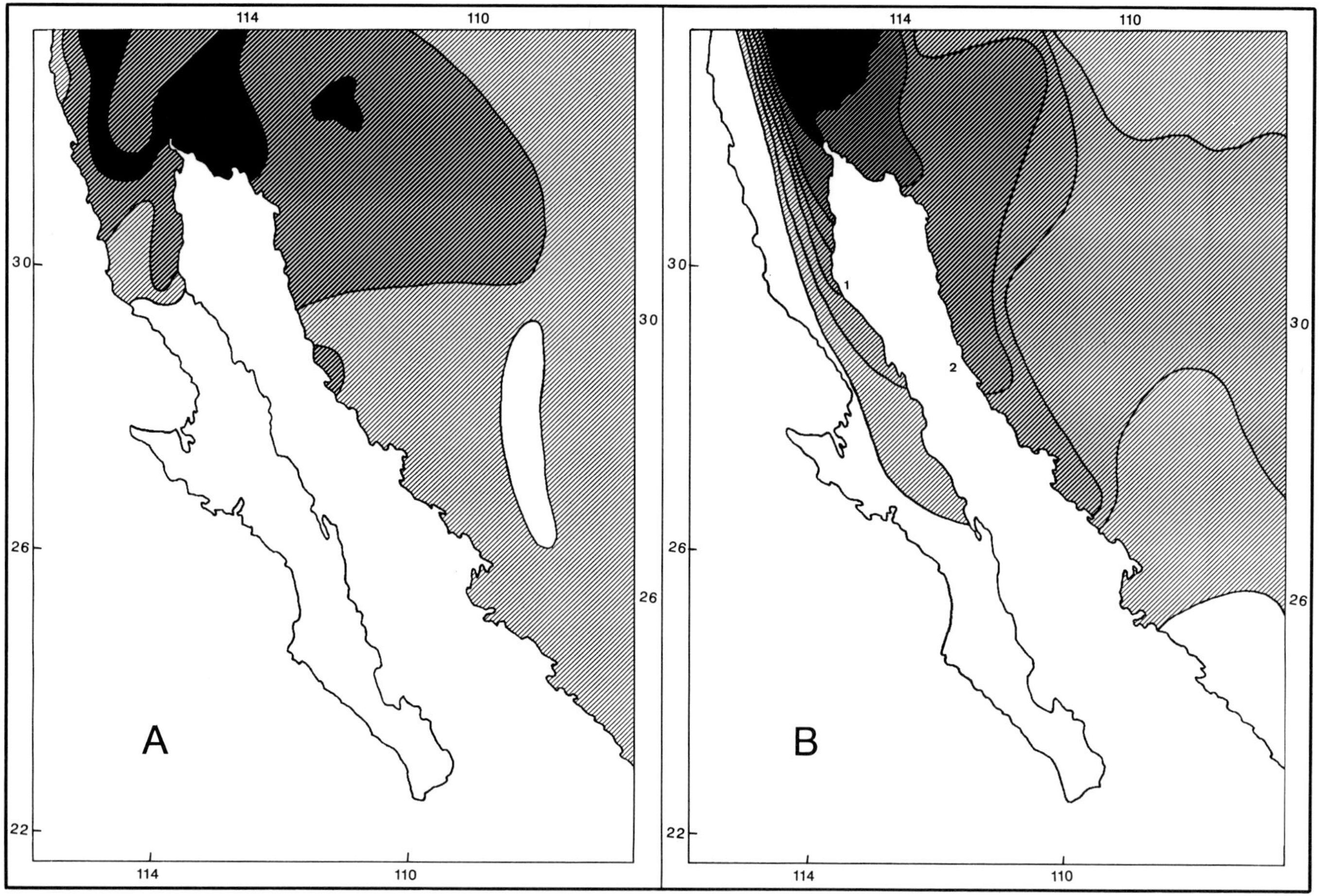

Fig. 3. Combined distributions of arid-adapted small mammals that have recently arrived to Baja California: a) three species of indeterminate origin (*Perognathus longimembris, Onychomys torridus,* and *Reithrodontomys megalotis*); b) six species of the San Felipe desert region (*Spermophilus tereticaudus*, *Chaetodipus formosus, C. penicillatus, Dipodomys deserti, Peromyscus crinitus*, and *Neotoma albigula*. Species density (contour interval = 1) indicated by darker shading; numbers off-shore indicate insular species.

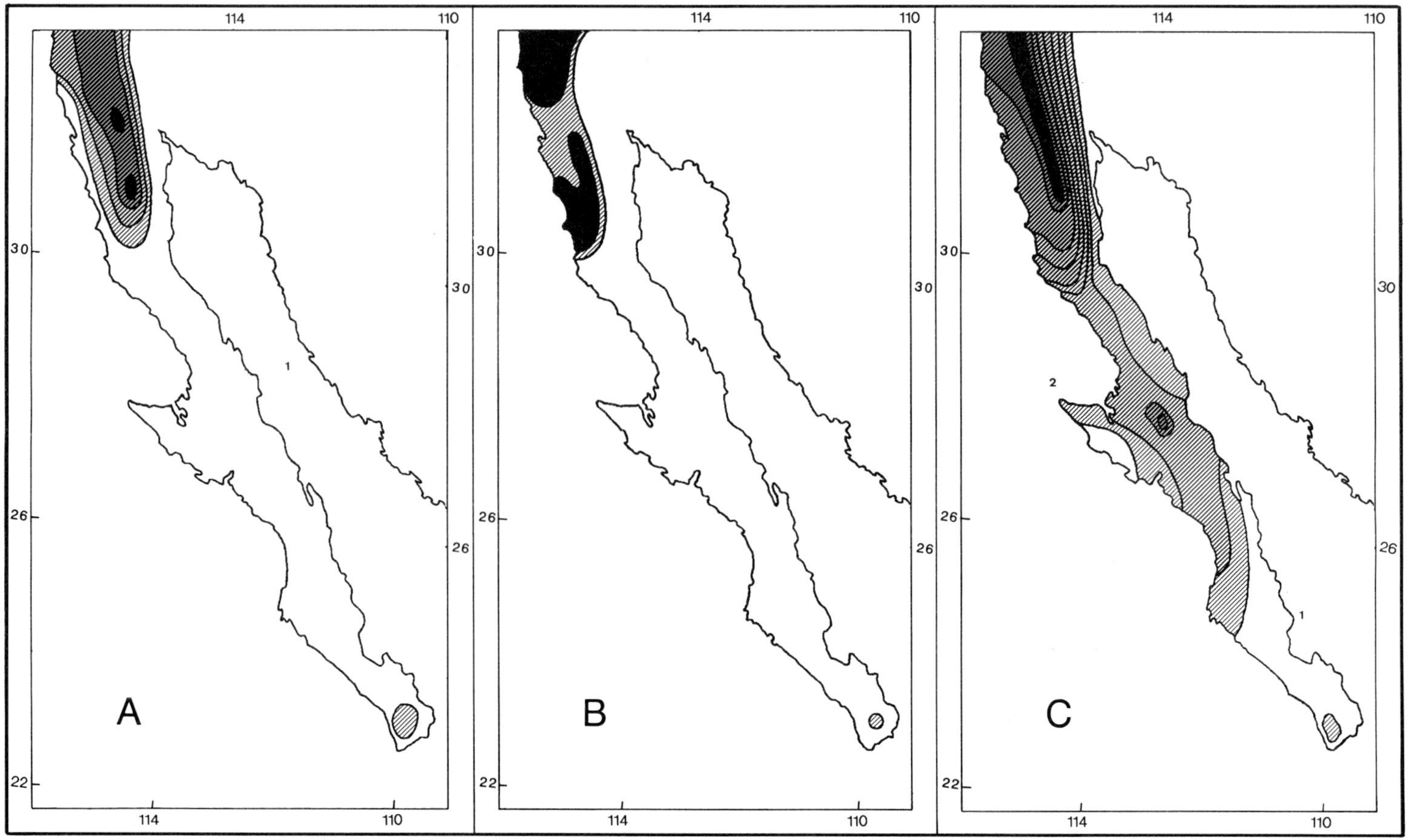

Fig. 4. Combined distributions of mesic-adapted small mammals with pluvial relicts on the Baja California peninsula: a) seven montane species (*Scapanus latimanus, Tamias merriami, T. obscurus, Sciurus griseus, Tamiasciurus mearnsi, Peromyscus boylii* and island derivative, and *Peromyscus truei*); b) two coastal species (*Microtus californicus* and *Sorex ornatus* [including *S. juncensis*]); c) nine chaparral species (*Sylvilagus bachmani* and island derivatives, *Spermophilus beecheyi, S. atricapillus, Chaetodipus fallax* and island derivative, *C. californicus, Dipodomys agilis, D. gravipes, Peromyscus californicus*, and *Neotoma fuscipes*). Species density (contour interval = 1) indicated by darker shading; numbers off-shore indicate insular species.

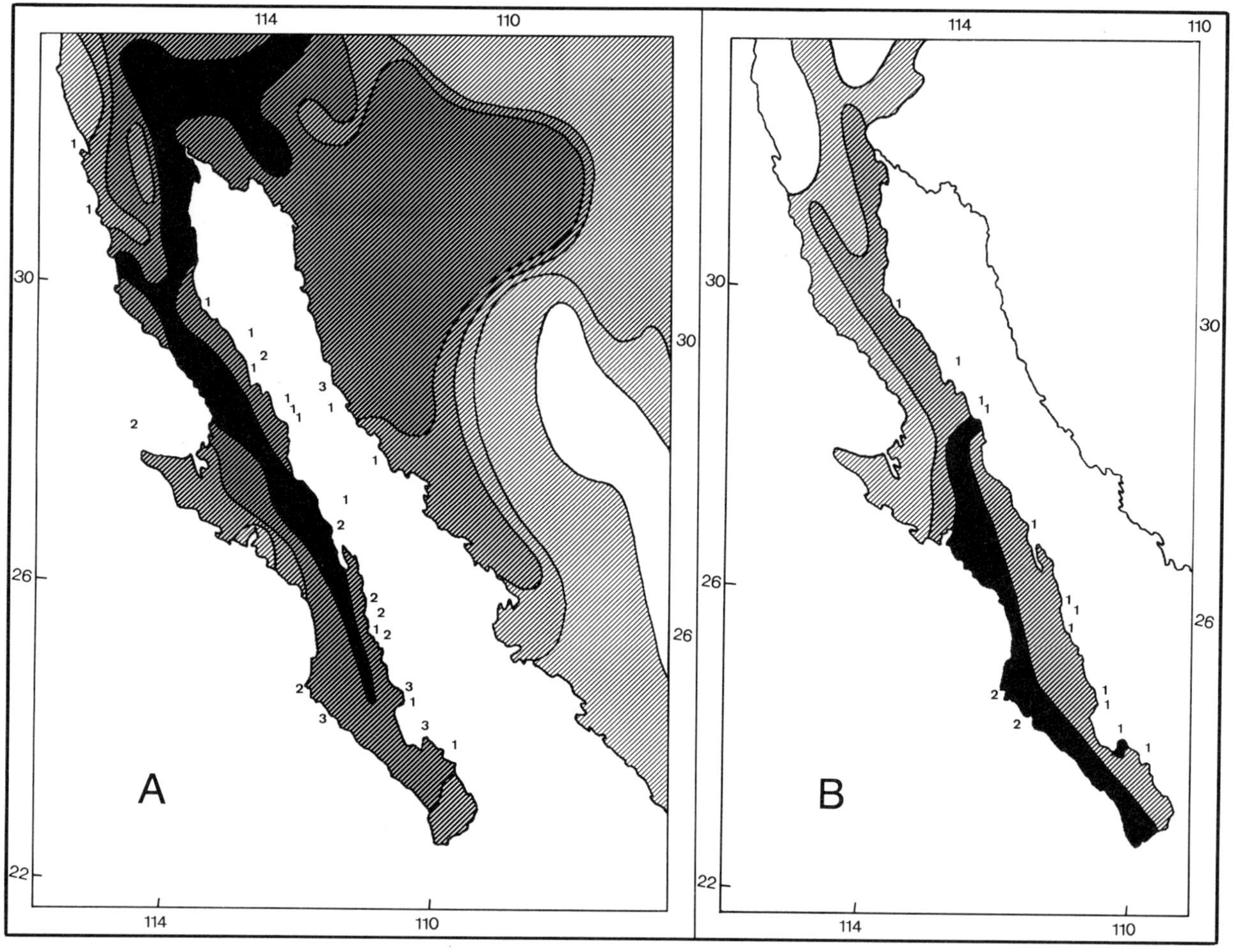

Fig. 5. Combined distributions of: a) five mammals with a circum-Gulf distribution (*Chaetodipus baileyi, Dipodomys merriami, Peromyscus* (*Haplomylomys*), *Neotoma lepida* species-group, *Ammospermophilus* spp., and island derivatives); b) three endemic (or near-endemic) Peninsular Desert mammals (*Chaetodipus spinatus, C. arenarius, Peromyscus eva*, and island derivatives). Species density (contour interval = 1) indicated by darker shading; numbers off-shore indicate insular species.

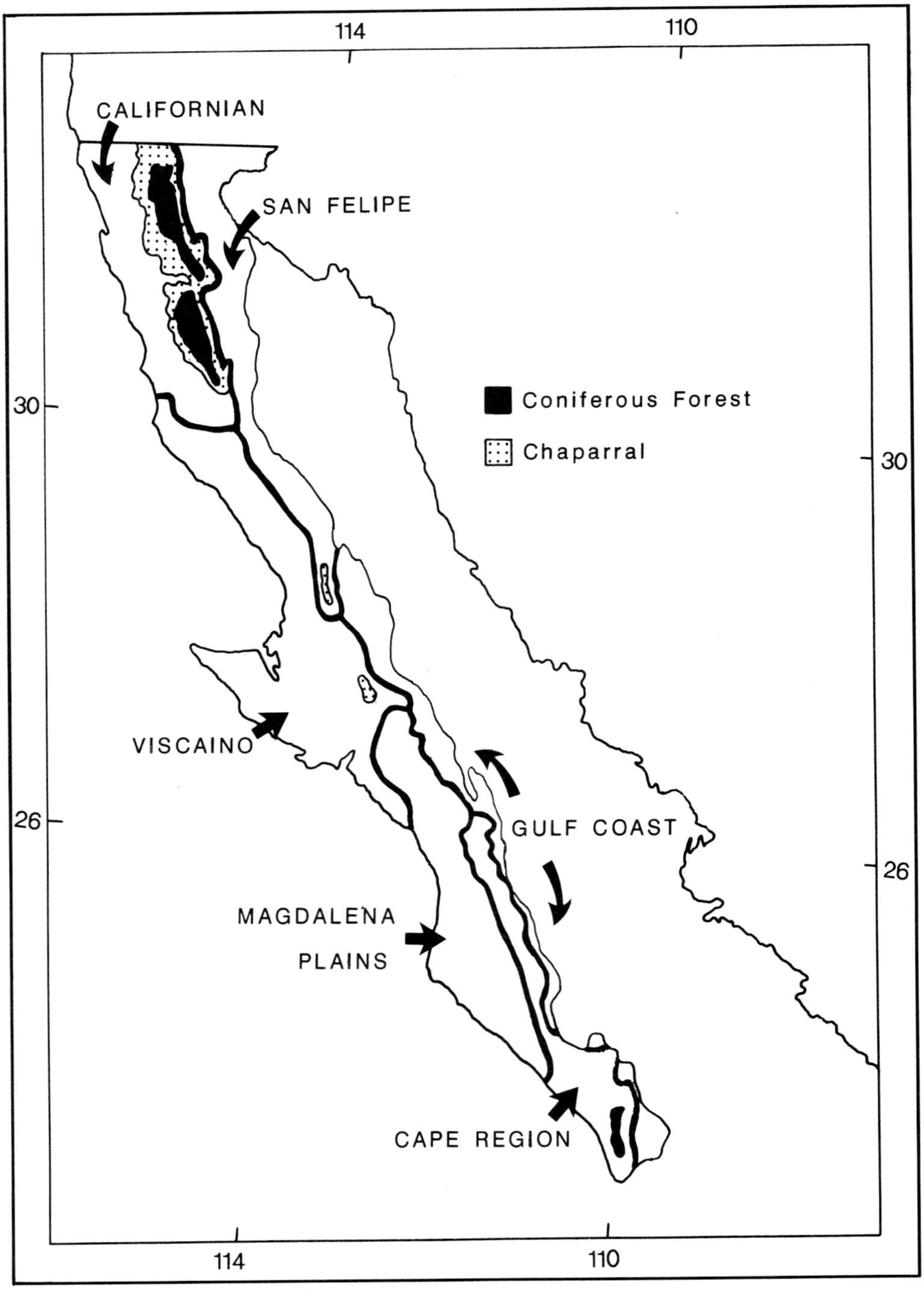

Fig. 6. Phytogeographic regions of Baja California.

DISCUSSION

Peninsular Desert Regional Biota

Recognition of a regional biota should apply to a wide array of flora and fauna. The biota of the Baja California peninsula has been relatively well described, often in relation to the geologic history of the peninsula. We summarize here the pertinent geologic history of the peninsula and major features of the peninsular biota relative to that of the adjacent Sonoran mainland.

Geologic History of the Peninsula

It has been predicted on the basis of geologic and paleoclimatic events (Hafner, 1981; Murphy, 1983a; 1983b; Grismer, 1994) that the Baja California peninsula should support a distinct regional desert biota. The Cape Region of the Baja California peninsula and the northern Peninsular Ranges were torn away from the Sinaloan and Sonoran mainland (respectively) during middle Miocene rifting that opened a shallow proto-Gulf of California (= Sea of Cortéz) 12-5.5 million years ago (Fig. 7a; Gastil et al., 1983; Grismer, 1994; Karig and Jensky, 1972; Lonsdale, 1989; Spencer and Normark, 1989). Continued plate-boundary expansion during the late Miocene and Pliocene (5.5-3 mya) simultaneously enlarged and extended the Gulf of California (Fig. 7b; Buising, 1990), elevated the Baja California peninsula and California's Tehachapi and Coast Ranges, and expelled marine waters from the Central Valley (Norris and Webb, 1976). The emergent Baja California peninsula was isolated from the neighboring Mojave and Sonoran regions by the San Gorgonio and Bouse extensions of the Gulf of California (Buising, 1990; Murphy, 1983a; Grismer, 1994), while the San Joaquin desert began to form in the southern Central Valley (Friesen, 1979; Norris and Webb, 1976). During accelerated uplift of the Colorado Plateau, Sierra Nevada, and Sierra Madres (Lucchitta, 1979), the northern Gulf extensions retracted (Buising, 1990) and the region gained essentially its modern geography. Late Pleistocene glacial-interglacial climatic cycles, which became markedly longer and more extreme about 0.7 mya (Webb and Bartlein, 1992), subsequently caused extensive range contraction and expansion of arid-adapted taxa (Fig. 7c).

Peninsular endemics could date from one of three distinct time intervals: Miocene, Pliocene, or late Pleistocene. Sinaloan thornscrub or Mexican highland taxa that were present on the Cape Region when it was torn from the mainland and survived to spread north along the emergent Baja California peninsula would have begun genetic divergence from their mainland counterparts at least 5.5 mya. Taxa that entered the emergent peninsula from the north would have been isolated from their Sonoran sister taxa by the San Gorgonio and Bouse embayments (3 mya). Finally, taxa that entered the peninsula (again from the north) subsequent to the retreat of the embayments would have been isolated from their Sonoran sister taxa in peninsular desert refugia during the lengthy and extreme full-glacial intervals of the last 700,000 years.

Peninsular Flora

The unique characteristics of the flora of the Baja California peninsula have been obscured by the traditional inclusion of the region within the Sonoran Desert. Shreve (1942: 235) stated that the "flora of Baja California has long been noted for its strong endemism." Wiggins (1960) stated that "Endemism is high among the plants of Baja California" but later (1980) noted that "it supports relatively fewer endemics than does the adjacent California Floristic Province," adding that "in spite of the low percentage of endemism in Baja California, several of its endemic genera are of considerable interest." Following Axelrod's (1979) arguments that "Sonoran Desert" (including the Peninsular Desert) endemics have been incorporated into developing regional deserts continuously since the Cretaceous, it is likely that at least some of the Peninsular endemics date from the initial Miocene separation of the peninsula from the mainland.

Peninsular Mammals

Huey (1964) summarized the status of Baja California mammals as of 1960, the same year that Orr (1960) attempted a broad review of the biogeographic origin of the peninsula's mammalian fauna. Unfortunately, Orr's review was

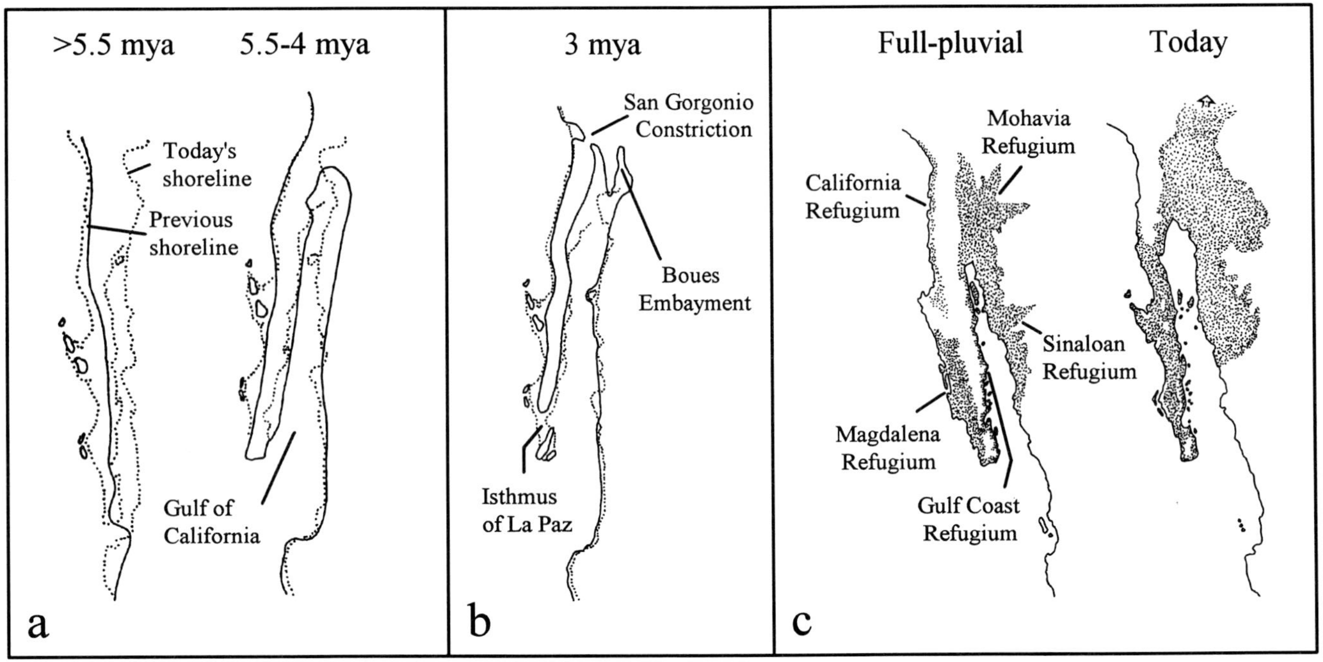

Fig. 7. Tectonic evolution of the Baja California peninsula and Gulf of California: a) origin of the peninsula and gulf; b) marine transgressions; and c) distribution of regional deserts and refugia (stippled) during Pleistocene pluvials and interpluvials.

handicapped by the prevailing static view of continental configurations; the concept of plate tectonics to explain continental drift had not yet been proposed. Consequently, Orr (1960) emphasized Pleistocene climatic fluctuations as the major factors promoting endemism on the peninsula. Findley and Caire (1977) identified mammalian faunal provinces in southwestern North America, naming a separate province (Bajan) for Baja California. Lawlor (1983a) analyzed the historical biogeography of mammals on the islands surrounding Baja California, and briefly summarized the development of regional deserts.

In terms of recognized species of arid-adapted mammals, the Peninsular Desert is the second most distinct regional desert after the Chihuahuan Desert, and has as many endemic mammal species (exclusive of insular forms) as the other recognized regional deserts. Whether the distinctiveness of the Peninsular Desert mammalian fauna arose during the Miocene rifting of the Cape Region, the Pliocene marine transgressions, during isolation in Pleistocene full-pluvial refugia, or a combination of all three remains to be determined from genetic and paleontological studies of individual taxa.

Peninsular Reptiles

Savage (1960) undertook a biogeographic analysis of the evolution of the peninsular herpetofauna. As his study predated acceptance of plate tectonics and continental drift, he was unable to incorporate these processes into his study. Savage therefore emphasized the importance of Pleistocene glacial events in producing the peninsular herpetofauna, even concluding that "waves of immigrants from the north . . . have all but replaced the early Tertiary fauna of the area." Savage did note, however, the distinctiveness of the peninsular herpetofauna, recognizing the Peninsular Desert District as a subdivision of the Sonoran Desert Subprovince of the Desert and Plains herpetofaunal province.

Seib (1980) described three current distributional patterns for reptiles of the Baja California peninsula: 1) ecological isolates, most of which "are probably relics [sic], having dispersed southward under moister climatic conditions"; 2) mainland disjuncts and/or peninsular endemics; and 3) northern dispersers. These would correspond to our pluvial relicts, peninsular endemics, and recent arrivals. In his thorough review of the origin and evolution of the Baja California reptiles, Murphy (1983a,b) incorporated studies by Atwater (1970), Atwater and Molner (1973), Gastil and Jensky (1973), and Gastil et al. (1975) that documented the Miocene-age fault-displacement of the peninsula from the mainland. Murphy (1983a) concluded that his Transgulfian Vicariance Model was sufficient to explain the origin of most of the peninsular reptilian fauna, and that Pleistocene glacial events had little effect on peninsular distributional patterns. Although Murphy (1983a) identified Pleistocene immigrants to the peninsula, he noted that they have all failed to disperse far southward on the peninsula, and that none have been reported from islands surrounding Baja California. Thus, Murphy underscored the considerable age of the Peninsular Desert reptilian fauna, and provided evidence from reptiles that is congruent with the major features of mammalian biogeography reviewed here. Grismer (1994) summarized advances in geotectonic models concerning the Gulf region, and presented phylogenetic hypotheses for the Peninsular herpetofauna constructed within an explicitly cladistic paradigm. The major features of Murphy's (1983a) Transgulfian Vicariance Model were corroborated by Grismer (1994).

Peninsular Avifauna

Cody (1983) stated that the "Sonoran [including Peninsular] Desert avifauna is at best an indistinct entity"; this is perhaps due to inclusion of two regional deserts in this avifauna as well as the great vagility of birds. He did note, however, three patterns of endemism on the peninsula. The Cape Region supports remnants of an original Mexican highlands avifauna (three species) as well as more numerous relicts (including one endemic species) of the Pleistocene expansion of chaparral and pine-oak woodland down the length of the peninsula. Six species are characteristic of the Peninsular Desert: four endemics, one of which presumably originated in the Cape Region and spread north, and two California chaparral species that have shifted to desert habitat only in Baja California. Eleven other species exhibit subspecific boundaries at or near 30°N latitude (the northern margin of the Peninsular Desert), and two species have a disjunct distribution on the Peninsula south of 27°N latitude. Thus, even within the highly mobile birds there is evidence for both early (Miocene–Pliocene) and more recent (Pleistocene) origin of Peninsular Desert taxa, and a distinct northern boundary of Peninsular Desert avifauna at about 30°N latitude.

Peninsular Invertebrates

Truxal (1960) reported origins and affinities for Baja California entomofauna that appear very similar to those of birds, and probably reflects the high vagility of both groups. Truxal concluded that there are three insect faunal areas on the peninsula: 1) a northern area of Californian and Vancouveran elements; 2) the Cape Region, species of which share affinities with the Mexican highlands; and 3) the central deserts, species of which are closely associated with Sonoran Desert taxa. Thus, endemism is restricted to the Cape Region and is presumably of Miocene–Pliocene age, and the Peninsular Desert entomofauna is not considered by Truxal as distinct from that of the Sonoran Desert. A study on Baja California spiders by Chamberlin (1924), referred to by Truxal (1960) as reporting "somewhat the same distribution pattern," instead states that "This fauna, in the main, has been long established, as indicated by the large percentage of forms peculiar to the area." Brown (1987) described species patterns among butterflies of Baja California as representing a transition between two major faunas: Neotropical (Cape Region) and Nearctic (Californian). Brown (1987) believed that the isolation of much of the current Neotropical biota of the Cape Region may be traced to the Miocene separation of the Cape Region from the Mexican mainland.

Scorpions of Baja California and adjacent islands display patterns of distribution and diversity that parallel closely patterns observed for both mammals and reptiles. According to Williams (1980), the peninsular scorpion fauna "is one of the richest . . . of any place of comparable size in the world" and is "rich in endemics." Of 36 peninsular mainland species, 21 (58.3%) are endemic. Williams attributes this diversity to the complex geological history of the region, variety of available habitats, and multiple origins of the fauna. Specifically, the three largest sources of the scorpion fauna are: 1) an archaic element that may have been present prior to the separation of the peninsula from the mainland; 2) invasion of more northern species during Pleistocene climatic oscillations; and 3) autochthonous species that arose on the peninsula *in situ* following separation from the Mexican mainland.

Peninsular Desert Biota

The Peninsular Desert should be recognized as a regional biota independent of the other recognized regional deserts but similar in climate and floral growth form to the adjacent Sonoran Desert. Gross ecological similarities of the flora of the two regions (Axelrod, 1979; Shreve, 1942; Shreve, 1951; Shreve and Wiggins, 1964) probably resulted from parallel evolution of long-separated floras in similar climatic regimes. Existence of Peninsular Desert endemics that may have originated in the Miocene, Pliocene, or Pleistocene and the degree of differentiation of peninsular plants, mammals, reptiles, and scorpions (at least) underscores the long-separate evolutionary trajectory of the peninsular biota.

Historical Biogeography of Peninsular Desert Mammals

Peripheral Influences

Prior to Pleistocene climatic fluctuations, the arid-adapted mammalian fauna probably consisted of species that dated from the middle Miocene separation of the Cape Region from the Sinaloan–Sonoran mainland, those that originated in isolation during the Pliocene Gulf transgressions, and those that moved onto the peninsula subsequent to the retreat of the Gulf embayments. As pluvial–interpluvial cycles became increasingly pronounced about 700,000 years ago (Webb and Bartlein, 1992), the Peninsular fauna experienced sequential invasion of mesic-adapted and arid-adapted species. Relictual isolates of mesic-adapted forms (Fig. 4) and direct fossil evidence (Axelrod, 1979, 1983; Miller, 1977; Van Devender and Spaulding, 1979) document the full-pluvial spread of chaparral and pine-oak forest down the length of the peninsula, which provided a dispersal corridor for mesic-adapted mammals. Following the Wisconsinan pluvial, arid-adapted forms entered the northern end of the peninsula, where most have not penetrated south of 30°N latitude. It is not clear, based on distribution alone, whether endemic, circum-Gulf, or widespread desert taxa evolved in intermittent isolation on the peninsula since the Miocene separation from the mainland, the Pliocene marine transgressions, or during isolation in pluvial refugia. We consider it unlikely that all circum-Gulf and widespread desert taxa represent post-Wisconsinan immigrants that spread throughout the length of the peninsula. Rather, at least some of these species probably diverged in isolation but did not speciate on the peninsula, or are unrecognized cryptic species that have experienced secondary contact with their Sonoran counterparts.

Full-Pluvial Distributions

Desert species that were present on the Peninsula prior to the last pluvial maximum must have been forced to follow depressed vegetative zones down in elevation and latitude and onto exposed continental shelves (Betancourt et al., 1990). Gastil et al. (1983), Murphy (1983b), Porter (1989), and Van Devender and Spaulding (1979) have provided approximate figures for the extent of life-zone depression (ca. 1,000 m) and sea level drop (ca. 150 m). Postulated full-pluvial elevational limits of deserts (Fig. 7c) indicate that at least two full-pluvial desert refugia may have been present on the peninsula: a broad Pacific-slope (Magdalena Plains) refugium and a smaller Gulf Coast refugium. The two peninsular refugia may have been divided by a continuous corridor of more-mesic vegetation through the narrow spine of the peninsula west of present-day La Paz in southern Baja California. Peninsular refugia may have been partially or fully isolated from the adjacent Mohavia refugium (Axelrod, 1950) by inhospitable and poorly vegetated salt and mud flats of newly exposed land surrounding the head of the Sea of Cortéz, such as that found today near San Felipe. Vegetation in the present-day San Felipe region is extremely sparse; this is the driest region on the peninsula (annual precipitation < 5 cm), and it also receives the highest annual percentage of sunshine, producing one of the highest evaporation rates (Roberts, 1989).

The full-pluvial latitudinal distribution of taxa on the peninsula may be estimated based on the occurrence of relict populations on landbridge islands surrounding the peninsula. Species that do not occur on oceanic islands may be regarded as poor across-water dispersers. For these species, occurrence of relictual populations on landbridge islands documents their previous occurrence on the adjacent mainland. All landbridge islands were connected to the peninsular mainland during the Wisconsinan pluvial, and were separated between 12,500 and 5,600 years ago during post-Wisconsinan eustatic sea level rise (Milliman and Emery, 1968; Wilcox, 1978). Efforts by Banks (1964a, 1964b, 1967), Brand and Ryckman (1969), Dickey (1983), Huey (1964), Lawlor (1971, 1983a), O.J. Reichman (pers. comm.), and Schultz et al. (1970) have revealed 101 insular populations of 40 species on 36 Baja California islands. These include at least 76 populations of 31 species of small, non-volant mammals on 33 islands, including 21 landbridge islands (Appendix I). Landbridge island populations document the full-pluvial latitudinal distribution of desert rodents that presumably colonized the islands across landbridges during restriction to full-pluvial refugia. Indirect evidence for the partial disjunction of these hypothesized refugia derives from their different presumed species compositions (Table 4).

A Model of Peninsular Desert Biogeographic History

Arid-adapted taxa may have experienced any or all of three alternating periods of isolation and dispersal between the peninsula and the mainland, depending on the time of their arrival to the region, whether they diverged in allopatry, whether they subsequently dispersed to the alternate region following isolation, and whether they left surviving lineages (Fig. 8). This vicariant model predicts the potential establishment of peninsular lineages of up to four disparate ages: Miocene (12-5.5 mya), Pliocene (3 mya), late Pleistocene (700 kya), or Holocene (< 10,000 years ago).

Alternatives to Vicariant Explanations

We have interpreted existing distributional patterns of mammals on the Baja California pen-

Table 4. Presumed distribution of *Ammospermophilus, Chaetodipus, Peromyscus (Haplomylomys), Dipodomys merriami* species-group, and *Neotoma lepida* species-group in full-pluvial refugia around the Sea of Cortéz, based on occurrence of relictual populations on landbridge islands and current mainland distribution. The majority of small, non-volant Peninsular Desert mammals are included in these taxa.

	Pleistocene Full-pluvial Refugia				
	California	Magdalena	Gulf Coast	Mohavia	Sinaloan
[1] *Chaetodipus fallax*	X				
[1] *Neotoma intermedia*	X	X	X		
[1-2] *Peromyscus eremicus*	X	X	X	X	X
[1-2] *Dipodomys merriami*		X	X	X	X
[1] *Chaetodipus arenarius*		X	X		
[1] *Chaetodipus spinatus*		X	X		
[1] *Peromyscus eva*			X		
[1] *Ammospermophilus leucurus*			X	X	
[1,2] *Chaetodipus baileyi*			X	X	X
Chaetodipus formosus				X	
Neotoma lepida				X	
Ammospermophilus harrisii				X	X
[2] *Chaetodipus penicillatus*				X	X
[2] *Chaetodipus intermedius*				X	X
[2] *Peromyscus merriami*					X
Neotoma devia					X
Chaetodipus artus					X
Chaetodipus goldmani					X
Chaetodipus pernix					X

[1] species and/or closely related insular endemics occur on island(s) of Baja California peninsula
[2] species and/or closely related insular endemics occur on island(s) of the Sonoran mainland

insula in terms of history and isolating events. However, distributional limits instead may reflect differing ecological requirements or peninsular effects (*sensu* Simpson, 1964). Certainly, each species has its own ecological requirements, and although we find the degree of coincidence among species ranges to be remarkable, exact coincidence should not be expected. The failure of certain taxa (e.g., *Spermophilus tereticaudus*, *Perognathus longimembris,* and *Onychomys torridus*) to penetrate the Peninsular Desert may reflect some ecological limitation rather than recency of arrival. Lawlor (1983b) suggested that stenophagous species such as heteromyid rodents and grasshopper mice (*Onychomys*) may be more subject to extinction than other rodents, such that a peninsular effect would be more likely to occur in these species.

Taylor and Regal (1978) considered three explanations for their observation of reduced species diversity with increased distance from the mainland base (peninsular effect *sensu lato*) in reptiles, birds, and mammals of the Baja California peninsula: 1) too little time has elapsed since the peninsula became desert (following Orr, 1960); 2) species are limited ecologically; and 3) a balance exists between extinction and colonization (the peninsular effect *sensu stricto*). Subsequent studies demonstrated that there is no general peninsular effect among Baja California reptiles (Seib, 1980) or mammals (Lawlor, 1983b), and Brown (1987) found that butterflies of Baja California also did not conform to that principle. Brown (1987), Lawlor (1983b), and Seib (1980) each described two simultaneous and countering clines in species diversity (north to south and south to north) that

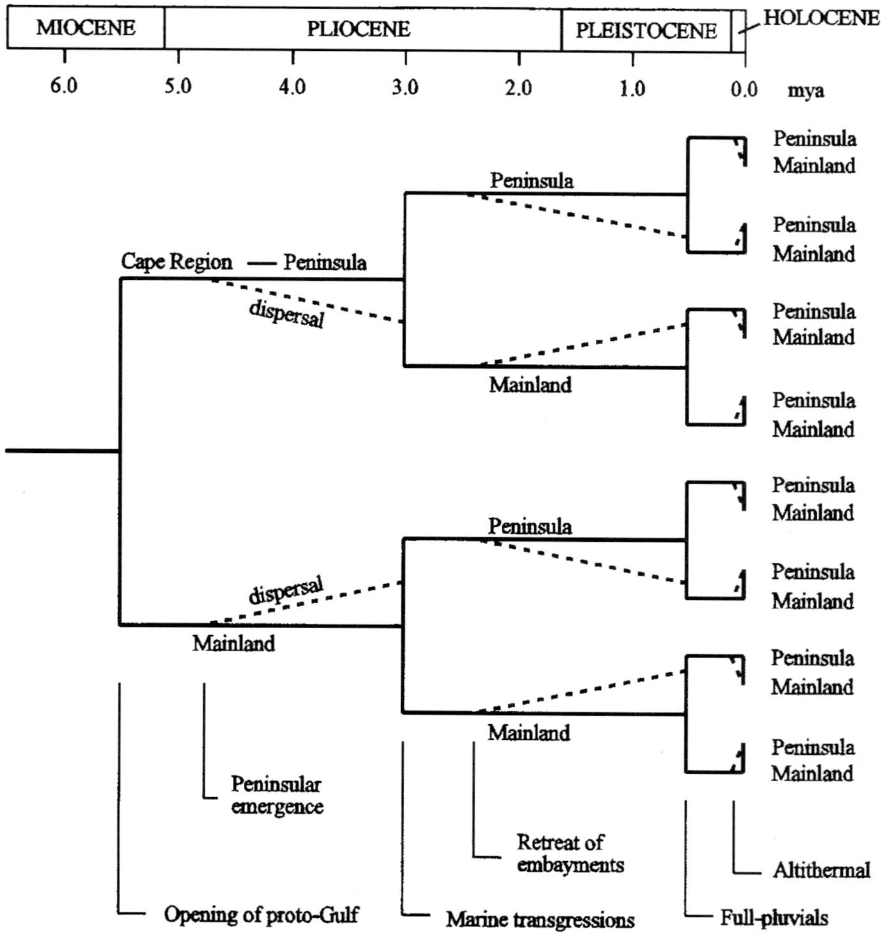

Fig. 8. Model depicting multiple possible Neogene origins of Peninsular Desert taxa. Dashed lines indicate dispersal following vicariant events.

indicate contact between northern dispersers and peninsular endemic reptiles (Seib, 1980), Californian and Cape Region butterfly faunas (Brown, 1987), and northern (or San Felipe) and Cape Region mammals (Lawlor, 1983b). These patterns of species diversity conform to our historical explanations rather than to ecological or peninsular-effect models.

Limitations of Species-Level and Faunal Similarity Analyses

Species-level analyses carry inherent limitations and biases for phylogenetic and biogeographic interpretation. Such analyses are dependent on taxonomic accuracy, yet it is reasonable to assume that some allopatric "species" would not be reproductively isolated if they were to experience secondary contact; that other allopatric conspecifics have speciated, but have not been recognized due to lack of a sympatric test; and that cryptic species remain undetected within other, particularly widespread, species. Taxonomic consideration of populations in allopatric desert regions is usually based on morphological criteria, which is likely to be affected by the ecological similarity of the regions: homoplasy or morphological stasis would be expected in ecologically similar regions, while allopatric taxa in dissimilar regions are likely to be more divergent morphologically, regardless of the sequence of divergence of the regions.

Finally, the criterion of reproductive isolation, while perhaps critical in terms of true separation of evolutionary lineages, is insufficient for phylogenetic and, hence, biogeographic analysis. Not only is this criterion non-hierarchical (based on attainment of a single grade of development), but widespread conspecifics may exhibit internally embedded biogeographic structure that is unrecognized at the species level. Speciation may arbitrarily preserve the record of an isolation event in a taxon that speciates while the same event is not preserved in a sympatric nonspeciating taxon. Thus, biogeographically informative genetic units, if not cryptic species, probably exist within both the widespread and circum-Gulf species, while identification of sister taxa of Baja California endemics would provide additional comparisons of genetic divergence with presumed isolating events.

Faunal similarity does not directly sort distributional data into historical versus ecological components. Even if all evolutionary differentiation were included in the species-level taxonomy (it is not), and eliminating the effects of species richness (e.g., by using Simpson's faunal similarity index), a similarity phenogram (e.g., Fig. 2) is not a biological area cladogram. Species unique to one region might include forms that were endemic to the region prior to isolation (paleoendemics) or autochthonous species (neoendemics). Taxa shared between regions could include those that occupied a common region prior to regional differentiation as well as forms that dispersed to the other region. Finally, absence of a taxon might reflect either historical absence or extinction. Thus, faunal similarity analysis provides a measure of the relative distinctiveness of regional biotas, but only an imperfect indication of the historical biogeographic affinities among biotas.

In short, limitations of species-level and faunal similarity analyses are largely the product of an ambiguous and unspecified mixture of units of analysis and the absence of an explicit hierarchical structure. Regional biotas are likely to include a temporally and spatially complex set of biogeographic relationships to the extent that taxa have responded differentially to geological and climatic events. Analysis of components of biogeographic structure (e.g., historical isolation and its approximate timing versus recent or ongoing dispersal) will therefore require: 1) an alternative conceptual framework that accommodates spatio-temporal hierarchical structure (i.e., phylogenetic systematics); 2) character data that are more likely to be reflective of shared common ancestry rather than of adaptation to particular environmental conditions; and 3) a character source that provides information on relative times of divergence and dispersal among populations and regional biota.

Phylogeographic Analysis of mtDNA

Recent advances in molecular genetic techniques, particularly analysis of mitochondrial DNA (mtDNA), provide powerful tools for revealing biogeographic history among regional populations (Avise, 1992, 1994; Avise et al., 1987; Riddle and Honeycutt, 1990). Avise

(1994) has found that phylogenetic gaps among different taxa revealed by mtDNA analysis may be geographically concordant, and that such gaps are likely to be associated with either boundaries between traditionally recognized biogeographic provinces or with boundaries postulated from geotectonic and paleoclimatic inferences. Riddle (1995), Riddle and Honeycutt (1990), and Riddle et al. (1993) have found these same qualities to hold in studies of mtDNA patterns among certain taxa of aridland rodents.

A molecular phylogeographic approach may resolve several of the disadvantages of reliance on a morphologically based, species-level data base. Specifically, the potential to identify relatively recent isolation and divergence among regional populations is enhanced using mtDNA. The overall evolutionary rate of this molecule is rapid relative to single-copy nuclear DNA (Wilson et al., 1985), with new mutations (i.e., nucleotide base substitutions) providing a pool of new character-states. Further, these mutations may more rapidly be fixed within a population, due to the smaller (female-only) initial effective population sizes for mtDNA relative to nuclear DNA. Given sufficient time for sorting of ancestral polymorphisms, the probability of discordance between gene genealogies and biogeographic area cladograms is minimized. Finally, the non-recombinant nature of mammalian mtDNA results in a prolonged signal of historical isolation and divergence following secondary contact of populations that are not reproductively isolated (e.g., phylogeographic category II of Avise, 1989; Riddle et al., 1993; Taberlet et al., 1992; Wayne et al., 1990).

In order to relate levels of mtDNA sequence divergence to known geologic events, it is necessary to rely on molecular clocks (Page, 1990); fossil evidence of adequate geographic and temporal sampling is rarely available. The concept of a universal molecular clock has been criticized due to heterogeneity in molecular evolutionary rates and errors involved with estimating those rates (Hillis and Moritz, 1990; Li et al., 1990). However, if generation time (Li et al., 1990) and/or metabolic rate (Martin and Palumbi, 1993; Rand, 1994) are primary sources of rate heterogeneity, then the error in estimating rates of divergence should be reduced substantially by calibrating the rate of divergence for a specified molecular sequence within a specified group of taxa (e.g., Bermingham and Lessios, 1993; Hafner et al., 1994; Irwin et al., 1991; Martin et al., 1992). For example, the temporally well-resolved split between *Mus* and *Rattus* (Jaeger et al, 1986) at 10 mya has been used to calibrate rates of molecular divergence in muroid rodents (Catzeflis et al., 1993).

Phylogeographic analysis of mtDNA from *Chaetodipus, Ammospermophilus,* and *Neotoma* of the Baja California peninsula provide support for both ancient and more recent isolation of the Peninsular Desert from the Mexican mainland. All are small rodents of similar body size, metabolic rate, and generation time as *Mus* and *Rattus*, permitting use of the *Mus–Rattus* split to calibrate rates of sequence divergence. Sequence data appropriate for calibrating times of divergence are already available for *Chaetodipus*, while sequence divergence estimates based on restriction fragment length polymorphisms (RFLP) are available for *Ammospermophilus* and *Neotoma*.

Riddle (1995) used mtDNA COIII and cytochrome b sequences to infer branching times among lineages in three genera of North American aridland rodents, *Onychomys, Perognathus,* and *Chaetodipus*. The sequence divergence rate (from transversion substitutions) of these genes based on the *Mus–Rattus* split was 0.8 to 1.1% per million years. Using this rate, divergence among most extant lineages in *Perognathus* and *Chaetodipus* appears to coincide with the late Tertiary transformation of savanna woodland to steppe and desert ecosystems in western North America. The emergence of "desertscrub" taxa in both pocket mouse genera coincide temporally with the postulated late Miocene–early Pliocene expansion of the regional deserts. Calibrated divergence times between the circum-Gulf *C. baileyi* and other *Chaetodipus* species are consistent with the middle Miocene separation of the Cape Region from the Sinaloan mainland (10.1-12.2% sequence divergence), and peninsular populations of *C. baileyi* appear to be a distinct phylogeographic group relative to Sonoran populations (based on mtDNA RFLP data; T.E. Lee, Jr., in litt.), in accord with allozymic and karyotypic data (Patton et al., 1981).

Although sequence data suitable for calibrating times of divergence are not yet available from *Ammospermophilus* or *Neotoma*, several lines of evidence point to historical isolation and divergence of endemic Peninsular lineages relative to mainland (Mojave and Sonoran) lineages. Preliminary RFLP examination of mtDNA haplotypes found in *A. leucurus* along the peninsula (J.W. Demastes, in litt.) revealed a 5.4% estimated sequence divergence that is coincident with the morphologically identified contact zone at about 30°N latitude (Hafner, 1981). Although this zone also represents the northern extent of a new genus of flea (*Tiptonella*) endemic to *A. leucurus* of the peninsula (T.D. Galloway, pers. comm.), it is unaccompanied by any concordant shift in allozyme complements of *A. leucurus* of the peninsula (K.K. Kleyboecker, in litt.). This overall pattern is consistent with mtDNA divergence unaccompanied by speciation, followed by secondary contact and introgressive hybridization (without transfer of the flea ectoparasite) between formerly allopatric populations. Planz (1992) has proposed recognition of peninsular *Neotoma* as a distinct species (*N. intermedia*) based on allozyme, bacular, and RFLP analysis of mtDNA. Mascarello (1978) had previously noted the distinction of peninsular *Neotoma* based on morphological (external, cranial, and bacular), allozyme, and chromosomal analyses. Both studies indicate that the peninsular form extends north into the California chaparral region, and contacts the Mojave Desert form in northeastern Baja California at about 30°N latitude.

In summary, estimates of genetic divergence based on sequence divergence of mtDNA haplotypes indicate that at least one mammal species (*C. baileyi*) may have originated on the Cape Region during its initial separation from the Sinaloan mainland. The origin of another species (the previously unrecognized *N. intermedia*) may date from Pliocene or Pleistocene isolation of the peninsula. Traces of these more-recent isolating events appear to have been retained in mtDNA patterns among populations of species that are distributed in both the Peninsular and Sonoran regional deserts (*A. leucurus* and *C. baileyi*). Fossil records of early Blancan age (Pliocene, 2-4 mya) are known from the Cape Region of the Peninsula for both *Ammospermophilus* and *Neotoma* (Miller, 1980). It would be of particular interest to investigate levels of divergence among populations of two other arid-adapted rodent species that are continously distributed between the two deserts (*Dipodomys merriami* and *Peromyscus* [*Haplomylomys*] *eremicus*) and among species of *Ammospermophilus, Chaetodipus, Haplomylomys,* and *Neotoma* that have a similar distribution. Avise et al. (1974) found that Peninsular populations of *P. eremicus* were more similar to Sonoran *P. merriami* than either were to Sonoran populations of *P. eremicus*, based on allozyme analysis. They suggested that eastern and western forms of *P. eremicus* may have evolved in isolation during a Pliocene or Pleistocene embayment of the Gulf of California. The rodent taxa that we consider most likely to have pre-Pleistocene lineages on the Peninsula each have fossil records in western North America that are of sufficient antiquity to place them in the geographic vicinity by at least the late Miocene: *Ammospermophilus* (Clarendonian, 8-12 mya; James, 1963); *Chaetodipus* (Hemingfordian, 16-20 mya; Wahlert, 1993); *Dipodomys* (Barstovian, 12-16 mya; Wahlert, 1993); *Peromyscus* (Hemphillian, 5-8 mya; Hibbard, 1964); and *Neotoma* (Hemphillian, 5-8 mya; Hibbard, 1968).

CONCLUSIONS

The Peninsular Desert of the Baja California peninsula south of 30°N latitude should be regarded as a separate regional desert of North America. For some floral and faunal elements, historical divergence may date from the middle Miocene separation of the Baja California peninsula from the Sonoran-Sinaloan mainland. For others, it began during the late Miocene and early Pliocene formation of regional semideserts, when the emergent peninsula was isolated from adjacent semideserts by marine transgressions. More recently, Peninsular Desert taxa were isolated in one or more desert refugia during late Pleistocene full-pluvial intervals. Subsequent to each period of isolation, taxa from adjacent regions had the opportunity to disperse between the Peninsular Desert and adjacent xeric regions. Despite limitations of species-level and faunal similarity analyses, traces of both ancient and more recent isolation and dispersal periods are evident in the plants, mammals, reptiles, birds, scorpions, and insects of the Peninsular Desert. Furthermore, phylogeographic patterns and divergence times estimated from analysis of mtDNA sequence data for small mammals thus far studied are consistent with those predicted from geological events.

ACKNOWLEDGMENTS

An earlier draft of this manuscript benefited greatly from the comments of J.H. Brown, G.L. Kirkland, and an anonymous reviewer. Laboratory analysis and access to unpublished data were generously provided by J.W. Demastes, K.K. Kleyboecker, and T.E. Lee, Jr.

LITERATURE CITED

Atwater, T. 1970. Implications of plate tectonics for the Cenozoic tectonic evolution of western North America. Geological Society of America Bulletin, 81:3513-3536.

Atwater, T., and P. Molner. 1973. Relative motion of the Pacific and North American plates deduced from sea-floor spreading in the Atlantic, Indian, and South Pacific Oceans. Pp. 136-148, *in* Proceedings of the conference on tectonic problems of the San Andreas Fault system (R.L. Kovach and A. Nur, eds.). Stanford University Publications in Geological Science, 13.

Avise, J.C. 1989. Gene trees and organismal histories: A phylogenetic approach to population biology. Evolution, 43:1192-1208.

Avise, J.C. 1992. Molecular population structure and the biogeographic history of a regional fauna: A case history with lessons for conservation biology. Oikos, 63:62-76.

Avise, J.C. 1994. Molecular markers, natural history, and evolution. Chapman and Hall, London, 511 pp.

Avise, J.C., M.H. Smith, R.K. Selander, T.E. Lawlor, and P.R. Ramsey. 1974. Biochemical polymorphism and systematics in the genus *Peromyscus*. V. Insular and mainland species of the subgenus *Haplomylomys*. Systematic Zoology, 23:226-238.

Avise, J.C., J. Arnold, R.M. Ball, E. Bermingham, T. Lamb, J.E. Neigel, C.A. Reeb, and N.C. Saunders. 1987. Intraspecific phylogeography: The mitochondrial DNA bridge between population genetics and systematics. Annual Review of Ecology and Systematics, 18:489-522.

Axelrod, D.I. 1950. The evolution of desert vegetation in Western North America. Carnegie Institute of Washington Publications, 590:215-306.

Axelrod, D.I. 1966. The Pleistocene Soboba flora of southern California. University of California Publications in Geological Science, 60:1-109.

Axelrod, D.I. 1967. Drought, diastrophism, and quantum evolution. Evolution, 21:201-209.

Axelrod, D.I. 1979. Age and origin of Sonoran Desert vegetation. Occasional Papers, California Academy of Sciences, 132:1-74.

Axelrod, D.I. 1983. Paleobotanical history of the western deserts. Pp. 113-129, *in* Origin and evolution of deserts (S.G. Wells and D.R. Haragan, eds.). University of New Mexico Press, 228 pp.

Banks, R.C. 1964a. The mammals of Cerralvo Island, Baja California. Transactions of the San Diego Society of Natural History, 13:397-404.

Banks, R.C. 1964b. Birds and mammals of the voyage of the "Gringa." Transactions of the San Diego Society of Natural History, 13:177-184.

Banks, R.C. 1967. The *Peromyscus guardia-interparietalis* complex. Journal of Mammalogy, 48:210-218.

Bermingham, E., and H.A. Lessios. 1993. Rate variation of proteins and mitochondrial DNA evolution as revealed by sea urchins separated by the Isthmus of Panama. Proceedings of the National Academy of Sciences USA, 90:2734-2738.

Betancourt, J.L., T.R. Van Devender, and P.S. Martin. 1990. Packrat middens: The last 40,000 years of biotic change. University of Arizona Press, Tucson, 467 pp.

Brand, L.R., and R.E. Ryckman. 1969. Biosystematics of *Peromyscus eremicus, P. guardia,* and *P. interparietalis*. Journal of Mammalogy, 50:501-513.

Brown, D.E., and C.H. Lowe. 1980. Biotic communities of the Southwest. United States Department of Agriculture, General Technical Report, RM-78 (map).

Brown, J.W. 1987. The peninsular effect in Baja California: An entomological assessment. Journal of Biogeography, 14:359-365.

Buising, A.V. 1990. The Bouse Formation and bracketing units, southeastern California, and western Arizona: Implications for the evolution of the proto-Gulf of California and the lower Colorado River. Journal of Geophysical Research, 95:20,111-20,132.

Catzeflis, F.M., A.W. Dickerman, J. Michaus, and J.A.W. Kirsch. 1993. DNA hybridization and rodent phylogeny. Pp. 159-172, *in* Mammal phylogeny: Placentals (F.S. Szalay, M.J. Novacek, and M.C. McKenna, eds.). Springer-Verlag, New York, 321 pp.

Chamberlin, R.V. 1924. Expedition of the California Academy of Sciences to the Gulf of California in 1921—The spider fauna of the shores and islands of the Gulf of California. Proceedings of the California Academy of Sciences, 12:561-694.

Cody, M.L. 1983. The land birds. Pp. 210-245, *in* Island biogeography in the Sea of Cortéz (T.J. Case and M.L. Cody, eds.). University of California Press, 508 pp.

Cody, M.L., R. Moran, and H. Thompson. 1983. The plants. Pp. 49-97 *in* Island biogeography in the Sea of Cortéz (T.J. Case and M.L. Cody, eds.). University of California Press, 508 pp.

Dickey, K.J. 1983. A natural history guide to Baja California. Published by the author, 226 pp.

Findley, J.S. 1969. Biogeography of southwestern boreal and desert mammals. Pp. 113-128, *in* Contributions in Mammalogy (J.K. Jones, ed.). University of Kansas Miscellaneous Publications of the Museum of Natural History, 51:1-428.

Findley, J.S. and W. Caire. 1977. The status of mammals in the northern region of the Chihuahuan Desert. Pp. 127-139, *in* Transactions of the symposium on the biological resources of the Chihuahuan Desert region, United States, and Mexico (R.H. Wauer and D.H. Riskind, eds.). National Park Service Proceedings and Transactions Series, 3:658 pp.

Friesen, R.D. 1979. Aspects of systematics and biogeography of *Dipodomys* (Rodentia: Heteromyidae) in the Tehachapi Mountain Region of California. Unpubl. Ph.D. dissertation, University of Southern California, 621 pp.

Gastil, R.G., and W. Jensky. 1973. Evidence for strike-slip displacement beneath the trans-Mexican volcanic belt. Pp. 171-180, *in* Proceedings of the conference on tectonic problems of the San Andreas Fault system (R.L. Kovach and A. Nur, eds.). Stanford University Publications in Geological Science, 13.

Gastil, R.G., R.P. Phillips, and E.C. Allison. 1975. Reconnaissance geology of the state of Baja California. Geological Society of America Memoirs, 140:xiv + 170 pp.

Gastil, R.G., J. Minch, and R.P. Phillips. 1983. The geology and ages of the islands. Pp. 13-25, *in* Island biogeography in the Sea of Cortéz (T.J. Case and M.L. Cody, eds.). University of California Press, 508 pp.

Grismer, L.L. 1994. The origin and evolution of the Peninsular herpetofauna of Baja California, Mexico. Herpetological Natural History, 2:51-106.

Hafner, D.J. 1981. Evolutionary relationships and historical zoogeography of antelope ground squirrels, genus *Ammospermophilus* (Rodentia: Sciuridae). Unpubl. Ph.D. dissertation, University of New Mexico, Albuquerque, 224 pp.

Hafner, D.J. 1992. Speciation and persistence of a contact zone in Mojave Desert ground squirrels, subgenus *Xerospermophilus*. Journal of Mammalogy, 73:770-778.

Hafner, M.S., P.D. Sudman, F.X. Villablanca, T.S. Spradling, J.W. Demastes, and S.A. Nadler. 1994. Disparate rates of molecular evolution in cospeciating hosts and parasites. Science, 265:1087-1090.

Hall, E.R. 1981. The mammals of North America. Second ed. John Wiley and Sons, New York, xv + 1181 + *90*.

Hibbard, C.W. 1964. A contribution to the Saw Rock Canyon local fauna of Kansas. Papers of the Michigan Academy of Science, Arts, and Letters, 49:115-127.

Hibbard, C.W. 1968. Paleontology, Pp. 6-26, *in* Biology of *Peromyscus* (Rodentia) (J.A. King, ed.). Special Publication, American Society of Mammalogists, 2:1-593.

Hillis, D.M, and C. Moritz. 1990. An overview of applications of molecular systematics. Pp. 502-515, *in* Molecular systematics (D.M. Hillis and C. Moritz, eds.). Sinauer Associates, Sunderland, Massachusetts, 588 pp.

Hubbard, J.P. 1973. Avian evolution in the aridlands of North America. Living Bird, 12:155-196.

Huey, L.M. 1964. The mammals of Baja California, Mexico. Transactions of the San Diego Society of Natural History, 13:85-168.

Humphries, C.J., and L.R. Parenti. 1986. Cladistic biogeography. Clarendon, Oxford, U.K.

Irwin, D.M., T.D. Kocher, and A.C. Wilson. 1991. Evolution of the cytochrome *b* gene of mammals. Journal of Molecular Evolution, 32:128-144.

Jaeger, J.-J., H. Tong, and C. Denys. 1986. The age of *Mus-Rattus* divergence: Paleontological data compared with the molecular clock. Comptes Rendus Academie des Sciences Paris (ser. 2), 302:917-922.

James, G.T. 1963. Paleontology and nonmarine stratigraphy of the Cuyama Valley Badlands, California. Part I. Geology, faunal interpretations, and systematic descriptions of Chiroptera, Insectivora, and Rodentia. University of California Publications in Geological Sciences, 45:1-170.

Karig, D.E., and W. Jensky. 1972. The proto-Gulf of California. Earth and Planetary Science Letters, 17:169-174.

Küchler, A.W. 1970. Potential natural vegetation (revised 1985). Pp. 89-92, *in* National atlas of the United States of America. U.S. Department of the Interior, U.S. Geological Survey.

Lawlor, T.E. 1971. Evolution of *Peromyscus* on northern islands in the Gulf of California, Mexico. Transactions of the San Diego Society of Natural History, 16:91-124.

Lawlor, T.E. 1983a. The mammals. Pp. 265-289, *in* Island biogeography in the Sea of Cortéz (T.J. Case and M.L. Cody, eds.). University of California Press, 508 pp.

Lawlor, T.E. 1983b. The peninsular effect on mammalian species diversity in Baja California. The American Naturalist, 121:432-439.

Li, W-H., M. Gouy, P.M. Sharp, C. O'Huigin, and Y-W. Yang. 1990. Molecular phylogeny of Rodentia, Lagomorpha, Primates, Artiodactyla, and Carnivora and molecular clocks. Proceedings of the National Academy of Sciences USA, 87:6703-6707.

Lonsdale, P. 1989. Geology and tectonic history of the Gulf of California. Pp. 499-521, *in* The Eastern Pacific Ocean and Hawaii (E.L. Winterer, D.M. Hussong, and R.W. Decker, eds.). The Geological Society of America, Vol. N:1-563.

Lucchitta, I. 1979. Late Cenozoic uplift of the southwestern Colorado Plateau and adjacent lower Colorado River region. Tectonophysics, 61:63-95.

Martin, A.P., G.J. P. Naylor, and S.R. Palumbi. 1992. Rates of mitochondrial DNA evolution in sharks are slow compared with mammals. Nature, 357:153-155.

Martin, A.P., and S.R. Palumbi. 1993. Body size, metabolic rate, generation time, and the molecular clock. Proceedings of the National Academy of Sciences USA, 90:4087-4091.

Mascarello, J.T. 1978. Chromosomal, biochemical, mensural, penile, and cranial variation in desert woodrats (*Neotoma lepida*). Journal of Mammalogy, 59:477-495.

Miller, W.E. 1977. Pleistocene terrestrial vertebrates from southern Baja California. Geological Society of America, Abstract with Programs, 9:468.

Miller, W.E. 1980. The late Pliocene Las Tunas local fauna from southernmost Baja California, Mexico. Journal of Paleontology, 54:762-805.

Milliman, J.D., and K.O. Emery. 1968. Sea levels during the past 35,000 years. Science, 162:1121-1123.

Munz, P.A., and D.D. Keck. 1949. California plant communities. El Aliso, 2:87-105.

Munz, P.A., and D.D. Keck. 1950. California plant communities. El Aliso, 2:199-202.

Murphy, R.W. 1983a. Paleobiogeography and patterns of genetic differentiation of the Baja California herpetofauna. Occasional Papers, California Academy of Sciences, 137:1-48.

Murphy, R.W. 1983b. The reptiles: Origins and evolution. Pp. 130-158, *in* Island biogeography in the Sea of Cortéz (T.J. Case and M.L. Cody, eds.). University of California Press, 508 pp.

Norris, R.M., and R.W. Webb. 1976. Geology of California. John Wiley and Sons, New York, 365 pp.

Orr, R.T. 1960. An analysis of the recent land mammals. Systematic Zoology, 9:171-179.

Page, R.D.M. 1990. Temporal congruence and cladistic analysis of biogeography and cospeciation. Systematic Zoology, 39:205-226.

Patton, J.L., S.W. Sherwood, and S.Y. Yang. 1981. Biochemical systematics of chaetodipine pocket mice, genus *Perognathus*. Journal of Mammalogy, 62:477-492.

Planz, J.V. 1992. Molecular phylogeny and evolution of the American woodrats, genus *Neotoma* (Muridae). Unpubl. Ph.D. dissertation, University of North Texas, Denton, 164 pp.

Porter, S.C. 1989. Some geological implications of average Quaternary glacial conditions. Quaternary Research, 32:245-261.

Rand, D.M. 1994. Thermal habit, metabolic rate, and the evolution of mitochondrial DNA. Trends in Ecology and Evolution, 9:125-131.

Riddle, B.R. 1995. Molecular biogeography in the pocket mice (*Perognathus* and *Chaetodipus*) and grasshopper mice (*Onychomys*): The late Cenozoic development of a North American aridlands rodent guild. Journal of Mammalogy, 76:283-301.

Riddle, B.R., and R.L. Honeycutt. 1990. Historical biogeography in North American arid regions: An approach using mitochondrial-DNA phylogeny in grasshopper mice (genus *Onychomys*). Evolution, 44:1-15.

Riddle, B.R., R.L. Honeycutt, and P.L. Lee. 1993. Mitochondrial DNA phylogeny in northern grasshopper mice (*Onychomys leucogaster*)—The influence of Quaternary climatic oscillations on population dispersion and divergence. Molecular Ecology, 2:183-193.

Roberts, N.C. 1989. Baja California plant field guide. Natural History Publishing Co., La Jolla, California, 309 pp.

Savage, J.M. 1960. Evolution of a Peninsular herpetofauna. Systematic Zoology, 9:184-212.

Schmidly, D.J., K.T. Wilkins, and J.N. Derr. 1994. Biogeography. Pp. 319-356, *in* Biology of the Heteromyidae (H.H. Genoways and J.H. Brown, eds.). Special Publication, American Society of Mammalogists, 10:1-719.

Schultz, T.A., F.J. Radovsky, and P.D. Budwiser. 1970. First insular record of *Notiosorex crawfordi*, with notes on other mammals of San Martin Island, Baja California, Mexico. Journal of Mammalogy, 51:148-150.

Seib, R.L. 1980. Baja California: A peninsula for rodents but not for reptiles. The American Naturalist, 115:613-620.

Shreve, F. 1940. The edge of the desert. Association of Pacific Geographers Yearbook, 6:6-11.

Shreve, F. 1942. The desert vegetation of North America. Botanical Review, 8:195-246.

Shreve, F. 1951. The vegetation of the Sonoran Desert. Carnegie Institute of Washington Publications, 591:1-192.

Shreve, F., and I.L. Wiggins. 1964. Vegetation and flora of the Sonoran Desert. Stanford University Press, 1740 pp.

Simpson, G.G. 1960. Notes on the measurement of faunal resemblance. American Journal of Science, 258A:300-311.

Simpson, G.G. 1964. Species density of North American recent mammals. Systematic Zoology, 13:57-73.

Spencer, J.E., and W.R. Normark. 1989. Neogene plate-tectonic evolution of the Baja California Sur continental margin and the southern Gulf of California, Mexico. Pp. 489-498, *in* The Eastern Pacific Ocean and Hawaii (E.L. Winterer, D.M. Hussong, and R.W. Decker, eds.). The Geology of North America, Vol. N:1-563.

Steinhauser, F. 1979. Climatic atlas of North and Central America. Volume I. Maps of mean temperature and precipitation. World Meteorological Organization, Geneva, 28 maps.

Taberlet, P., A. Meyer, and J. Bouvet. 1992. Unusual mitochondrial DNA polymorphism in two local populations of blue tit *Parus caeruleus*. Molecular Ecology, 1:27-36.

Taylor, R.J., and P.J. Regal. 1978. The peninsular effect on species diversity and the biogeography of Baja California. The American Naturalist, 112:583-593.

Truxal, F.S. 1960. The entomofauna with special reference to its origin and affinities. Systematic Zoology, 9:165-170.

Van Devender, T.R., and W.G. Spaulding. 1979. Development of vegetation and climate in the southwestern United States. Science, 240:701-710.

Wahlert, J.H. 1993. The fossil record. Pp. 1-37, *in* Biology of the Heteromyidae (H.H. Genoways and J.H. Brown, eds.). Special Publication, American Society of Mammalogists, 10:1-719.

Wayne, R.K., A. Meyer, N. Lehman, B. Van Valkenburgh, P.W. Kat, T.K. Fuller, D. Girman, and S.J. O'Brien. 1990. Large sequence divergence among mitochondrial DNA genotypes within populations of eastern African black-backed jackals. Proceedings of the National Academy of Sciences, USA, 87:1772-1776.

Webb, T., III, and P.J. Bartlein. 1992. Global changes during the last three million years: Climatic controls and biotic responses. Annual Review of Ecology and Systematics, 23:141-173.

Wiggins, I.L. 1960. The origin and relationships of the land flora. Systematic Zoology, 9:148-165.

Wiggins, I.L. 1980. Flora of Baja California. Stanford University Press, 1025 pp.

Wilcox, B.A. 1978. Supersaturated island faunas: A species-age relationship for lizards on post-Pleistocene land-bridge islands. Science, 199:996-998.

Williams, D.F., H.H. Genoways, and J.K. Braun. 1993. Taxonomy and systematics. Pp. 38-196, *in* Biology of the Heteromyidae (H.H. Genoways and J.H. Brown, eds.). Special Publication, American Society of Mammalogists, 10:1-719.

Williams, S.C. 1980. Scorpions of Baja California, Mexico, and adjacent islands. Occasional Papers of the California Academy of Sciences, 135:1-127.

Wilson, D.E., and D.M. Reeder (eds.). 1992. Mammal species of the world: A taxonomic and geographic reference. Second ed. Smithsonian Institution, 1206 pp.

Wilson, A.C., R.L. Cann, S.M. Carr, M. George, Jr., U.B. Gyllensten, K.M. Helm–Bychowski, R.G. Higuchi, S.R. Palumbi, E.M. Prager, R.D. Sage, and M. Stoneking. 1985. Mitochondrial DNA and two perspectives on evolutionary genetics. Biological Journal of the Linnean Society, 26:375-400.

Woodburne, M.O. (ed.). 1987. Cenozoic mammals of North America: Geochronology and biostratigraphy. University of California Press, 336 pp.

David J. Hafner, New Mexico Museum of Natural History, 1801 Mountain Road NW, Albuquerque, New Mexico 87104, and Brett R. Riddle, Department of Biological Sciences, University of Nevada–Las Vegas, 4505 Maryland Parkway, Las Vegas, Nevada 89154-4004

Appendix I. **Non-volant small mammals recorded from islands of the Baja California Peninsula, Mexico. Oceanic (versus landbridge) islands are listed in italics.**

Island	°N latitude	Species
PACIFIC OCEAN:		
Los Coronados	32°25'	*Peromyscus maniculatus assimilis*
Todos Santos	31°48'	*Peromyscus maniculatus dubius*
		Neotoma anthonyi
San Martín	30°29'	*Notiosorex crawfordi*
		Peromyscus maniculatus exiguus
		† *Neotoma martinensis*
San Gerónimo	29°47'	*Peromyscus maniculatus geronimensis*
Cedros	28°10'	*Sylvilagus bachmani cerrosensis*
		Chaetodipus fallax anthonyi
		Peromyscus eremicus cedrosensis
		Neotoma bryanti
Natividad	27°53'	*Peromyscus maniculatus dorsalis*
San Roque	27°09'	† *Peromyscus maniculatus cineritius*
Asunción	27°06'	*Peromyscus maniculatus*
Magdalena	24°50'	*Lepus californicus magdalenae*
		Thomomys bottae magdalenae
		Chaetodipus arenarius albulus
		Chaetodipus spinatus magdalenae
		Peromyscus eremicus
		Peromyscus maniculatus magdalenae
		Neotoma lepida pretiosa
Santa Margarita	24°25'	*Lepus californicus magdalenae*
		Chaetodipus spinatus margaritae
		Chaetodipus arenarius ammophilus
		Dipodomys merriami margaritae
		Peromyscus eremicus polypolius
		Peromyscus maniculatus margaritae
		Neotoma lepida pretiosa
SEA OF CORTÉZ		
Willard (San Luís Gonzaga)	29°50'	*Peromyscus crinitus pallidissimus*
		Peromyscus maniculatus hueyi
Mejia	29°34'	*Chaetodipus spinatus evermanni*
		Peromyscus guardia mejiae

Appendix I. Continued.

Granito	29°34'	*Peromyscus guardia harbisoni*
Angel de la Guarda	29°15'	*Chaetodipus spinatus guardiae* *Peromyscus guardiae guardiae* *Neotoma lepida insularis*
Coronado (Smith)	29°03'	*Chaetodipus baileyi* *Peromyscus maniculatus*
Salsipuedes	28°45'	*Peromyscus interparietalis ryckmani*
San Lorenzo Norte (Las Animas)	28°42'	*Chaetodipus spinatus lorenzi* *Peromyscus interparietalis lorenzi*
San Lorenzo (San Lorenzo Sur)	28°38'	*Chaetodipus spinatus lorenzi* *Peromyscus interparietalis interparietalis*
San Esteban	28°41'	*Peromyscus stephani*
Tortuga	27°26'	*Peromyscus dickeyi*
San Marcos	27°13'	*Ammospermophilus leucurus* *Chaetodipus spinatus marcosensis* *Neotoma lepida marcosensis*
San Ildefonso	26°37'	"two small rodent species" (Dickey, 1983)
Coronado	26°06'	*Chaetodipus spinatus pullus* *Peromyscus pseudocrinitus* † *Neotoma bunkeri*
Carmen	25°55'	*Lepus californicus sheldoni* *Chaetodipus spinatus occultus* *Peromyscus eva carmeni* *Neotoma lepida nudicauda*
Danzante	25°48'	*Chaetodipus spinatus seorsus* *Neotoma lepida latirostra*
Monserrat	25°41'	*Chaetodipus baileyi fornicatus* *Peromyscus caniceps*
Santa Catalina	25°39'	*Peromyscus slevini pallescens*
Santa Cruz	25°17'	*Peromyscus sejugis*
San Diego	25°13'	*Peromyscus sejugis*
San José	25°00'	*Sylvilagus mansuetus* *Chaetodipus spinatus bryanti* *Dipodomys merriami insularis* *Peromyscus eremicus cinereus* *Neotoma lepida perpallida*

Appendix I. Continued.

San Francisco	24°51'	*Chaetodipus spinatus latijugularis* *Neotoma lepida abbreviata*
Espíritu Santo	24°28'	*Lepus insularis* *Ammospermophilus insularis* *Chaetodipus spinatus lambi* *Peromyscus eremicus insulicola* *Neotoma lepida vicina*
Cerralvo	24°13'	*Chaetodipus arenarius siccus* *Peromyscus eremicus avius*

† extinct

Annotated Checklist of the Recent Land Mammals of Sonora, Mexico

WILLIAM CAIRE

Abstract

This annotated checklist describes the distribution of 124 native species of mammals known to occur in Sonora, Mexico. I have also included three introduced species of rodents, eight domesticated species, and 28 species of mammals which occur close to the political boundaries of Sonora and ultimately may be found in Sonora.

INTRODUCTION

Since Burt's (1938) account of Sonoran mammals, various species that occur there have been discussed in treatments of mammals in adjacent states: Arizona (Hoffmeister, 1986), Chihuahua (Anderson, 1972), Sinaloa (Armstrong and Jones, 1971; Armstrong et al., 1972), and Baja California (Huey, 1964). However, a current checklist of the recent mammals of Sonora and its islands is not available and has long been needed by a variety of biologists. Much of the current information about what species occur in Sonora and their distribution is scattered in the literature and in natural history museums containing collections of mammals from Sonora.

My aim is to provide a list that is as complete as possible through the end of 1994. It serves as a progress report of a larger monograph being prepared on the mammals of mainland Sonora, Mexico, and in adjacent waters. Burt (1938) listed 91 native and two introduced mammal species as occurring in Sonora. The list presented below contains 124 species. It lists three species which have been introduced, eight domesticated species, and 28 (indicated by an asterisk preceding the scientific name) which could possibly occur in Sonora because presently known distributional limits are near the political boundaries of the state.

METHODS

The list presented here was compiled from a search of the literature, visits to museums containing specimens from Sonora, collecting trips to Sonora, and correspondence with taxonomists and systematists familiar with the mammals of the area. The taxa are presented in sequence following Wilson and Reeder (1993) through species. Species within a genus are listed in alphabetical order. Vernacular names for most follow Jones et al. (1992) or Hall (1981).

A brief description of the distribution of each species in Sonora is provided. In summarizing the distributions, not every locality from which specimens are known is cited. I chose to use more familiar localities and selected some of these in order to guide "the mind's eye" geographically across Sonora. For example, the use of "near Alamos" should imply to the reader the very southern portions of the state. In several cases, specific localities are listed and these refer to places from which the few specimens of a particular species are known. Many, but not all, of the museum specimens and literature citations used to compile this checklist are listed by Caire (1978). Included is a map (Fig. 1) of Sonora which indicates the majority of the localities mentioned in the text.

Life Among the Muses: Papers in Honor of James S. Findley. Edited by T.L. Yates, W.L. Gannon & D.E. Wilson.

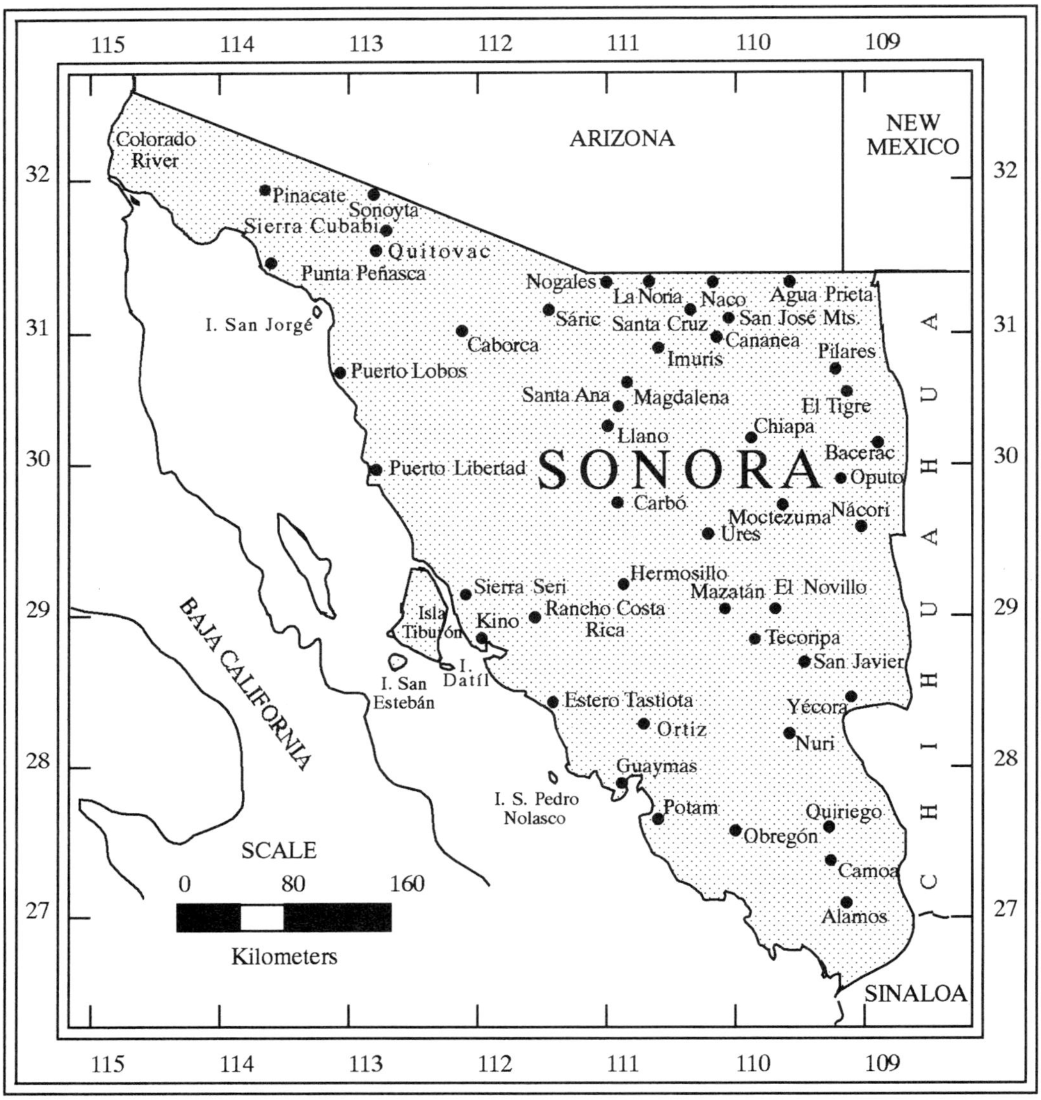

Fig. 1. Localities in Sonora, Mexico, listed in the species accounts.

RESULTS AND DISCUSSION

ORDER DIDELPHIMORPHA-Opossums
Family Didelphidae (Opossums)

Didelphis virginiana (Virginia Opossum) is known from near Alamos north through the foothill region to Ures, Oputo, and Llano.

Marmosa canescens (Grayish Mouse-opossum) is known from a single specimen captured near Alamos.

ORDER XENARTHRA–Xenarthrans
Family Dasypodidae (Armadillos)

Dasypus novemcinctus (Nine-banded Armadillo) is not known by specimens from Sonora although Leopold (1959) mapped the range of armadillos as including the southern part of the state. Inhabitants near Alamos speak of armadillos as frequenting the area and this species does occur in northern Sinaloa (Armstrong and Jones, 1971).

ORDER INSECTIVORA–Insectivores

Family Soricidae (Shrews)

Notiosorex crawfordi (Desert Shrew) is known from near Guaymas, Nogales, and Alamos, but probably occurs across the state in suitable habitats.

**Sorex arizonae* (Arizona Shrew) is not known from Sonora; however, it has been collected in southeastern Arizona (Hoffmeister, 1986) and in Chihuahua (Caire et al., 1978).

**Sorex monticolus* (Dusky or Montane Shrew) is not known from Sonora; however, records exist from southeastern Arizona (Hoffmeister, 1986), western Chihuahua (Anderson, 1972), and southwestern New Mexico (Findley et al., 1975), suggesting this shrew could be found in eastern Sonora.

ORDER CHIROPTERA–Bats

Family Emballonuridae (Emballonurid Bats)

Balantiopteryx plicata (Peter's bat) occurs in southern Sonora as far north as Ciudad Obregón.

Family Mormoopidae (Mormoopid Bats)

Mormoops megalophylla (Ghost-faced Bat) is known from southern Sonora. Northernmost records are near Carbó in central Sonora; however, two specimens captured in southern Arizona (Hoffmeister, 1986) suggest it probably occurs north of Carbó.

Pteronotus davyi (Davy's Naked-backed Bat) occurs at lower elevations from southern regions north to near Carbó in central Sonora.

Pteronotus parnellii (Parnell's Mustached Bat) ranges from southern regions of the state north to near Carbó in central Sonora.

Pteronotus personatus (Wagner's Mustached Bat) records are from near Alamos.

Family Phyllostomatidae (Phyllostomid Bats)

Macrotus californicus (California Leaf-nosed Bat) occurs throughout most of northern Sonora, south to Quiriego, and is known from Isla Tiburón.

Macrotus waterhousii (Waterhouse's Leaf-nosed Bat) occurs in southern Sonora to as far north as Quiriego.

Choeronycteris mexicana (Mexican Long-tongued Bat) occurs at low to medium elevations from near Alamos north through the central foothills into southeastern Arizona. No records exist from the arid northwestern areas or from higher elevations in the mountains of eastern Sonora.

Glossophaga soricina (Pallas' Long-tongued Bat) ranges from near Alamos north to El Novillo in central Sonora and Guaymas in the west.

Leptonycteris curasoae (Southern Long-nosed Bat) occurs over most of the state at low to moderate elevations and is absent from higher elevations of the mountains of eastern Sonora and the extremely arid areas of northwestern Sonora. It occurs on Isla Tiburón.

**Leptonycteris nivalis* (Mexican Long-nosed Bat) is not known by specimens from Sonora. It probably occurs in northeastern Sonora and possibly farther south because specimens exist in southwestern New Mexico and southeastern Arizona (Hoyt et al., 1994).

Artibeus hirsutus (Hairy Fruit-eating Bat) occurs from near Alamos north to Guaymas and El Novillo.

**Chiroderma salvini* (Salvin's White-lined Bat) is not known from Sonora, but two specimens captured in extreme southwestern Chihuahua (Anderson, 1972) suggest the possibility of its occurrence.

Sturnira lilium (Yellow-shouldered Bat) occurs at lower elevations from near Alamos to as far north as El Novillo.

Desmodus rotundus (Vampire Bat) is known from near Alamos and Pótam.

Family Natalidae (Funnel-eared Bats)

Natalus stramineus (Mexican Funnel-eared Bat) occurs from the Alamos area north through central regions to near Carbó, Ures, and Yécora.

Family Vespertilionidae (Vespertilionid Bats)

Antrozous pallidus (Pallid Bat) occurs over most of Sonora and on Isla Tiburón, but is less abundant in southern regions and at higher elevations of the mountains of eastern Sonora.

Eptesicus fuscus (Big Brown Bat) occurs over most of the state. Abundance declines toward the south and through the arid northwestern areas. It occurs on Isla Tiburón and probably in the Colorado River area.

**Euderma maculatum* (Spotted Bat) is not known by specimens from Sonora. Villa (1962) indicated natives near Nogales made references to a "pinto bat," possibly indicating the spotted bat.

**Idionycteris phyllotis* (Allen's Big-eared Bat) is not known by specimens from Sonora. Occurrences in southeastern Arizona (Hoffmeister, 1986) and northwestern Chihuahua (Anderson, 1972) suggest its occurrence in northeastern Sonora.

**Lasionycteris noctivagans* (Silver-haired Bat) occurs in southeastern Arizona (Hoffmeister, 1986) and southwestern New Mexico (Findley et al., 1975), but has not been found in Sonora. It could eventually be found in northeastern areas of the state.

Lasiurus borealis (Western Red Bat) ranges from near Alamos north through eastern and central areas to the United States' border. It is not known from coastal or western areas of Sonora.

Lasiurus cinereus (Hoary Bat) records of occurrence are from the Alamos area, north through the eastern half of Sonora into Arizona and New Mexico. A mummified specimen was found on Isla Datíl. It might also occur in the Colorado River area and at other localities in western Sonora where water sources exist.

Lasiurus xanthinus (Western Yellow Bat) occurs through the central region of Sonora from the Alamos area north into Arizona and New Mexico. It is absent from the relatively treeless areas of the west, but may occur in the Colorado River area because specimens are known to the north in Arizona (Hoffmeister, 1986).

Myotis auriculus (Southwestern Myotis) is known from near Alamos north to Santa Ana. It probably occurs north to the United States border at intermediate and higher elevations of the mountains of eastern Sonora because specimens exist from southeastern Arizona (Hoffmeister, 1986), southwestern New Mexico (Findley et al., 1975), and northwestern Chihuahua (Anderson, 1972).

Myotis californicus (California Myotis) occurs statewide at low to intermediate elevations and on Isla Tiburón.

Myotis fortidens (Cinnamon Myotis) occurs from near Alamos northward at lower elevations to Nácori and Moctezuma.

**Myotis ciliolabrum* (Western Small-footed Myotis) is not known by specimens from Sonora. It probably occurs in northeastern Sonora because specimens exist from southeastern Arizona (Hoffmeister, 1986), northwestern Chihuahua (Anderson, 1972), and southwestern New Mexico (Findley et al., 1975). Specimens reported as *M. leibii* by Ingles (1958) from Bahía San Carlos near Guaymas are *M. californicus*.

**Myotis lucifugus* (Little Brown Myotis) is not known by specimens from Sonora. It could occur in northeastern Sonora as records exist from northwestern Chihuahua (Anderson, 1972).

Myotis thysanodes (Fringed Myotis) occurs across the northeastern corner of the state, but possibly ranges farther south through the western foothills of the Sierra Madre.

Myotis velifer (Cave Myotis) range in Sonora closely parallels the foothill region from near Alamos north into Arizona. Higher elevations of the Sierra Madre and extremely arid western areas seemingly exclude this species.

Myotis vivesi (Fish-eating Bat) is becoming uncommon in Sonora. Former occurrence was along the Gulf of California coast of Sonora and on the adjacent islands. Records exist from Isla San Jorgé south to the Guaymas area.

**Myotis volans* (Long-legged Myotis) is not known by specimens from Sonora. However, its occurrence in southeastern Arizona (Hoffmeister, 1986), southwestern New Mexico (Findley et al., 1975), and northwestern Chihuahua (Anderson, 1972) suggests it might eventually be found in northeastern Sonora.

Myotis yumanensis (Yuma Myotis) occurs at lower and intermediate elevations from near Alamos north through central areas. It is absent from the arid western regions although a record exists from the Colorado River area.

Pipistrellus hesperus (Western Pipistrelle) ranges throughout Sonora at lower and medium elevations. It occurs on Isla Tiburón.

Plecotus mexicanus (Mexican Big-eared Bat) is known only from northeastern Sonora. Because specimens exist from southwestern

Chihuahua (Anderson, 1972), it might occur farther south in the mountains of eastern Sonora.

Plecotus townsendii (Townsend's Big-eared Bat) occurs from the Alamos region north through central and eastern areas. Records exist from Isla Tiburón.

Rhogeessa parvula (Little Yellow Bat) occurs in the Alamos area north to near Mazatán.

Family Molossidae (Molossid Bats)

Eumops perotis (Western Mastiff Bat) occurs from near Alamos north through central parts of the state in the foothills of the mountains of eastern Sonora to the United States border.

Eumops underwoodi (Underwood's Mastiff Bat) specimens exist from the Guaymas and Rancho Costa Rica areas. Because specimens exist from Sinaloa (Jones et al., 1972), southwestern Chihuahua (Anderson, 1972), and southern Arizona (Hoffmeister, 1986), this species probably occurs across the central length of Sonora at low to moderate elevations. No specimens are known from northwestern Sonora.

Nyctinomops aurispinosa (Peale's Free-tailed Bat) is known by two specimens from near Alamos.

Nyctinomops femorosaccus (Pocketed Free-tailed Bat) occurs at low to moderate elevations in central and eastern Sonora from the Alamos area north to the United States border. A record exists from near the Colorado River.

Nyctinomops macrotis (Big Free-tailed Bat) occurs near Alamos and Guaymas, and probably to the north since specimens exist from Arizona (Hoffmeister, 1986) and northwestern Chihuahua (Anderson, 1972).

Tadarida brasiliensis (Brazilian Free-tailed Bat) occurs throughout the state at low to moderate elevations and on Isla Tiburón.

ORDER CARNIVORA–Carnivores

Family Canidae (Canids)

Canis latrans (Coyote) occurs throughout the state and on Isla Tiburón.

Canis lupus (Gray Wolf) is known by only a few records from Sonora. Now extirpated, wolves probably occurred throughout the mountainous eastern portions of the state.

Urocyon cinereoargenteus (Common Gray Fox) occurs throughout the state and on Isla Tiburón.

Vulpes velox (Kit or Swift fox) records of occurrence are few and are from northern parts of the state south to west of Hermosillo.

Family Felidae (Cats)

**Herpailurus yaguarondi* (Jaguarundi) is not known by records in Sonora. This rare and endangered species is known to occur in Sinaloa (Armstrong et al., 1972) and a questionable sight record is listed by Hoffmeister (1986) for Arizona. It might occur in southern Sonora.

Leopardus pardalis (Ocelot) is rare and endangered in Sonora. Specimens exist from Camoa and Guirocoba (near Alamos) in southern Sonora. In former times, it probably ranged along the western slopes of the mountains of eastern Sonora.

**Leopardus wiedii* (Margay) is not known by specimens from Sonora; however, the close proximity of occurrences in Chihuahua (Anderson, 1972) and in Sinaloa (Armstrong et al., 1972) suggest it might occur in Sonora.

Lynx rufus (Bobcat) occurs throughout the state.

Puma concolor (Mountain Lion) occurred throughout the state in former times, but numbers are declining and it is now probably restricted to more remote regions.

Panthera onca (Jaguar) is rare and endangered in Sonora. It has been reported from the foothills near Nogales, and the northeastern and southern areas of the state. A sighting was recorded along the Colorado River delta region (Mearns, 1907).

Family Mustelidae (Mustelids)

Lontra canadensis (Northern River Otter) is recorded by Leopold (1959) from near where the international boundary intersects the Colorado River. I do not know of any preserved specimens from Sonora. Hoffmeister (1986) records specimens from Arizona along the Colorado River.

Lontra longicaudis (Southern River Otter) occurs in eastern Sonora from the Alamos area north to Los Pilares. It might occur farther north because specimens are known from western

Chihuahua near Pacheco (Anderson, 1972) near the Sonoran border.

Conepatus mesoleucus (Common Hog-nosed Skunk) probably ranges across the eastern part of the state from the Camoa area in the south, north to the Nogales area throughout the foothills, and at moderate elevations in the mountains of eastern Sonora. It probably occurs farther west.

Mephitis macroura (Hooded Skunk) occurs from the Alamos area north to Guaymas, Hermosillo, and to the westernmost record near Sierra Cubabi. It probably occurs in northeastern Sonora since records exist from southwestern New Mexico (Findley et al., 1975) and northwestern Chihuahua (Anderson, 1972).

Mephitis mephitis (Striped Skunk) is known only from the northeastern area of the state, but it might occur in the Colorado River area since records exist to the north from Arizona (Hoffmeister, 1986).

Spilogale gracilis (Western Spotted Skunk) probably occurs throughout the state, but there are only a few widely scattered records.

**Mustela frenata* (Long-tailed Weasel) is not known from Sonora; however, due to the close proximity of occurrence in southeastern Arizona (Hoffmeister, 1986) and Chihuahua (Anderson, 1972), it seems probable weasels do or at least in former times did occur in Sonora.

Taxidea taxus (American Badger) occurs throughout the state, but is absent from higher elevations in the east.

Family Procyonidae (Procyonids)

Bassariscus astutus (Ringtail) occurs across Sonora in areas of suitable rocky habitats and on Isla Tiburón.

Nasua narica (White-nosed Coati) occurs from southern parts of the state northward in the foothill region into Arizona and New Mexico.

Procyon lotor (Common Raccoon) occurs along waterways over much of the state. Occurrence is sparse in arid northwestern portions of the state.

Family Ursidae (Bears)

Ursus americanus (Black Bear) records are from northeastern Sonora. It probably is extant in remote areas of the mountains of the Sierra Madre.

Ursus arctos (Grizzly or Brown Bear), now extirpated, is known from a single record near Nogales in northeastern Sonora. Former distribution probably included the eastern mountainous areas of Sonora.

ORDER ARTIODACTYLA
Even-toed Ungulates

Family Tayassuidae (Peccaries)

Pecari tajacu (Collared Peccary) occurs statewide, but is rare in extreme northwestern regions west of Río Sonoyta and at higher elevations in the mountains of eastern Sonora.

Family Cervidae (Cervids)

Cervus elaphus (Wapiti or Elk) is known in Sonora by a sight record on San José Mountain in 1892 (Mearns, 1907). It could have been more plentiful in former times (Ceballos and Navarro L., 1991).

Odocoileus hemionus (Mule or Black-tailed Deer) occurs throughout the northern and western parts of the state and on Isla Tiburón, but is seemingly absent from eastern mountainous regions.

Odocoileus virginianus (White-tailed Deer) occurs statewide, but is rare in areas west of Río Sonoyta.

Family Antilocapridae (Pronghorn)

Antilocapra americana (Pronghorn) used to number in the thousands in Sonora. However, numbers have declined to about 100 individuals which occur in the Pinacate region and in extreme northeastern Sonora. Records of former occurrence exist from western plains of Sonora as far south as Hermosillo.

Family Bovidae (Bovids)

Bison bison (American Bison) is not known by specimens from Sonora, although Brand (1937) cites an account from northeastern Sonora. This species could have ranged into northern Chihuahua (Anderson, 1972), but it is not known from Arizona (Hoffmeister, 1986) or western New Mexico (Findley et al., 1975).

Ovis canadensis (Mountain or Bighorn sheep) still exists in small numbers in the Pinacate region of northwestern Sonora. Formerly, these sheep occurred in the mountains of northeastern Sonora and as far south as Sierra Seri in the west. They have been introduced to Isla Tiburón, and occur in mountains on the adjacent mainland.

ORDER RODENTIA–Rodents

Family Sciuridae (Squirrels)

Ammospermophilus harrisi (Harris' Antelope Squirrel) occurs across the northwestern region of the state south through central areas near Hermosillo, Ortiz, and Guaymas. Northeasternmost records are from near Agua Prieta.

Cynomys ludovicianus (Black-tailed Prairie Dog) is not known by preserved specimens from the state. A record is listed by Mearns (1907) from the San Pedro River, but the exact locality is unknown and could be in Arizona. Specimens are known from southeastern Arizona (Hoffmeister, 1986), southwestern New Mexico (Findley et al., 1975), and northwestern Chihuahua (Anderson,1972), suggesting prairie dogs probably occurred in northeastern Sonora in former times.

Sciurus aberti (Abert's Squirrel) is recorded by Leopold (1959) as occurring in northeastern Sonora. I examined a skin from Huachinera collected in 1935 by O.E. Muller and tentatively identified it as an Abert's squirrel (skull is lacking). It was originally deposited in the Cleveland Museum of Natural History and is now housed in the University of Michigan. Mearns (1907) reports that J.A. Allen recorded a capture of a specimen on the upper Río Bavispe in northeast Sonora by members of the Lumholtz Archaeological Expedition of 1890-1892.

Sciurus arizonensis (Arizona Gray Squirrel) is known from the northcentral region of Sonora near Chinapa, Magdalena, Nogales, and Santa Cruz.

Sciurus colliaei (Collie's Squirrel) occurs in southern Sonora north to San Javier.

Sciurus nayaritensis (Mexican Fox Squirrel) occurs at higher elevations of the mountains of eastern Sonora along the length of the state.

**Spermophilus madrensis* (Sierra Madre Mantled Ground Squirrel) has not been recorded from Sonora; however, it does occur in mountainous areas of western Chihuahua (Anderson, 1972) near Yécora, in Sonora.

Spermophilus spilosoma (Spotted Ground Squirrel) is known from extreme northeastern Sonora at La Noria and Naco.

Spermophilus tereticaudus (Round-tailed Ground Squirrel) range in Sonora is at lower elevations west of the foothills of the mountains of eastern Sonora from Camoa in the south, north to near Nogales, and west to the Colorado River. It occurs on Isla Tiburón.

Spermophilus variegatus (Rock Squirrel) occurs in rocky habitats nearly statewide and on Isla Tiburón.

Tamias dorsalis (Cliff Chipmunk) is known from northeastern Sonora near Bacerac and El Tigre at higher elevations of the mountains of eastern Sonora. It probably occurs to the south in the mountains of eastern Sonora since specimens are known from southwestern Chihuahua (Anderson, 1972). An isolated population occurs in the vicinity of Guaymas.

**Glaucomys volans* (Southern Flying Squirrel) is not known to occur in Sonora; however, two specimens are recorded in the literature (Goodwin, 1961) from the mountains of western Chihuahua (Anderson, 1972). Anderson (1972) indicated neither specimen had a field tag and that until more are found it should be regarded as unverified in Chihuahua.

Family Castoridae (Beaver)

Castor canadensis (American Beaver) is uncommon in Sonora. Records exist from the Colorado River area and riparian areas in the northeast.

Family Geomyidae (Pocket Gophers)

Thomomys bottae (Botta's Pocket Gopher) is common over most of Sonora at low to moderate elevations.

Thomomys umbrinus (Southern Pocket Gopher) occurs at moderate to higher elevations in eastern Sonora.

Family Heteromyidae (Heteromyids)

Dipodomys deserti (Desert Kangaroo Rat) occurs in western parts of Sonora from the Colorado River area east to near Caborca, then south along the coast to areas west of Hermosillo and Estero Tastiota.

Dipodomys merriami (Merriam's Kangaroo Rat) occurs statewide at lower elevations in suitable habitats and on Isla Tiburón.

Dipodomys ordii (Ord's Kangaroo Rat) occurs in the northeastern part of Sonora. Records are from near Magdalena, Cananea, Naco, and Nogales.

Dipodomys spectabilis (Banner-tailed Kangaroo Rat) occurs in the northcentral and eastern areas of Sonora and south to near Hermosillo. The westernmost record is near Quitovac.

Liomys pictus (Painted Spiny Pocket Mouse) occurs from southern Sonora northward in a narrow strip through central regions at lower and moderate elevations to near Nogales.

Chaetodipus artus (Narrow-skulled Pocket Mouse) occurs in the Alamos area.

Chaetodipus baileyi (Bailey's Pocket Mouse) occurs at low to moderate elevations from southern regions through the central foothills, the northeastern area, and over much of the western areas of the state. It occurs on Isla Tiburón.

**Chaetodipus formosus* (Long-tailed Pocket Mouse) is not known by specimens from Sonora. However, records exist from southeastern California and northern Baja California (Hall, 1981). The Colorado River might have been a barrier to dispersal into Sonora.

Chaetodipus goldmani (Goldman's Pocket Mouse) occurs from southern regions through the Río Yaqui and Río Bavispe valley drainage systems of eastern Sonora.

Chaetodipus hispidus (Hispid Pocket Mouse) is confined to a narrow strip across the northeastern part of the state. Specimens are known from the Cananea area.

Chaetodipus intermedius (Rock Pocket Mouse) occurs across the northern portion of the state, along the foothills, and south to the vicinity of Guaymas. It occurs on Isla Tiburón and on Isla Datíl (Hoffmeister, 1974; Lawlor, 1983).

Chaetodipus penicillatus (Desert Pocket Mouse) occurs statewide at low to moderate elevations and is known from Isla Tiburón.

Chaetodipus pernix (Broad-nosed Pocket Mouse) occurs in southern Sonora north to Guaymas and Tecoripa.

**Chaetodipus spinatus* (Spiny Pocket Mouse) is not known by specimens from Sonora and Hoffmeister (1986) reports none from southwestern Arizona. However, specimens are known from across the Colorado River in northern Baja California.

Perognathus amplus (Arizona Pocket Mouse) occurs in the arid western areas of the state from the Colorado River and along the coast south to near Hermosillo, and possibly farther south to near Guaymas.

Perognathus flavus (Silky Pocket Mouse) occurs at lower elevations in northeastern Sonora west to the Caborca region and south to Hermosillo and Guaymas.

Perognathus longimembris (Little Pocket Mouse) occurs in the arid western portion of the state from the Colorado River area east to Puerto Peñasco, then south along the coast to Kino and possibly to Guaymas.

Family Muridae (Mice, Rats and Voles)

**Microtus mexicanus* (Mexican Vole) is not known from Sonora, but a series from Colonia García in Chihuahua (Anderson, 1972) suggest it could occur at higher elevations in the mountains of northeastern Sonora.

**Ondatra zibethicus* (Common Muskrat) is not known by specimens preserved from Sonora. If it occurs, it is scarce and is probably limited to the Colorado River area and northern river systems. The source of Leopold's (1959) mapped occurrence in northeastern Sonora is unknown to me. Mearns (1907) indicated the Yuma Indians knew of muskrat around the Colorado River.

Mus musculus (House Mouse), an introduced species, probably occurs in most places inhabited by humans, including buildings and agricultural fields. Few specimens have been preserved.

**Rattus norvegicus* (Norway Rat), an introduced species, has not been collected in Sonora. I know of no specimens from Chihuahua and only one is reported from southern Arizona (Hoffmeister, 1986). It might eventually be found to occur around human habitations. Mearns (1907) indicated he saw dead specimens in the streets of Nogales.

Rattus rattus (Black Rat), an introduced species, probably occurs near all human habitations in Sonora, but few specimens have been preserved. A population exists on Isla San Estebán (Lawlor, 1983).

Baiomys taylori (Northern Pygmy Mouse) is known from agricultural regions around Ciudad Obregón in southern Sonora. Another possible area of occurrence is northeastern Sonora since specimens exist from southeastern Arizona (Hoffmeister, 1986), southwestern New Mexico (Findley et al., 1975), and northwestern Chihuahua (Anderson, 1972).

Neotoma albigula (White-throated Woodrat) occurs throughout the state and on Isla Tiburón, but is less abundant in eastern areas at higher elevations.

Neotoma devia (Arizona Woodrat) is known from extreme northwestern Sonora from the Pinacate and Puerto Peñasco areas westward.

Neotoma lepida (Desert Woodrat) May occur in extreme western Sonora near the Colorado River.

Neotoma mexicana (Mexican Woodrat) occurs at moderate to higher elevations in the mountains of eastern Sonora.

Neotoma phenax (Sonoran Woodrat) occurs from near Alamos north to Guaymas.

Neotoma varia (Isla Datíl; Lawlor, 1983) occurs only on Isla Datíl (see also Bogan, this volume).

**Onychomys arenicola* (Mearns' Grasshopper Mouse) is not known from Sonora, but because of the close proximity of occurrence in southwestern New Mexico (Findley, 1987), it might occur in northeastern Sonora.

Onychomys leucogaster (Northern Grasshopper Mouse) is known only from the northeastern part of Sonora near the international border. Westernmost specimens are from near La Noria.

Onychomys torridus (Southern Grasshopper Mouse) occurs statewide at low to moderate elevations. It is absent from higher elevations of the mountains of eastern Sonora.

Oryzomys couesi (Marsh Rice Rat) is known only from the vicinity of Guaymas. It probably occurs to the south and possibly farther north to the Kino region.

Peromyscus boylii (Brush Mouse) occurs at middle to higher elevations in the mountains of eastern Sonora and across northern parts of the state in the broken foothill terrain west to Sáric. An insular race, *P. b. glasselli*, is reported from Isla San Pedro Nolasco (Lawlor, 1971).

Peromyscus crinitus (Canyon Mouse) occurs in the extreme northwestern area of the state. Easternmost records are from near Sonoyta.

**Peromyscus difficilis* (Rock Mouse) is not known by specimens from Sonora, although it occurs in southwestern Chihuahua (Anderson, 1972). It might eventually be found at higher elevations in eastern Sonora.

Peromyscus eremicus (Cactus Mouse) occurs statewide at low to moderate elevations, and it occurs on Isla Tiburón and Isla Datíl.

Peromyscus leucopus (White-footed Mouse) occurs across the northeastern part of the state near the international border as far west as Sáric.

Peromyscus maniculatus (Deer Mouse) occurs across the northeastern region of Sonora in a narrow strip to near Nogales. It is rare in the arid desert areas of the west (May, 1973), but does occur near the Colorado River.

**Peromyscus melanotis* (Black-eared Mouse) is not known from Sonora. This species occurs in western Chihuahua (Anderson, 1972) near Pacheco and could eventually be found in isolated areas of the mountains of eastern Sonora.

Peromyscus merriami (Merriam's Mouse) ranges through central Sonora from Sonoyta and Nogales south to the Alamos region.

**Peromyscus nasutus* (Northern Rock Mouse) is not reported from Sonora; however, specimens are known from southeastern Arizona (Hoffmeister, 1986).

Peromyscus pembertoni (Pemberton's Mouse) is known only from Isla San Pedro Nolasco. Recent surveys to find it on the island have failed (Lawlor, 1983).

**Peromyscus polius* (Chihuahuan Mouse) is not known from Sonora. However, Anderson (1972) recorded specimens from near the northeastern Sonora–Chihuahua border in the higher elevations of the Sierra Madre. It will probably be found to occur in Sonora west of Colonia Garcia in Chihuahua.

Peromyscus stephani (Isla San Estebán Mouse) is known only from Isla San Estebán.

**Peromyscus truei* (Piñon Mouse) is not recorded from Sonora. However, records exist from northwestern Chihuahua (Anderson, 1972) and southwestern New Mexico (Findley et al., 1975) near the Sonoran border. Isolated areas in the mountains of eastern Sonora might support this species.

Reithrodontomys burti (Sonoran Harvest Mouse) records exist on the coastal plain from near Ciudad Obregón in the south to near Hermosillo farther north.

Reithrodontomys fulvescens (Fulvous Harvest Mouse) occurs in the eastern half of the state from the Alamos area north to near Nogales.

Reithrodontomys megalotis (Western Harvest Mouse) occurs across the northern parts of the state from the Colorado River to the Chihuahua–Sonora border. Southernmost record is near Puerto Lobos.

Reithrodontomys montanus (Plains Harvest Mouse) is known in Sonora by two specimens collected in the northeast near Agua Prieta and Nogales.

Sigmodon arizonae (Arizona Cotton Rat) occurs from the Alamos region north through the central portions of the state into Arizona. The westernmost specimens are from the vicinity of Caborca. It is not present at higher elevations in the mountains of eastern Sonora.

Sigmodon fulviventer (Tawny-bellied Cotton Rat) is known from only the northeastern parts of Sonora near the international border.

Sigmodon hispidus (Hispid Cotton Rat) is known from the vicinity of the Colorado River in northwestern Sonora and might be found in extreme northeastern Sonora as it occurs in southeastern Arizona (Hoffmeister, 1986) and southwestern New Mexico (Findley et al., 1975).

**Sigmodon ochrognathus* (Yellow-nosed Cotton Rat) is not reported from Sonora, but its occurrence in the northeastern parts of the state is almost certain. Specimens exist north of Nogales in Arizona (Hoffmeister, 1986), in southwestern New Mexico (Findley et al., 1975), and in northwestern Chihuahua (Anderson, 1972).

Family Erethizontidae (New World Porcupines)

Erethizon dorsatum (Common Porcupine) is rare in Sonora and is known by three specimens from near Imuris and Kino. It might occur in northeastern regions of the state because records exists from southwestern New Mexico (Findley et al., 1975) and northwestern Chihuahua (Anderson, 1972).

ORDER LAGOMORPHA–Lagomorphs
Family Leporidae (Hares and Rabbits)

Lepus alleni (Antelope Jackrabbit) occurs across the central portion of Sonora from southern parts of the state north into Arizona. It is less common in the desert areas of the extreme northwest and the higher elevations in the mountains of eastern Sonora. It occurs on Isla Tiburón.

Lepus californicus (Black-tailed Jackrabbit) occurs across the northern portion of the state south to near Hermosillo and Guaymas. It is not common farther south in neotropical areas nor at higher elevations.

**Lepus callotis* (White-sided Jackrabbit) is not known by specimens from Sonora. However, it does occur very close to the Sonoran border in extreme northwestern Chihuahua (Anderson, 1972) and southwestern New Mexico (Findley et al., 1975). A possible sight record exists from the base of the Huachuca Mountains in Arizona (Hoffmeister, 1986). I suspect it will be found in northeastern Sonora.

Sylvilagus audubonii (Desert Cottontail) occurs at moderate and low elevations over most of Sonora.

Sylvilagus floridanus (Eastern Cottontail) occurs at higher elevations in the mountains of eastern Sonora.

Domesticated Mammals: The following domesticated mammals occur in Sonora:

Canis familiaris (Dog), *Felis catus* (Cat), *Equus asinus* (Burro), *Equus caballus* (Horse), *Sus scrofa* (Pig), *Bos taurus* (Cow), *Capra hircus* (Goat), and *Ovis aries* (Sheep).

ACKNOWLEDGMENTS

This publication is dedicated to Jim Findley. Under his direction, I wrote my dissertation "The Distribution and Zoogeography of the Mammals of Sonora, Mexico," but, because of other involvement, I never published the results of this research. His continued interest and gentle prodding will, I'm sure, result in its eventual publication. During the time I was a student at the University of New Mexico, he helped shape my understanding of research. I will always admire his ability to question the events of nature that the rest of us often blindly accept. For all of this, I thank you, Jim.

LITERATURE CITED

Anderson, S. 1972. Mammals of Chihuahua, Mexico, taxonomy and distribution. Bulletin, American Museum Natural History,148:151-410.

Armstrong, D.M., and J.K. Jones, Jr. 1971. Mammals from the Mexican state of Sinaloa. I. Marsupialia, Insectivora, Edentata, Lagomorpha. Journal of Mammalogy, 52:747-757.

Armstrong, D.M., J.K. Jones, Jr., and E.C. Birney. 1972. Mammals from the Mexican state of Sinaloa. III. Carnivora and Artiodactyla. Journal of Mammalogy, 53:48-61.

Brand, D.D. 1937. The natural landscape of northwestern Chihuahua. University of New Mexico Bulletin, 316:1-74.

Burt, W.H. 1938. Faunal Relationships and geographic distribution of mammals in Sonora, Mexico. Miscellaneous Publication Museum of Zoology University of Michigan, 39:1-77.

Caire, W. 1978. The distribution and zoogeography of the mammals of Sonora, Mexico. Unpublished dissertation, The University of New Mexico, Albuquerque, New Mexico, 613 pp.

Caire, W., J.E. Vaughan, and V.E. Diersing. 1978. First record of *Sorex arizonae* (Insectivora: Soricidae) from Mexico. Southwestern Naturalist, 23:529-530.

Ceballos, G., and D. Navarro L. 1991. Diversity and conservation of Mexican mammals Pp. 167-198, *in* Latin American mammalogy, history, biodiversity and conservation (M.A. Mares and D.J. Schmidley, eds.). University of Oklahoma Press, Norman, Oklahoma, 408 pp.

Findley, J.S. 1987. The natural history of New Mexico mammals. University of New Mexico Press, Albuquerque, New Mexico, 164 pp.

Findley J.S., A.H. Harris, D.E. Wilson, and C. Jones. 1975. Mammals of New Mexico. University of New Mexico Press, Albuquerque, New Mexico, 360 pp.

Goodwin, G.G. 1961. Flying squirrels *(Glycomys volans)* of Middle America. American Museum Novitates, 2059:1-22.

Hall, E.R. 1981. The Mammals of North America. Second ed. John Wiley and Sons, New York, 1:1-600 + 90 and 2:601-1181 + 90.

Hoffmeister, D.F. 1974. The taxonomic status of *Perognathus penicillatus minimus* Burt. Southwestern Naturalist, 19:213-214.

Hoffmeister, D.F. 1986. Mammals of Arizona. University of Arizona Press, Tucson, Arizona, 602 pp.

Hoyt, R.A., J.S. Altenbach, and D.J. Hafner. 1994. Observations on long-nosed bats (*Leptonycteris*) in New Mexico. Southwestern Naturalist, 39:175-179.

Huey, L.M. 1964. The mammals of Baja California. Transactions of the San Diego Society of Natural History, 13:85-165.

Ingles, L.G. 1958. Notas acerca de los mamiferos mexicanos. Anales de la Instituto Biologicas, Mexico, 29:379-407.

Jones, J.K., Jr., J.R. Choate, and A. Cadena. 1972. Mammals from the Mexican state of Sinaloa. II. Chiroptera. Occasional Papers of the Museum of Natural History, University of Kansas, 6:1-29.

Jones, J. Knox, Jr., R.S. Hoffman, D.W. Rice, C.J. Jones, R.J. Baker, and M.D. Engstrom. 1992. Revised checklist of North American Mammals north of Mexico, 1991. Occasional Papers The Museum, Texas Tech University, 146:1-23.

Lawlor, T.E. 1971. Distribution and relationships of six species of *Peromyscus* in Baja California and Sonora, Mexico. Occasional Paper of the Museum of Zoology, University of Michigan, 661:1-29.

Lawlor, T.E. 1983. The Mammals. Pp. 265-289, *in* Island Biogeography in the Sea of Cortez (T.J. Case and M.L. Cody, eds.). University of California Press, Berkeley, California, 508 pp.

Leopold, A.S. 1959. Wildlife of Mexico, The game birds and mammals. University of California Press, Berkeley and Los Angeles, California, 568 pp.

May, L.A. 1973. Resource reconnaissance of the Gran Desierto region, northwestern Sonora, Mexico. M.S. thesis, The University of Arizona, Tucson, Arizona, 173 pp.

Mearns, E.A. 1907. Mammals of the Mexican boundary of the United States: A descriptive catalogue of the species of mammals occurring in that region; with a general summary of the natural history, and a list of trees. United States National Museum Bulletin, 56:1-530.

Villa, R.B. 1962. Nota acerca de la distribution

de los murcielagos *Euderma maculatum* (J.A. Allen) y *Chiroderma isthmicum* en Mexico. Anales de la Instituto Biologicas, Mexico, 33:379-384.

Wilson, D.E., and D.M. Reeder. 1993. Mammal species of the world: A taxonomic and geographic reference. Smithsonian Institution Press in association with The American Society of Mammalogist, Washington, D.C., 1207 pp.

William Caire, Biology Department, University of Central Oklahoma, Edmond, Oklahoma 73034

On the Status of *Neotoma varia* from Isla Datil, Sonora

Michael A. Bogan

Abstract

Neotoma varia Burt, 1932 from Isla Datil (= Turner's Island), Sonora, previously known only by the type specimen, is reduced to subspecific status under *Neotoma albigula*, based on comparisons with additional material. Specimens of *N. varia* were compared with specimens of other subspecies of *N. albigula* from Sonora using discriminant and principal component analyses.

INTRODUCTION

During the winter of 1931-1932, William Henry Burt accompanied an expedition to islands in the Gulf of California (Sea of Cortez). The objective of the expedition was the continuation of the work Donald R. Dickey and others had started, namely the discovery and elaboration of the kinds of small mammals inhabiting the Gulf islands. Among several new taxa found by them, one of the more enigmatic was a single specimen of a woodrat, later named *Neotoma varia* by Burt (1932), from a small island just south of Isla Tiburon. The island, called Turner's Island by Burt but known generally as Isla Datil (Gastil et al., 1983), has been visited by other mammalogists since Burt worked there. For example, Richard Van Gelder procured the jaw of a woodrat (AMNH 180715) from the east side of the island in 1957. However, no other complete specimens of this putative endemic species were obtained until 1975 when Edward Roth trapped three woodrats, two adults and one juvenile, on the north end of the island at an elevation of 200 feet (61 m). Roth's specimens are in the Collection of Mammals at the University of Arizona.

Burt (1932:179) thought it advisable "to give full specific designation to the Turner's Island Woodrat because of the relative constancy in tooth pattern in the genus *Neotoma* and because of this marked divergence [of the type and only specimen of *N. varia*] from specimens of *Neotoma albigula*." Later in the same paper he commented (p. 182), however, "that the primary purpose of systematic studies—the outlining of true relationships and natural groupings—is defeated when slightly differentiated insular forms are given full specific rank chiefly because there is no present geographic intergradation with other insular forms or with those on the mainland." In keeping with this approach, he then treated three previously described insular species of woodrats from the peninsular side of the Gulf as subspecies of *N. lepida* and described two additional subspecies (of *lepida*) from other islands. As noted by Lawlor (1983), *albigula* is the species found on Gulf islands nearer the Sonoran side, *lepida* is found on islands closer to Baja California.

Burt (1938) recorded five subspecies of *albigula* in the state of Sonora, Mexico, and described the ranges as follows: *N. a. albigula* Hartley, 1894 from the desert area of north–central Sonora south as far as Hermosillo; *N. a. sheldoni* Goldman, 1915 from the Pinacate Mountain region east to Saric; *N. a. venusta* True, 1894 from a narrow strip along the gulf coast of Sonora south to Costa Rica Ranch; *N. a. melanura* Merriam, 1894 from the lowlands of southern Sonora north as far as Ortiz; and *N. a. seri* Townsend, 1912 from Isla Tiburon. Burt (1938) retained *N. varia* as a full species.

Hall and Kelson (1959) concurred with these distributions, although their range map implies the possible occurrence of *N. a. mearnsi*

Life Among the Muses: Papers in Honor of James S. Findley. Edited by T.L. Yates, W.L. Gannon & D.E. Wilson.

Goldman, 1915 in northwestern Sonora. Hall and Genoways (1970) and subsequently Hall (1981) adjusted the depiction of the range of *mearnsi* but otherwise maintained the same distributions. Hoffmeister (1986) discussed the status of several of these taxa in Arizona and noted that up to six subspecies of *albigula* might occur in Arizona. After careful scrutiny, Hoffmeister chose to recognize only three: *albigula*, *mearnsi*, and *venusta*.

Given the unknown status of *varia*, other than as a relative of *albigula* (e.g., Hall, 1981), and the availability of two additional adult specimens of woodrats from Isla Datil, I re-examined the status of this taxon.

MATERIALS AND METHODS

I examined 58 woodrats from Sonora in the collections of the National Museum of Natural History (USNM), San Diego Museum of Natural History (SDNHM), Los Angeles County Museum (LACM), Donald R. Dickey (DRD, located at UCLA), and the University of Arizona Museum of Natural History (AU). I did not examine the jaw in the American Museum of Natural History (AMNH 180715). Using adult animals, as indicated by presence of adult pelage and permanent dentition at occlusal level, I made notes on pelage coloration, external details, dental and cranial configuration, and recorded the following measurements from the skin label or with dial calipers to the nearest 0.05 mm: total length, tail length, hindfoot length, ear length, greatest skull length, length of nasal bone, interorbital breadth, mastoid breadth, condyloincisive length, maxillary toothrow length, and maxillary breadth. The listing of specimens examined provides numbers of animals from each locality.

Discriminant function analyses (BMD07M) were performed on various groupings of specimens to distinguish among them and to explore possible classification possibilities. Means and variances were generated as a portion of the analysis. Principal component analyses (BMD09M) were conducted on the correlation matrix of characters to discern patterns of variation among individual animals not *a priori* assigned to groups. Analyses were conducted on the U.S. Department of Interior IBM computer in Washington, D.C.

RESULTS AND DISCUSSION

The type specimen of *N. varia* is an adult female, the characteristics of which are detailed in Burt (1932). Although he mentioned external characters of size and color that distinguished the type from specimens of *N. a. albigula*, *N. a. seri*, *N. a. melanura*, and *N. a. sheldoni*, Burt seemed most impressed with details of toothrow structure and form. He noted (1932:179) that, "the one specimen of *varia* available differs also from specimens representing the races mentioned above in the shape of the maxillary tooth rows and in the structure of the last upper molar. In all the specimens of *albigula*, *seri*, *melanura*, and *sheldoni* examined the outer border of the maxillary tooth row forms a nearly straight line whereas in *varia* the outer border is distinctly convex. The last upper molar in *varia* differs from that of all the others examined in having but two lobes with one external reentrant angle instead of three lobes with two external reentrant angles. Whether this is an abnormal condition present in the one specimen available or whether it is a constant character will not be known until more specimens are obtained from Turner's Island." He further noted that the tooth conditions described are not the result of age or wear. I examined the type of *varia* and concur with Burt's description in all essential details.

The woodrats collected by Roth from Isla Datil do not exhibit the tooth characters of the type. The toothrows of these specimens are straight, as in other *Neotoma* I have examined (although USNM 513991, *N. a. seri* from Isla Tiburon, has slightly convex upper toothrows). With respect to Burt's emphasis on shape of M3, both adults taken by Roth possess the typical complement of three lobes with two external reentrant angles. The first anterior lobe is small, but I have observed this feature on other specimens of *albigula* as well, especially among woodrats from Isla Tiburon (*N. a. seri*). Thus, I question the constancy of the dental features Burt ascribed to the woodrat population represented by the type of *N. varia*.

Other characteristics mentioned by Burt (1932) seem within the range of individual variation. Unlike the type, Roth's two adult specimens do not possess especially heavier rostra with more incurved incisors, broader nasals, shorter palatal bridges, lighter pterygoids, or different coronoid processes. Burt noted the skull of the type is more arched anteroposteriorly on the dorsal surface than that of *N. albigula seri*; I fail to discern such a difference in the more recent specimens. The skins of the Roth specimens are slightly darker (blacker) compared to the type of *N. varia* and to specimens of *N. albigula seri* (the latter known to me by a series from Isla Tiburon in the USNM); however, the differences are subtle.

To compare *N. varia* with subspecies of *N. albigula* other than *seri*, within a context of regional variation, I conducted discriminant function analyses using additional specimens from mainland Sonora. These analyses used only cranial measurements because external measurements are missing for several specimens. The analyses revealed considerable overlap among the subspecies in discriminant space, suggesting that the taxa are very similar. *N. a. albigula* occupied a central position in the plot (Fig. 1a); most specimens of *venusta* and *seri* are placed to the right, whereas *sheldoni* and *melanura* occupy positions that may reflect slight shape differences among these taxa and *albigula*. Roth's two adult specimens of *varia* are plotted closest to specimens of *seri* and *sheldoni*. The type of *varia* falls outside the cluster of specimens. Carleton (*in litt.*) noted that just such an extreme placement would be expected from a discriminant function analysis as it exaggerates the singularity of the *varia* holotype given the *a priori* definition of groups and the weighting factors assigned to the mensural variables to achieve separation of the groups.

In a similar analysis, specimens of *varia* were entered as unknowns, and the discriminant analysis program calculated probabilities of the specimens belonging to the other five target groups (subspecies). Roth's two (adult) specimens were assigned to *sheldoni* (V2) and *albigula* (V3) with probabilities of 58.3% and 54.7%, respectively. The type was assigned to *albigula* with a probability of 61.8%, an anomalous figure when the two-dimensional plot of this analysis is viewed, as the type is almost off (lower left corner) the computer plot (Fig. 1b). Roth's specimens are positioned near the polygons of those taxa to which they were assigned by the program (Fig. 1b).

The data also were analyzed with a principal component analysis (Fig. 2). Again, there was considerable overlap among the taxa. The first two principal components (PCs) accounted for 83% of the variation. Specimens were arranged generally by size with more specimens of *venusta* and *albigula* (the largest subspecies) to the right and specimens of *sheldoni* and *seri* to the left of the plot. *N. a. melanura* occupied a more central position on the first two PCs. All three adult specimens of *varia* were plotted within the overlapping polygons containing specimens of *seri* and *sheldoni*. The scatter of specimens of *venusta* tended to overlap all other groups. In Arizona, *venusta* is the largest subspecies whereas *mearnsi* is the smallest (Hoffmeister, 1986). An investigation of the contact zone between *mearnsi* and *sheldoni* would be of interest for, as Hoffmeister points out, the type localities for these two subspecies are less than 80 km apart.

Isla Datil, with an area of 4 km^2, is 2 km from the southern shore of Isla Tiburon, separated by a channel less than five fathoms deep. According to Gastil et al. (1983), Datil consists of basement rock and was probably isolated from Isla Tiburon through erosion during the Holocene. Tiburon itself, about 2 km from the mainland and separated by a minimum channel depth of three fathoms, parted from mainland Sonora in the Pleistocene, primarily by faulting. Exact times of separation of the islands from the mainland or each other may never be known; probable times of separation for Datil (from Isla Tiburon) and for Tiburon from the mainland range between 5,000–10,000 ybp (Wilcox, 1978; Lawlor, 1983).

Only two other species of land mammals are known from Isla Datil, a *Chaetodipus* and a *Peromyscus*. The pocket mouse was believed by Burt (1932) to represent an endemic subspecies (*minimus*) of *Chaetodipus penicillatus*. He compared his single specimen with specimens of *C. p. seri* from Isla Tiburon and *C. p. pricei* from

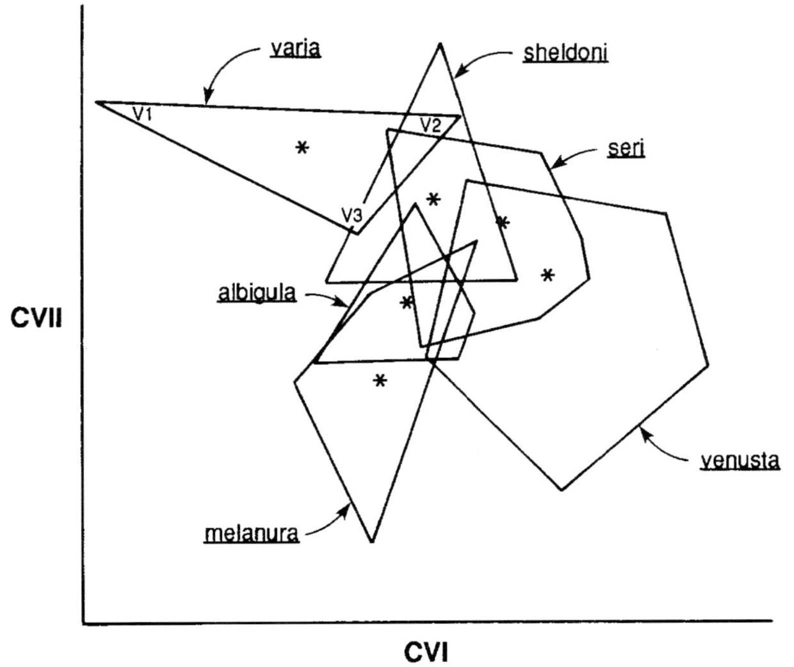

a.

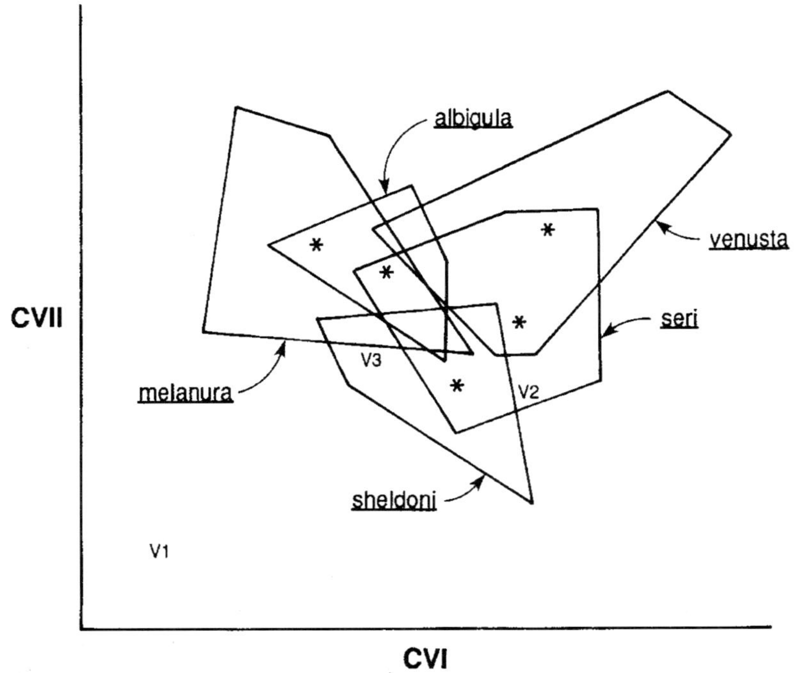

b.

Fig. 1a. Discriminant function analysis of six groups (subspecies) of *Neotoma albigula* from Sonora, Mexico. Polygons encompass all specimens from each group and asterisks represent group centroids.

Fig. 1b. Discriminant function analysis of five groups (subspecies) of *Neotoma albigula* from Sonora, Mexico. For this analysis specimens of *N. varia* were assigned by the analysis. Polygons encompass all specimens from each group. For both analyses, V1 is the type of *N. varia* and V2 and V3 are the adult specimens taken by Roth.

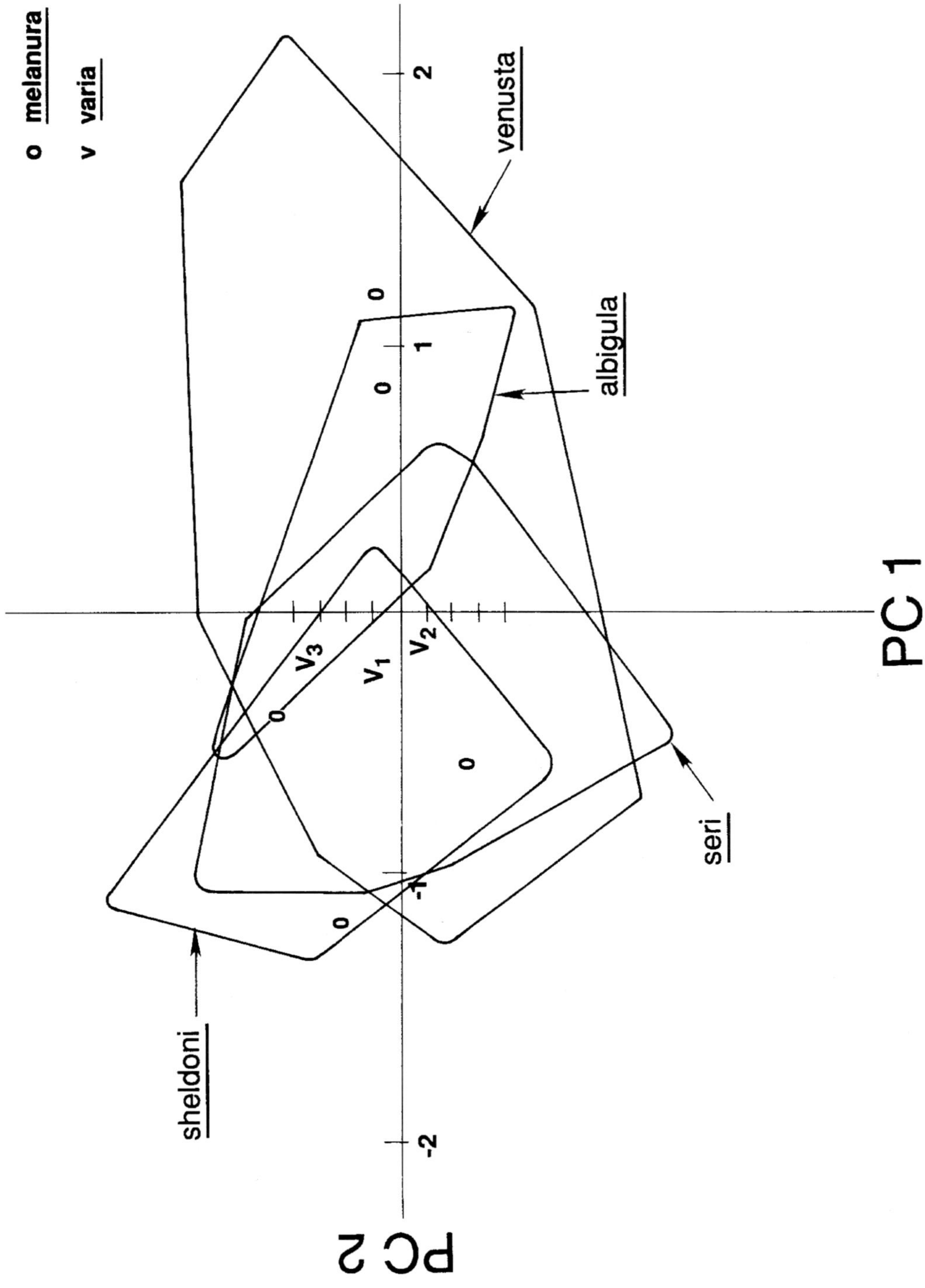

Fig. 2. Principal component analysis of woodrats from Sonora, Mexico. Polygons enclose all specimens from each group subjected to analysis.

the mainland. Hoffmeister and Lee (1967), in their study of variation in *C. penicillatus*, referred *minimus* to *C. intermedius*; Hoffmeister (1974) later formally recognized this pocket mouse as *C. intermedius minimus*. Burt (1932) thought the *Peromyscus* was specifically distinct and named it *P. collatus*. Lawlor (1971) reduced *collatus* to a subspecies of *P. eremicus*, and noted that, except for the metacentric (rather than submetacentric) X chromosome, the differences separating *collatus* and *eremicus* are subtle and reminiscent of geographic variation exhibited by mainland populations of *eremicus*. Hall (1981) retained *collatus* as a full species. Both *C. intermedius* and *P. eremicus* occur on Isla Tiburon and on the mainland. Although Hall's (1981) map (p. 552) does not show *C. intermedius* on Isla Tiburon, Lawlor (1971) mentions the occurrence of, and I have trapped, *C. intermedius* on Isla Tiburon.

Given the relatively short time of isolation of woodrats on Isla Datil from those on Tiburon, the variable nature of the diagnostic characters of *varia*, and the generally subtle variation among the specimens and subspecies of *N. albigula*, I recommend that woodrats from Isla Datil, Sonora, should henceforth be known as *Neotoma albigula varia* Burt, 1932. This usage conforms to Burt's own apparent philosophy regarding insular species and to the current subspecific status of the other two species on Isla Datil. Moreover, this recommendation acknowledges the systematic relationship with the widespread and variable *N. albigula*, while still conveying a level of morphological distinctness appropriate to the isolation of woodrats on Isla Datil. An alternative view, and one supported by principal component analysis, would be to recognize only one subspecies in Sonora: *N. a. albigula*. I do not believe this treatment would accurately convey the extent of geographic variation in characters such as size and color of woodrats in Sonora.

The current biological status of Isla Datil woodrats is unknown to me at this writing. During a one-night visit to the island in 1976, we deployed 80 traps, including both Museum Specials and Victor rat traps. We captured small numbers of *Chaetodipus intermedius minimus* and *Peromyscus eremicus collatus*, saw fresh woodrat sign, and picked up two small skull fragments of *Neotoma*; however, we did not capture any woodrats.

ACKNOWLEDGMENTS

This paper is affectionately dedicated to Mateo Aguayo, Heriberto Ortega, Manuel Palacios, Margarito Rodriguez, and Vicente Velasquez, Mexican co-workers who, following an unexpectedly high tide on Isla Datil, diligently and successfully salvaged both our boat and motor from the sea, thus allowing our timely return to Isla Tiburon. Robert Fisher, Robert Reynolds, Richard Smartt, and Bruce Woodward assisted with field work on Isla Datil. Support for these studies was provided by the (then) Direccion General de la Fauna Silvestre of Mexico, which also provided permits for work on the islands, and the U.S. Fish and Wildlife Service under the terms of the Joint Committee Agreement on Wildlife Conservation. Clyde Jones was instrumental in arranging myriad details of the trip and Norm Scott provided logistical support and coordination. Yar Petryszyn facilitated the loan of specimens of *Neotoma* from the Collection of Mammals at the University of Arizona; curators at other institutions granted access to specimens in their collections. Michael Carleton, Kenneth Geluso, Clyde Jones, Timothy Lawlor, Rodrigo Medellin, and an anonymous reviewer commented on earlier versions of the manuscript.

SPECIMENS EXAMINED (ALL MEXICO: SONORA)

Neotoma a. albigula (5): Oputo, 2 (USNM); Costa Rica Ranch, 3 (USNM); *N. a. melanura* (6): Camoa, 1 (USNM); Batamotal, 2 (USNM); Alamos, 3 (USNM); *N. a. seri* (23): Isla Tiburon (various localities), 23 (USNM); *N. a. sheldoni* (6): Pinacate Mountains, 6 (USNM); *N. a. varia* (4): Isla Datil (= Turner's Island), 1 (DRD), 3 (AU); *N. a. venusta* (14): Bahia Kino, 6 (SDNHM); Carrizo, 7 km N Bahia Kino, 3 (LACM); 4 mi N Bahia Kino, 2 (USNM); Punta Chueca, 11.5 mi N Bahia Kino, 2 (USNM); 13 mi E Bahia Kino, 1 (LACM).

LITERATURE CITED

Burt, W.H. 1932. Descriptions of heretofore unknown mammals from islands in the Gulf of California, Mexico. Transactions San Diego Society of Natural History, 7:161-182.

Burt, W.H. 1938. Faunal relationships and geographic distribution of mammals in Sonora, Mexico. Miscellaneous Publications, Museum of Zoology, University of Michigan, 39:1-77.

Gastil, G., J. Minch, and R.P. Phillips. 1983. The geology and ages of islands. Pp. 13-25 *in* Island biogeography in the Sea of Cortez (T.J. Case and M.L. Cody, eds.). University of California Press, Berkeley, Los Angeles and London, 508 pp.

Hall, E.R. 1981. The mammals of North America. Second ed. John Wiley and Sons, New York, 2:601-1181 + 90.

Hall, E.R., and H.H. Genoways. 1970. Taxonomy of the *Neotoma albigula*-group of woodrats in central Mexico. Journal of Mammalogy, 51:504-516.

Hall, E.R., and K.R. Kelson. 1959. The mammals of North America. Ronald Press Co., New York, 2:547-1083 + 79.

Hoffmeister, D.F. 1974. The taxonomic status of *Perognathus penicillatus minimus* Burt. Southwestern Naturalist, 19:213-214.

Hoffmeister, D.F. 1986. Mammals of Arizona. University of Arizona Press and Arizona Game and Fish Department, Tucson, xx + 1-602 pp.

Hoffmeister, D.F., and M.R. Lee. 1967. Revision of the pocket mice, *Perognathus penicillatus*. Journal of Mammalogy, 48:361-380.

Lawlor, T.E. 1971. Evolution of *Peromyscus* on northern islands in the Gulf of California, Mexico. Transactions San Diego Society of Natural History, 16:91-124.

Lawlor, T.E. 1983. The mammals. Pp. 265-289 and 482-500 *in* Island biogeography in the Sea of Cortez (T.J. Case and M.L. Cody, eds.). University of California Press, Berkeley, Los Angeles and London, 508 pp.

Wilcox, B.A. 1978. Supersaturated island faunas: A species-age relationship for lizards on post-Pleistocene land-bridge islands. Science, 199:996-998.

Michael A. Bogan, Biological Resources Division, U.S. Geological Survey, Museum of Southwestern Biology, The University of New Mexico, Albuquerque, New Mexico 87131-1091

Systematics, Distribution, and Ecology of the Mammals of Catamarca Province, Argentina

MICHAEL A. MARES, RICARDO A. OJEDA, JANET K. BRAUN, AND RUBEN M. BARQUEZ

Abstract

The systematics, distribution, and ecology of the Didelphimorphia, Xenarthra, Chiroptera, Carnivora, Artiodactyla, Rodentia, and Lagomorpha of Catamarca Province, Argentina, are summarized. Observations for 81 species belonging to 20 families are presented; 10 additional species of probable occurrence are also discussed. A checklist is provided with information on occurrence and status. Information on taxonomy, specimens examined and additional records, standard external and cranial morphological measurements, reproduction, molting, habitat preferences, general natural history, and other information are presented in individual species accounts. The history, topography, climate, and vegetation of the province are given. Demographic, historical, and socioeconomic factors, as they relate to mammal populations, are discussed.

INTRODUCTION

Catamarca Province is located in northwestern Argentina between 25° and 30°S latitude and 65° and 69°W longitude (Fig. 1). It includes an area of 102,602 km^2, 3.7% of the territory of Argentina, excluding Antarctica and offshore islands (Anon., 1988; much of the information that follows is from the 1988 Atlas de la República Argentina). Human population density is low in the province, with the entire province supporting only about 207,000 inhabitants, or two people per km^2. Because the province includes primarily arid and semi-arid habitats, the standard of living of the people is low. There are four principal urban centers: San Fernando del Valle de Catamarca (the capital city), with 78,000 inhabitants; Tinogasta, with 8,000; Belén, with 7,400; and Andalgalá, with 7,000. There is some light industry, but most economic activity revolves about mining, agriculture, and ranching, with some significant production in wood and leather. Cattle are important in the more mesic parts of the province, with total head numbering 250,000. Sheep and goats are also important for meat, wool, and leather and about 134,000 head are maintained in the province.

Catamarca was first crossed by Europeans in 1535, when the Conquistador, Diego de Almagro, traveled from Cuzco, Perú, to the Kingdom of Chile (Anon., 1982; much of the historical information is from this publication) and described native tribes living in the province and mineral riches. This led to colonization of portions of the province as early as 1552, with the capital city being founded in 1683. Human activities within the present-day boundaries of Catamarca date from about 6000 BC. Pre-European Catamarca Province was undoubtedly a biotically rich area, with several species of deer, vicuña, guanacos, peccaries, tapirs, diverse carnivores, and other big game that would have attracted early hunters (e.g., Roig, 1991). Four archaeological periods are recognized within the province, with the last, the Inca Period, ending in 1535. There is evidence that a system of united tribes existed as early as 850 AD. By 1600, however, native tribes were in retreat and cattle, horses, sheep, goats, cotton, and grapes had been introduced. Livestock have thus been associated with, and have affected, the habitats of the province for four centuries.

Life Among the Muses: Papers in Honor of James S. Findley. Edited by T.L. Yates, W.L. Gannon & D.E. Wilson.

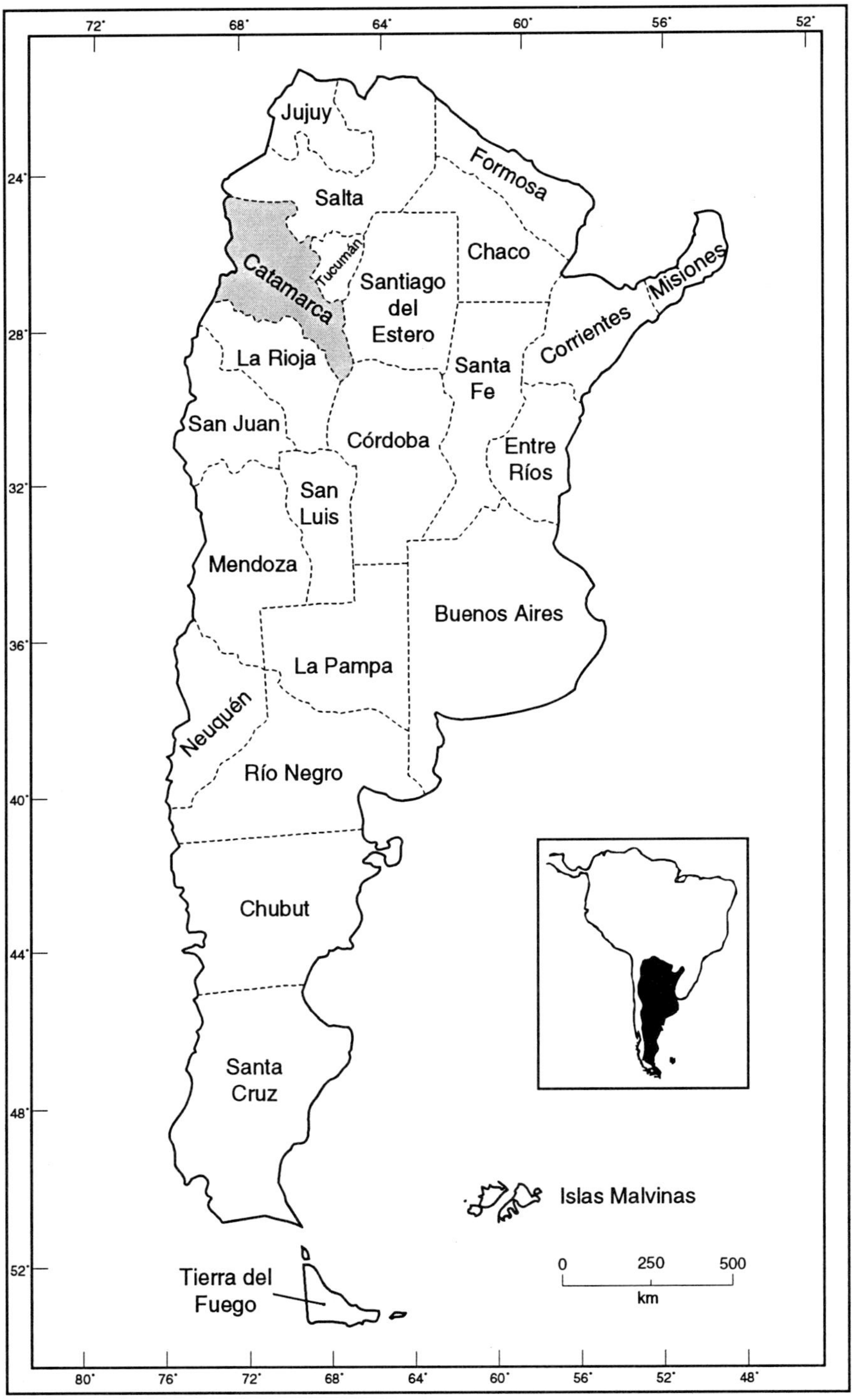

ARGENTINA

Fig. 1. Map showing the location of Catamarca Province in Argentina.

These demographic, historical, and socioeconomic factors are of considerable importance when examining the distribution of mammals within Catamarca, because they have played, and continue to play, an important role in how the habitats have been affected by farming, ranching, timber cutting, mining, and other activities. This is especially true in a province that contains such a low human population (two people/km^2). By comparison, the population densities (Hoffman, 1992) of Colorado, Arizona, Idaho, and New Mexico, are 12.6, 12.7, 4.9, and 4.9 inhabitants/km^2, respectively, 2-6 times the density of Catamarca. Only two states, Wyoming, with a population density of 1.8 people/km^2, and Alaska, with 0.4 inhabitant/km^2, are comparable in density. However, in the largest political department in the province, Antofagasta de la Sierra (28,097 km^2), a high-elevation desert, population density is only 0.03 inhabitant/km^2, more than an order of magnitude less than the population density of Alaska (Anon., 1982). Like Alaska, it might be supposed that the effects of human activities on the habitats of the province would be slight, but such is not the case. Human impact on Alaska has been minimal, with only limited intensive agricultural and ranching activies over its fairly recent history of European contact. In Catamarca, on the other hand, ranching, agriculture, mining, woodcutting, and other activities have been conducted for centuries within this largely desertic province and have had a pronounced impact on habitat integrity.

Catamarca Province is interesting mammalogically for a number of reasons. It contains great topographic, climatic, and vegetational diversity. It is also geographically located in a portion of northwestern Argentina where several major faunal regions abut (Ojeda and Mares, 1989). In an examination of the mammals of Salta Province, which borders Catamarca on the north, Ojeda and Mares concluded that earlier biogeographic hypotheses delimiting faunal regions in southern South America have suffered from a lack of precise data on mammalian distribution, especially in the critical northwestern portion of Argentina. Familial ranges, upon which some of these subregional classification schemes were based, failed to include at least seven families of mammals that extend into Salta, Tucumán, and Catamarca provinces. In this report, we will examine the mammal fauna of Catamarca Province, a political entity encompassing some areas of sweeping floristic and faunistic transitions.

TOPOGRAPHY

Catamarca offers enormous topographic relief (Fig. 2). At its eastern limits, the province supports lowland Chacoan thorn scrub with elevations generally below 300 m, whereas some mountains along the Chilean border have peaks exceeding 6,600 m. The province's highest mountain is Monte Pissis at 6,779 m in extreme western Catamarca on the border with La Rioja Province. Furthermore, pre-Andean chains, generally running north to south, produce numerous alternating montane regions with intermediate low valleys (Fig. 3). In some areas, mountain chains with an east-west axis form isolated valleys when they connect with other mountains that extend north to south. In general, the province is strongly mountainous and deeply dissected, rising rapidly in elevation from east to west.

VEGETATION AND CLIMATE

Catamarca Province extends from northwest to southeast. Human population density is greatest south and east of the Sierra de Ambato. The eastern slopes of these mountains support grasslands and fairly moist woodlands, while the small part of the province that is southeast of these mountains includes lowlands that are more mesic than the rest of the province. As might be expected, ranching, agriculture, and woodcutting are concentrated in this relatively small section of the province.

The major phytogeographic zones of Catamarca Province are shown in Fig. 4. There are seven principal zones, including the Chacoan thorn scrub, montane chaco, lowland evergreen forest, montane woodlands and grasslands, the prepuna, the monte desert, and the puna.

The most extensive of these habitats is the puna, or Andean desert, a sparse grass-shrub desert found between 3,400 and 4,500 m elevation. We also include in the puna the

Fig. 2. Landsat photo showing the isolated Campo Arenal (upper center) and a portion of the Bolsón de Pipanaco (lower center) in central Catamarca Province.

Andean plain occurring above 4,500 m (A.L. Cabrera, 1957; see also Vervoorst, 1982). The puna in Catamarca occupies the northwestern half of the province. Dominant plants include shrubs of the genera *Fabiana*, *Adesmia*, *Tetraglochin,* and *Pseudobaccharis*, and grasses and annuals of the genera *Stipa*, *Conyza*, *Hoffmanseggia*, and *Heliotropium*. Soils in the puna vary from sandy to rocky, and vegetation is generally less than 1 m high. Within the puna are extensive salt flats that may contain water for part, or all, of the year, and that often support verdant grass and sedge habitats along their margins. Because of the high elevation, insolation in the puna is intense, winds are strong and often incessant, and temperatures are cold (A.L. Cabrera, 1957; Mares et al., 1985)—mean monthly temperatures may reach 18°C in summer and fall to 7°C in winter. Precipitation may be received as winter snow or summer rain and hail; most rain falls in the summer months of December–February. Annual precipitation is less than 150 mm. Daily temperature fluctuations are ex-

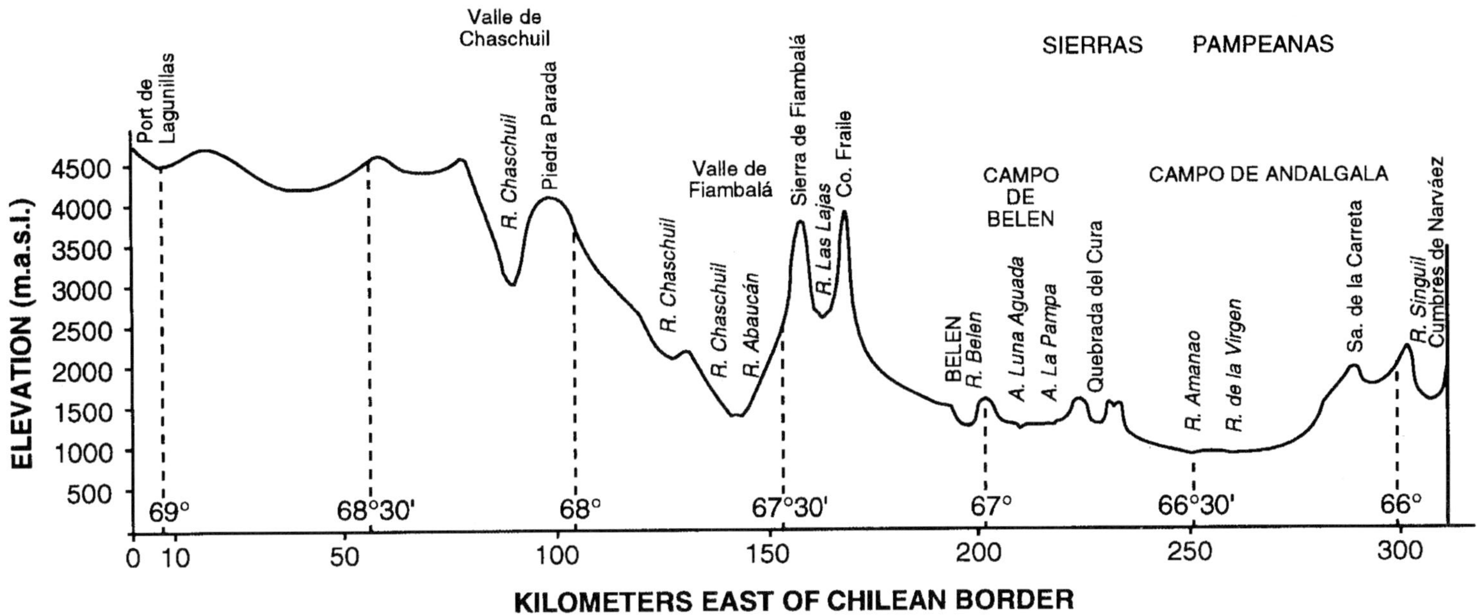

Fig. 3. An east-west topographic profile taken along latitude 27° 40' (after Anon., 1981). This profile corresponds roughly to a line drawn across the bottom one-third of Figure 2. Height is in meters above sea level; degrees west longitude are shown. Rivers are abbreviated by the letter R; arroyos by the letter A (both rivers and arroyos are shown in italics); mountains by the abbreviation Sa (for sierra); peaks by the abbreviation Co (for cerro).

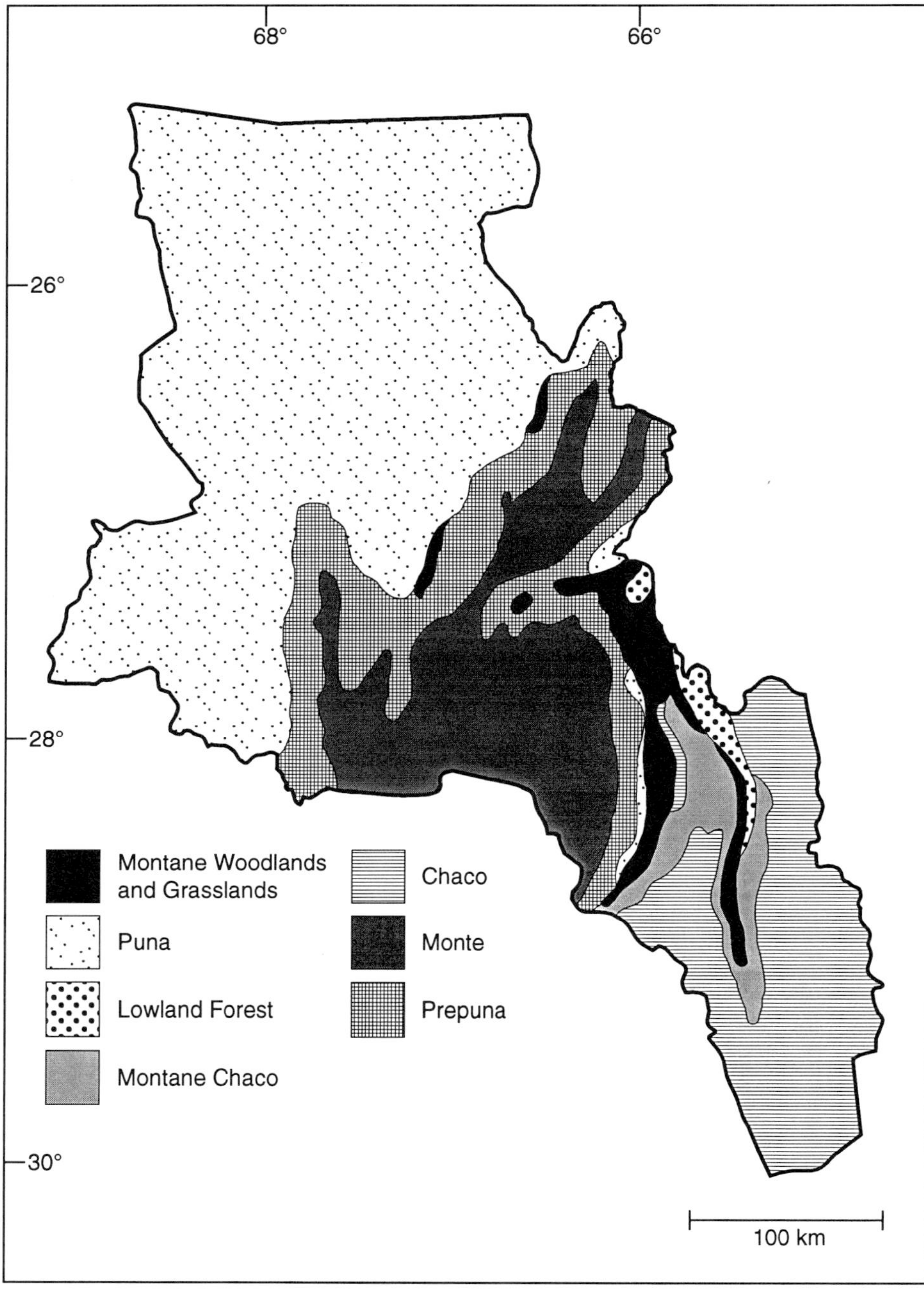

Fig. 4. The principal vegetation zones of Catamarca Province (after Vervoorst, 1982).

treme and exceed mean seasonal differences; frost days may occur up to half of the year.

The second most extensive habitat in Catamarca is the chaco, or lowland thorn scrub. This is a xerophytic woodland with numerous cacti and small shrubs present (e.g., Mares et al., 1985; Morello and Adámoli, 1968; Morello and Toledo, 1959a, 1959b; Papadakis, 1974). Principal trees include species in the genera *Prosopis*, *Acacia*, *Cercidium*, *Geoffroea*,

Ziziphus, *Schinopsis*, *Aspidosperma*, and *Tabebuia*. Palms (*Trithrinax campestris*) and numerous cacti (e.g., *Opuntia*, *Cereus*, *Cleistocactus*, *Trichocereus*) are common, as are many shrubs (e.g., *Larrea*, *Suaeda*, *Vallesia*). The chaco of Catamarca occurs between 250 and 500 m elevation. It is a hot scrubland with maximum summer temperatures exceeding 42°C and a frost-free period extending from October–April. Rain falls principally in summer (November–March) and, for most of the lowland chaco in the province, yearly precipitation is approximately 400 mm.

West of the Sierra de Ambato, east of the western highlands, and at elevations generally below 2,000 m lies the monte desert (Fig. 4). This desert extends for approximately 2,000 km in western Argentina, from Salta Province in the north to Chubut Province in the south (Morello, 1958). The vegetation of the monte has been well studied in the province because the International Biological Program Structure of Ecosystems Subprogram centered its research in the early 1970s in central Catamarca (Mabry et al., 1977; Orians and Solbrig, 1977; Simpson, 1977). Information on the vegetation of this zone is available in the preceding references as well as in Morello (1958), Vervoorst (1951, 1954, 1966, 1967, 1982), and Mares et al. (1985). The monte is a hot desert, where summer temperatures may exceed 40°C. Snow may fall in winter; however, almost all precipitation is received in summer (November–March). Annual rainfall is generally less than 250 mm, but total precipitation may vary. Andalgalá, in central Catamarca, for example, is one of the most mesic monte localities (293 mm/yr), whereas Tinogasta, at the southwestern edge of the same valley as Andalgalá and only 130 km distant, is one of the drier sites in the monte (130 mm/yr).

The monte is a low scrub desert with interspersed trees and cacti. Common trees are in the genera *Prosopis*, *Acacia*, *Cercidium*, and *Bulnesia*. Cacti include *Opuntia* and *Trichocereus*. Common shrubs are *Larrea*, *Atriplex*, *Cassia*, and *Suaeda*. Extensive *Prosopis* and *Acacia* forests can be found in sandy depressions, along ephemeral watercourses and surrounding the salt flats. On mid- to lower bajadas, creosote bush (*Larrea cuneifolia*) is the dominant shrub. On the rocky upper bajadas and lower montane slopes, ground bromeliads (*Dyckia*, *Deuterocohnia*) and cacti predominate. Some of these habitats are illustrated in Mares et al. (1985).

The remaining vegetation zones are small compared to the first three, which together encompass about 75% of the area of the province. A rather extensive xeric habitat lying at higher elevations (1,000–3,400 m) than the monte, but contiguous both with it and with the puna, is the prepuna. The prepuna is a cactus-scrub-bromeliad desert that occurs primarily on upper slopes of desert mountains or on high, arid plateaus (A.L. Cabrera, 1957; Vervoorst, 1982). At the lower elevations, it is a hot, dry desert, with annual precipitation being usually less than 200 mm, and in some places less than 100 mm. Soils are poor and the substrate is generally very rocky. Dominant plants are cacti (*Trichocereus*), bromeliads (*Dyckia*, *Deuterocohnia* and *Abromeitiella*), and shrubs such as *Caesalpinia*, *Zuccagnia*, *Lycium*, *Baccharis*, and *Cercidium*. In aspect, it is a rocky, cactus scrubland on moderate to steep slopes. At higher elevations, temperatures are lower. The prepuna extends along the arid mountains of central Catamarca Province.

Montane woodlands and grasslands are found principally on the mesic eastern slopes of the Sierra de Graciana, Sierra de Ancasti o del Alto, Sierra de Ambato, and the Sierra de Manchao. These forests at 1,400–2,100 m are fairly mesic (600–800 mm) woodlands of alder (*Alnus acuminata*), with occasional gymnosperms (*Podocarpus parlatorei*) and other trees (e.g., *Polylepis*, *Prunus*, *Schinus*, *Azara*, and *Sambucus*) (A.L. Cabrera, 1976; Digilio and Legname, 1966; Vervoorst, 1966, 1982). They form a part of the Yungas Province (A.L. Cabrera, 1976), a mesic, complex subtropical forest of the lowlands and highlands of northwestern Argentina that extends northward into Bolivia and thence north along the eastern Andean slopes and foothills. Scattered throughout these mountains in Argentina (at 2,000–3,000 m) are bunchgrass prairies (*Stipa*, *Festuca*, *Paspalum*) that are less mesic than the woodlands (annual rainfall < 500 mm).

The montane chaco habitat or chaco serrano—also termed horco-quebracho forest—extends from the foothills of Jujuy Province, in far northwestern Argentina, as far south as the low mountains of San Luis Province. In Catamarca, this vegetation zone is found only on the mountains of the extreme southeastern part of the province, where it is intermediate between the lowland chaco and the montane woodlands. It is a thorny woodland with trees reaching 8–10 m in height; a dense shrub layer is also present. Dominant trees are *Schinopsis haenkeana* (the horco-quebracho), *Lithraea*, *Fagara*, *Celtis*, *Schinus*, *Acacia*, *Prosopis*, and *Aspidospermum*. Shrubs include *Atamisquea*, *Colletia*, *Vernonia*, and *Aloyssia*; cacti (*Cereus* and *Opuntia*) are common.

The last of the major macrohabitats in Catamarca, and also the least extensive, is the lowland forest. This is a mesic forest (800–1,500+ mm annual precipitation) occurring at the base and along the lower east-facing slopes of mountains in extreme eastern parts of the province. This is the southernmost limit of the complex yungas forest (A.L. Cabrera, 1976), also termed the Tucumán-Bolivian Forest. This forest is largely evergreen. Tree genera include *Bocconia*, *Carica*, *Cedrela*, *Fagara*, *Jacaranda*, *Juglans*, and *Tipuana*. Bromeliads are common on tree branches, and the canopy may exceed 30 m in height. The forest is somewhat stratified, with a subcanopy of trees about 20 m in height, another of shrubs up to 4 m high, and an herbaceous layer about 1 m high. This is a dense forest, rich in plant species, and is one of the richest habitats in South America for mammals (Mares, 1992). This forest, which extends northward for thousands of kilometers, reaches its southern limits in Catamarca between 28° and 29° S latitude (A.L. Cabrera, 1976; Morello, 1958; Vervoorst, 1982).

METHODS

General

Our research in Catamarca began in January 1970 when Mares worked in the province as part of the International Biological Program Structure of Ecosystems Subprogram. The first field trip lasted two years, with almost a year being spent in the province. Initial research was ecological and physiological in nature, thus extensive collections of specimens were not made. Later trips to the province by Mares and/or collaborators occurred in 1974, 1979, 1983, 1986, 1988, 1989, and 1990. Trips were generally made while conducting more extensive expeditions to other provinces as well; thus, an intensive survey of Catamarca remains to be undertaken.

Specimens were collected using standard Museum Special snap traps, Victor Rat Traps, gopher traps, leg-hold traps, Sherman live traps (7.6 x 7.6 x 25.4 cm and 12.7 x 12.7 x 38.1 cm), homemade live traps, mist nets, and .22 caliber rifles.

Specimens were prepared as standard skin and skull or skeleton, or were preserved in 10% formalin and 70% ethyl alcohol. Beginning in 1990, tissues have been extracted and frozen, and chromosomal preparations made. Standard external measurements (in mm, unless otherwise noted) were obtained, reproductive condition was noted, and molting patterns were also recorded. Cranial measurements, given in mm, were made to the nearest 0.1 mm. Ecological data were noted when specimens were collected ancillary to ecological or physiological research. Specimens collected by us are deposited in the following museums: Museum of Southwestern Biology, University of New Mexico; Carnegie Museum, Pittsburgh; Oklahoma Museum of Natural History, University of Oklahoma; Colección Lillo, Instituto Miguel Lillo, Tucumán; IADIZA, Colección de Mamíferos, Mendoza; frozen tissues are deposited in the Texas Cooperative Wildlife Collection at Texas A&M University.

Taxonomy

With minor exceptions, we have chosen to follow the new edition of "Mammal Species of the World" (Wilson and Reeder, 1993) for the taxonomic arrangement of families, subfamilies, genera, and species. These exceptions are summarized below; additional comments may be found in the individual species accounts. We recognize as distinct species the following: *Lasiurus blossevillii*, *Lasiurus salinae*, *Akodon alterus*, *Akodon caenosus*, *Akodon tucumanensis*, *Eligmodontia marica*, and *Ctenomys fochi*. We have recognized *Pediolagus* as a distinct genus. Olds (1988) was used for the taxonomy of *Calomys*; we recognize *Calomys venustus* as distinct from *C. callosus*, but do not recognize *C. murillus* as distinct from *C. musculinus*.

Collecting Localities

The principal collecting localities for Catamarca Province are shown in Fig. 5. The gazetteer lists localities for specimens collected by us, as well as by others. Gazetteer No. 103 (Argentina) of the United States Board of Geographic Names, prepared by the Office of Geography, Department of Interior, Washington, D.C. (1968), was utilized as a source for coordinates of longitude and latitude. Coordinates of south latitude and west longitude in degrees and minutes are given for all localities that could be located. Localities have been kept as they appeared on the original labels, except for obvious misspellings. Elevation is given in m above sea level. Additional maps utilized included: the 1972, 1974, and 1986 maps of Catamarca published by the Automóvil Club Argentino; the series of 1:500,000 provincial maps published the Ejército Argentino, Instituto Geográfico Militar in 1964; and the 1: 200,000 maps of the province published by the Dirección Nacional de Geología y Minería in the 1940s and 1950s.

CHECKLIST OF MAMMALS OF CATAMARCA PROVINCE

A list of the orders, families, genera, and species of mammals known, or expected to occur, in Catamarca Province, Argentina, is given below. The status of each species is indicated as being: (*) collected, seen, or examined by us as museum specimens; (1) reported in the literature; (p) of probable occurrence due to the presence of the species in adjacent provinces (or in Chile) in similar habitats. In addition, we note whether the species is common (c), uncommon (u), rare (r), in danger of extirpation (e), or probably extirpated from the province (x). These designations of relative abundance are largely subjective and are based on our long experience with the species of the province. We present them only to guide biologists as to which species might be of special concern for conservation and which might be relatively easy to study in the province. In some cases, we have actual mark-recapture data, or collecting data, to estimate commonness or rarity. In other cases, we have formed impressions as to whether or not a species is common or rare. Although subjective, we hope that such designations are useful for some purposes.

Order Didelphimorphia
- Family Didelphidae
  - *Didelphis albiventris* *(1, c)
  - *Lutreolina crassicaudata* (p)
  - *Thylamys elegans* *(u)
  - *Thylamys pallidior* *(1, u)

Order Xenarthra
- Family Dasypodidae
  - *Chaetophractus vellerosus* *(1, c)
  - *Chaetophractus villosus* (p)
  - *Chlamyphorus truncatus* *(e)
  - *Dasypus hybridus* (p)
  - *Euphractus sexcinctus* (p)
  - *Tolypeutes matacus* *(r)

Order Chiroptera
- Family Phyllostomidae
  - *Desmodus rotundus* *(1, u)
  - *Sturnira erythromos* *(r)
  - *Sturnira lilium* *(u)
- Family Vespertilionidae
  - *Eptesicus furinalis* *(u)
  - *Histiotus macrotus* *(c)
  - *Histiotus montanus* (p)
  - *Lasiurus blossevillii* *(c)
  - *Lasiurus cinereus* *(c)
  - *Lasiurus ega* *(u)
  - *Lasiurus salinae* *(r)
  - *Myotis levis* *(c)
- Family Molossidae
  - *Nyctinomops macrotis* *(u)
  - *Promops nasutus* *(r)
  - *Tadarida brasiliensis* *(c)

Order Carnivora
- Family Canidae
  - *Pseudalopex culpaeus* *(u)
  - *Pseudalopex griseus* *(u)
  - *Pseudalopex gymnocercus* *(c)
- Family Felidae
  - *Herpailurus yaguarondi* *(l, r)
  - *Oncifelis colocolo* *(l, r)
  - *Oncifelis geoffroyi* *(c)
  - *Oreailurus jacobita* *(l, r)
  - *Puma concolor* *(l, c)
- Family Mustelidae
  - *Conepatus chinga* *(l, c)

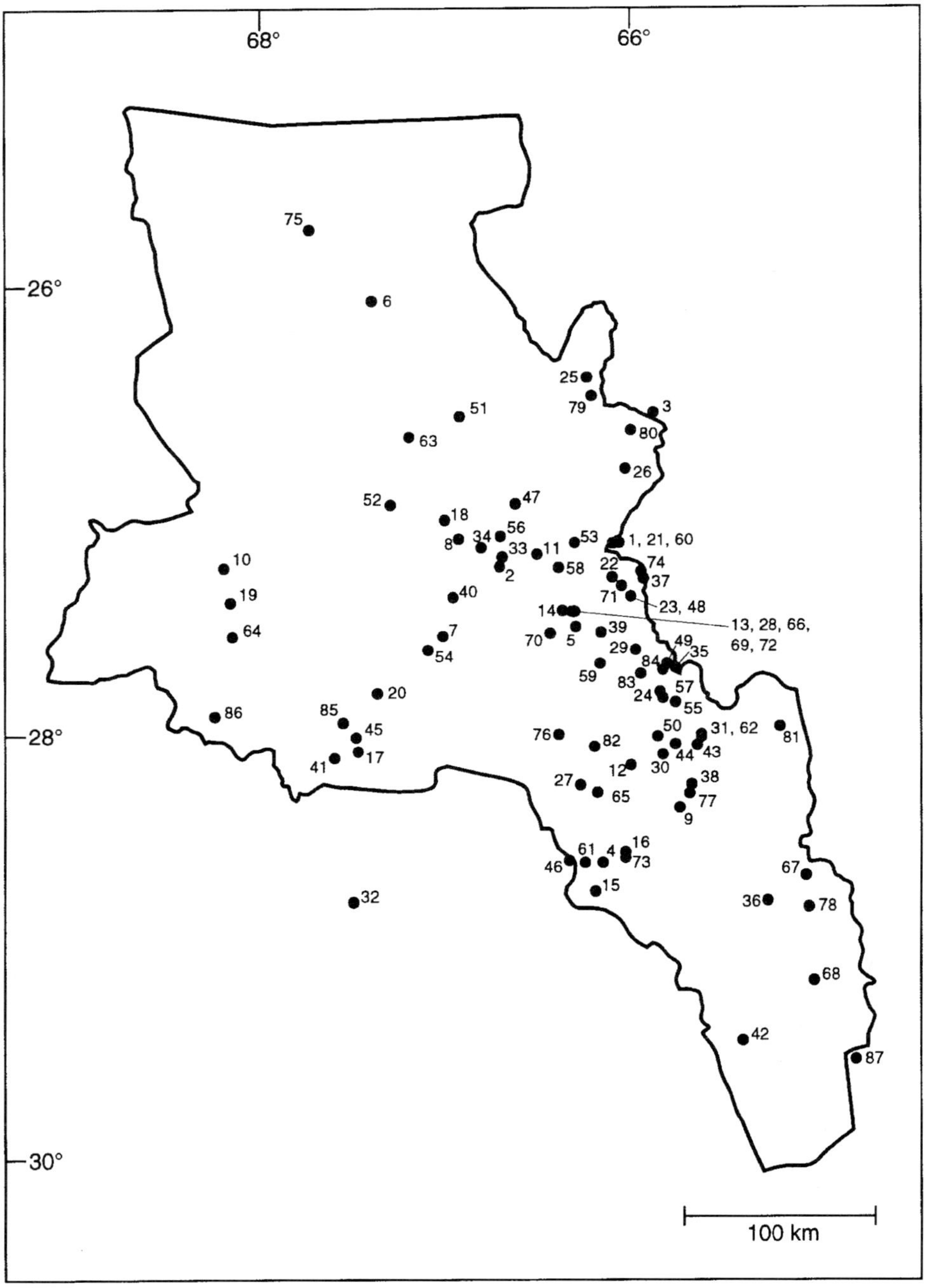

Fig. 5. The principal collecting localities for mammals in Catamarca Province. Numbers correspond to those listed in the gazeteer.

Eira barbara *(u)
Galictis cuja (p)
Lyncodon patagonicus *(l, r)
Family Procyonidae
Procyon cancrivorus *(u)
Order Artiodactyla
Family Tayassuidae
Pecari tajacu (l, u)
Family Camelidae
Lama guanicoe *(l, u)
Vicugna vicugna *(l, e)
Family Cervidae
Hippocamelus antisensis *(u)
Mazama americana (p)
Mazama gouazoupira *(u)
Ozotoceros bezoarticus *(u)
Order Rodentia
Family Muridae
Akodon alterus *(l, u)
Akodon caenosus *(u)
Akodon dolores *(l, u)
Akodon illuteus *(u)
Akodon simulator *(l, c)
Akodon spegazzinii *(c)
Akodon tucumanensis *(u)
Andalgalomys olrogi *(u)
Andalgalomys roigi *(u)
Andinomys edax *(l, u)
Bolomys lactens *(l, r)
Calomys musculinus *(l, c)
Calomys venustus (l, r)
Chroeomys andinus *(u)
Eligmodontia marica *(r)
Eligmodontia moreni *(l, c)
Eligmodontia puerulus *(u)
Eligmodontia sp. 1 *(r)
Eligmodontia sp. 2 *(r)
Graomys edithae *(r)
Graomys griseoflavus *(l, c)
Neotomys ebriosus (l, r)
Oligoryzomys flavescens *(l, u)
Oligoryzomys longicaudatus *(c)
Oxymycterus akodontinus *(r)
Phyllotis darwini *(l, c)
Phyllotis osilae *(l, c)
Reithrodon auritus (l, r)
Family Chinchillidae
Chinchilla brevicaudata (p)
Lagidium viscacia *(l, c)
Lagostomus maximus *(l, u)
Family Caviidae
Dolichotis patagonum *(l, c)
Galea musteloides (l, u)
Microcavia australis *(l, c)
Microcavia shiptoni *(r)
Pediolagus salinicola (l, u)
Family Ctenomyidae
Ctenomys coludo (l, c)
Ctenomys fochi *(l, c)
Ctenomys knightii (l, c)
Ctenomys opimus *(c)
Ctenomys sp. 1 *(r)
Ctenomys sp. 2 *(c)
Family Octodontidae
Octodontomys gliroides (p)
Octomys mimax *(r)
Family Abrocomidae
Abrocoma cinerea (l, r)
Family Myocastoridae
Myocastor coypus (p)
Order Lagomorpha
Family Leporidae
Lepus europaeus *(c)

ACCOUNTS OF SPECIES

In the following sections, we provide accounts of species known to occur in Catamarca Province. Order, family, and subfamily (where appropriate) are included, followed by the scientific name, authority for the species name and literature citation for the first recognized description. Genera within families or subfamilies are listed alphabetically, as are species within genera. Where appropriate, subspecies are mentioned within the species accounts. Specimens examined are in the following museums: (MACN) Museo Argentino de Ciencias Naturales "Bernardino Rivadavia," Buenos Aires, Argentina; (CMNH) Carnegie Museum, Pittsburgh, Pennsylvania; (TTU) The Museum, Texas Tech University, Lubbock, Texas; (MSB) Museum of Southwestern Biology, University of New Mexico, Albuquerque, New Mexico; (CML) Colección Lillo, Instituto Miguel Lillo, Tucumán, Argentina; (OMNH) Oklahoma Museum of Natural History, University of Oklahoma, Norman, Oklahoma; (IADIZA-CM) Colección de Mamíferos, Centro Regional de Investigaciones Científicas y Técnicas-CRICYT-Mendoza, Argentina; (RMB) personal collection of Rubén Barquez, Tucumán, Argentina; (BMNH) British Museum (Natural History), London; (Arg) denotes specimens housed in the OMNH whose disposition has not yet been determined.

Order Didelphimorphia
Family Didelphidae
Subfamily Didelphinae
Didelphis albiventris Lund

1840. *Didelphis albiventris* Lund, K. Dansk. Vid. Selsk. Afhandl., p. 20.

Specimens examined (3). Catamarca, 1 (MACN); Choya, 13 km NNW Andalgalá, 1219 m, 1 (OMNH); La Merced, 1 (MACN).

Additional records. Along Hwy 65, 7 km S Tucumán-Catamarca border; Andalgalá (Mares, 1973); 2 km S, 1 km W Choya (R.L. Humphrey, pers. obs.).

Measurements. External measurements (2 females): total length, 718, 432; tail, 330, 198; hindfoot, 53, 36; ear, 58, 45; weight, —, 210.0.

Remarks. Numerous individuals were seen in the town of Andalgalá; two were seen at about 2,000 m elevation near the Catamarca–Tucumán border in a lush forest situation. The species does not occur in the desert habitats of the province, but is found in a variety of other habitats, especially in agricultural areas where fruits are common. The specimen collected in Choya in December was caught in a Sherman live trap that had been placed outside of the town in tall, thick grass in an orchard surrounded by dry monte habitat. Opossums are omnivorous and nocturnal; they feed on the ripening fruits of plants found in irrigated verdant habitats and are frequently seen as road kills in and near towns. In Argentina, the species occurs as far south as Neuquén Province (Olrog and Lucero, 1981). The propensity of this species for non-xeric montane habitats and agricultural areas provides numerous natural corridors for colonization. Opossums have been heavily hunted as part of the legal fur trade in Argentina. Between 1972 and 1979, for example, 1.5 million skins of *Didelphis* were exported from the country (Mares and Ojeda, 1984; Ojeda and Mares, 1982, 1984); this number does not count the significant use of opossum skins within Argentina. There have been no extensive studies of opossums in Argentina, but Cajal (1981) examined home range in some individuals in neighboring Tucumán Province.

Thylamys elegans (Waterhouse)

1839. *Didelphis elegans* Waterhouse, Zool. H.M.S. "Beagle," Mammalia, p. 95.

Specimens examined (1). 5 km S Las Higuerillas on Hwy 9, 1,091 m, 1 (OMNH).

Measurements. External measurements (1 female): total length, 237; tail, 135; hindfoot, 18; ear, 26; weight, 22.0. Cranial measurements (1 female): greatest length of skull, 29.4; condylobasal length, 28.4; least interorbital breadth, 4.6; zygomatic breadth, 15.5; breadth of braincase, 10.9; length of maxillary toothrow, 11.5; palatal length, 15.3; length of mandibular toothrow, 12.0; greatest length of mandible, 12.3.

Remarks. This species has not been previously recorded for the province. The animal was captured along a wet gully in a pocket of yungas forest habitat just south of the Tucumán-Catamarca border, with large (> 10 m) trees supporting arboreal bromeliads and with a dense understory. The species may occasionally be abundant in mesic forest (yungas) in northern Argentina, particularly in Salta Province near the border with Bolivia (Mares et al., 1981). The specimen from Catamarca is the southernmost record for the species.

Thylamys pallidior (Thomas)

1902. *Marmosa pallidor* Thomas, Ann. Mag. Nat. Hist., ser. 7, 10:161.

Specimens examined (9). Immed. N Andalgalá, Río Andalgalá, 1 (CMNH); Belén, 1 (MACN); Chumbicha, 0.5 km E of Hwy 38 along Hwy 60, 457 m, 1 (OMNH); Chumbicha, 1.5 km E of Hwy 38 along Hwy 60, 457 m, 2 (1 CML, 1 IADIZA-CM); Minas Capillitas, 3,200 m, 2 (CMNH); off Hwy 62, 8 km E jct. Hwy 1 and 62, 1 (CMNH); Quirós, 1 (MACN).

Additional records. Río Andalgalá, 3 km N Andalgalá; off Hwy 62, 8 km E jct. Hwy 62 and 1, E of Andalgalá (Mares, 1973);

Chumbicha (Thomas, 1919a); Otro Cerro; Hualfin (Tate, 1933).

Measurements. External measurements (2 males, 5 females): total length, 180.5 (175–186), 190.0 (166–214); tail, 93.5 (89–98), 105.6 (92–119); hindfoot, 13.0 (13), 11.1 (9–13); ear, 21.5 (21–22), 22.4 (20–24); weight, 13.0 (13.0), 20.4 (12.1–32.0). Cranial measurements (1 male, 3 females): greatest length of skull, 24.7, 27.0 (24.5–28.5); condylobasal length, 23.9, 26.6 (23.8–28.0); least interorbital breadth, 3.6, 4.2 (4.1–4.3); zygomatic breadth, 13.7, 14.8 (13.4–15.6); breadth of braincase, 9.7, 10.8 (10.4–11.0); length of maxillary toothrow, 9.3, 10.2 (9.1–10.7); palatal length, 12.4, 14.2 (12.4–15.3); length of mandibular toothrow, 9.9, 10.6 (10.3–11.0); greatest length of mandible, 16.8, 19.3 (17.0–20.5).

Remarks. A male collected near Chumbicha in July was molting primarily on the dorsum. All three individuals from near Chumbicha were collected in disturbed creosote bush (*Larrea*) flats in the chaco. Two males collected near Chumbicha in July had testes 6 x 4 mm. A nonbreeding female was collected at the same time. This site was sampled five months later the same year, but no individuals were captured. *Thylamys pallidior* may appear in any dry habitat of Catamarca, including the high grasslands above 3,000 m, although it has not been taken in open stands of pure creosote bush in the driest portions of the monte. The species is never abundant and its occurrence can never be predicted with any degree of certainty. It can be captured at any time of year. In monte desert habitat in Mendoza Province, an individual was captured by Mares after a light snow had fallen. Fat is stored in the tail during the fall and winter and may permit the animals to spend greater amounts of time in their burrows and nests during inclement periods.

Order Xenarthra

Family Dasypodidae

Subfamily Dasypodinae

Chaetophractus vellerosus (Gray)

1865. *Dasypus vellerosus* Gray, Proc. Zool. Soc. London, 1865:376.

Specimens examined (14). 17 km S Andalgalá, along Hwy 1, 1 (OMNH); Belén, 1 (MACN); Catamarca, 2 (MACN); La Cienaguita, 3 (MACN); Laguna Blanca, 3,200 m, 2 (1 MACN, 1 IADIZA-CM); Pastos Largos, 4 (MACN); Singuil, 1 (MACN).

Additional records. Andalgalá Railroad Station, 4 km S Andalgalá; at km 99 on Hwy 1, 34 km S Andalgalá; at km 766, 23 km N Tinogasta; 1.5 km S, 150 m W jct. Hwy 1 and railroad track, S Andalgalá; 50 m N, 50 m E jct. Hwy 62 and Amanao Road, approx. 18 km E Andalgalá; Potrero River dike, Potrero, approx. 13 km N Andalgalá; Río Amanao, 15 km S jct. Hwy 1 and Río Amanao; Río Andalgalá, 3 km N Andalgalá; 6 km N Saujil, along Hwy 1 (Mares, 1973); near Andalgalá (Greegor, 1980a, 1980b, 1985).

Measurements. External measurements (2 males): total length, 218, 350; tail, 102, 105; hindfoot, 46.6, 52; ear, 31.8, 30; weight, 500.0, 610.0. Cranial measurements (1, sex unknown): greatest length of skull, 67.6; condylobasal length, 66.7; least interorbital breadth, 18.2; zygomatic breadth, 42.8; breadth of braincase, 28.0; length of maxillary toothrow, 29.9; length of mandibular toothrow, 32.0.

Remarks. This is the most common armadillo in the drier habitats of Catamarca at low and high elevations. It is widely hunted for food and its shell is used in fabricating a local stringed instrument, the charango. Animals are active primarily at night during the warmer months and may be found at any time of the day or night during the winter months. The species was studied in Catamarca by Greegor (1975, 1980a, 1980b, 1985), who examined both its ecology and behavior (see also Mares et al., 1977a). The animals are broadly omnivorous. The subspecies in Catamarca is *C. v. vellerosus* (Yepes, 1929a).

Tolypeutes matacus (Desmarest)

1804. *Loricatus matacus* Desmarest, Tabl. Meth. Hist. Nat., in Nouv. Dict. Hist. Nat., 24:28.

Specimens examined (1 shell). Catamarca, 1 (MACN).

Measurements. None.

Remarks. This species is extremely uncommon in Catamarca due to heavy hunting pressure by people. It is hunted with dogs and is used as food by the people of the chaco. The animal does not burrow (at least in Brazil), but rolls into a tight ball for protection (Schaller, 1983). It does not occur in the monte desert, but is primarily a Chacoan species (Barquez et al., 1991; Mares et al., 1989).

Chlamyphorinae

Chlamyphorus truncatus Harlan

1825. *Chlamyphorus truncatus* Harlan, Ann. Lyc. Nat. Hist., 1:235.

Specimens examined (1). La Guardia, 1 (MACN).

Measurements. External measurements (1 young female; from Yepes, 1932): total length, 152; tail length, 32; hindfoot, 32. Cranial measurements (1, sex unknown): greatest length of skull, 34.7; condylobasal length, 32.7; least interorbital breadth, 12.7; zygomatic breadth, 22.7; breadth of braincase, 20.5; length of maxillary toothrow, 14.0; palatal length, 17.4; length of mandibular toothrow, 13.8; greatest length of mandible, 24.0.

Remarks. This locality may be the northernmost known within the geographic range of the species. Additional records from central and southern provinces and a key for the identification of the subspecies were given by Yepes (1929a, 1929b, 1932). The subspecies in Catamarca is *C. t. patquiensis* (Yepes, 1932). A more recent revision was presented by Wetzel (1982, 1985), while Minoprio (1945) provided much information on the biology of this species. This species is cited as insufficiently known by IUCN's Red Data Book (1982, 1990) and is a protected species in Argentina (Fuller and Swift, 1984). It is never common anywhere in its range and its subterranean habits make it difficult to ascertain its abundance.

Order Chiroptera

Family Phyllostomidae

Subfamily Stenoderminae

Sturnira erythromos (Tschudi)

1844. *Phyllostoma erythromos* Tschudi, Fauna Peruana, p. 64.

Specimens examined (12). Cuesta del Clavillo, 5 km al S de La Banderita, 12 (RMB).

Measurements. External measurements (7 males, 5 females): total length, 56.9 (50–62), 52.2 (35–58); hindfoot, 9.6 (6–13.3), 9.2 (6–13.6); ear, 14.3 (12–17.5), 13.1 (8–17.5); weight, 14.4 (12.0–16.0), 15.7 (12.0–20.0); forearm, 40.1 (38.5–42.3), 40.3 (39.0–41.2).

Remarks. These are new records for the province.

Sturnira lilium (E. Geoffroy)

1810. *Phyllostoma lilium* E. Geoffroy, Ann. Mus. Hist. Nat. Paris, 15:181.

Specimens examined (3). Cuesta del Clavillo, 5 km al S de La Banderita, 1 (RMB); La Merced, 2 (TTU).

Measurements. External measurements (1 male, 1 female): total length, 62.5, 52; hindfoot, 15.2, 10; ear, 18.2, 14; weight, 25.0, 19.0; forearm, 45.9, 42.0. Cranial measurements (1 female, 1 sex unknown): greatest length of skull, 22.8, 22.0; condylobasal length, 20.9, 20.2; least interorbital breadth, —, 6.0; zygomatic breadth, 13.3, —; breadth of braincase, 10.5, 10.9; length of maxillary toothrow, 6.8, 6.4; palatal length, 10.0, 8.9; length of mandibular toothrow, 6.5, 6.1; greatest length of mandible, 15.4, 14.9.

Remarks. In western Argentina, this species reaches its southernmost limits in the lowland mesic forests of northern Catamarca; it occurs even further south along the east coast of Argentina (Barquez et al., 1993). In Catamarca, the species is uncommon, although its abundance increases rapidly northward into the yungas forests of Tucumán, Jujuy, and Salta provinces. The subspecies in Catamarca is *S. l. lilium* (Barquez, 1987; A. Cabrera, 1957).

Subfamily Desmodontinae

Desmodus rotundus (É. Geoffroy)

1810. *Phyllostoma rotundus* É. Geoffroy, Ann. Mus. Hist. Nat. Paris, 15:181.

Specimens examined (6). Andalgalá, 1 (MACN); Choya, 13 km NNW of Andalgalá, 1,219 m, 4 (1 CML, 1 IADIZA-CM, 2 OMNH); Dique El Potrero, 1 (MSB).

Additional records. Potrero River Dike, Potrero, approx. 13 km N Andagalá (Mares, 1973).

Measurements. External measurements (4 males): total length, 86.0 (82–88); hindfoot, 17.5 (16–19); ear, 20.5 (20–21); weight, 38.1 (36.0–42.5); forearm, 64.3 (63.0–67.0). Cranial measurements (1 male): greatest length of skull, 25.6; condylobasal length, 22.7; least interorbital breadth, 5.7; zygomatic breadth, 12.4; breadth of braincase, 13.0; length of maxillary toothrow, 3.5; palatal length, 9.3; length of mandibular toothrow, 4.7; greatest length of mandible, 14.6.

Remarks. Testes size (length x width) for the four males collected in mid-December from Choya ranged from 5 x 3 to 7 x 4 mm. These specimens were collected outside of the local police station. The vampire is very uncommon in the monte desert, occurring primarily along the foothills of the pre-Andean slopes along watercourses and in agricultural areas. It is more common in the chaco, although we have no records of the species in eastern Catamarca east of the pre-Andean chains. Nevertheless, it is common in neighboring Santiago del Estero Province. Although Mares netted the Potrero River Dike specimen in November 1972, efforts during the two previous years yielded no vampire bats (Mares, 1973). Neither did he observe evidence of feeding activity on area livestock. The subspecies in Catamarca is *D. r. rotundus* (Barquez, 1987; A. Cabrera, 1957). In Argentina, this species has the status of injurious wildlife (Fuller and Swift, 1984).

Family Vespertilionidae
Subfamily Vespertilioninae
Eptesicus furinalis (d'Orbigny)

1847. *Vespertilio furinalis* d'Orbigny, Voy. Amer. Merid., Atlas Zool., 4:13.

Specimens examined (15). Balneario El Caolín, 6 km NW Chumbicha, 8 (OMNH); Chumbicha, 1 km N and W of Balneario by road, 792 m, 4 (1 CML, 1 IADIZA-CM, 2 OMNH); Chumbicha, at Balneario, 732 m, 1 (OMNH); Dique El Potrero, 1 (MSB); San Antonio de la Falda, 1 (CML).

Measurements. External measurements (7 males, 5 females, unless otherwise noted): total length, 96.3 (80–105), 100.8 (83–107); tail, 42.0 (31–47), 42.4 (32–49); hindfoot, 9.4 (8–10), 9.2 (9–10); ear, 14.7 (12–16), 14.6 (13–17); weight (7 males, 4 females), 9.3 (5.0–11.0), 11.8 (10.5–14.0); forearm (1 male, 4 females), 37.0, 41.8 (38.0–44.0). Cranial measurements (3 males, 2 females, unless otherwise noted): greatest length of skull, 16.2 (15.8–16.4), 16.8 (16.5–17.0); condylobasal length, 15.7 (15.3–15.9), 16.3 (16.2-16.4); least interorbital breadth, 3.9 (3.8–3.9), 3.8 (3.7–3.9); zygomatic breadth (2 males, 2 females),11.1 (10.9–11.2), 11.2; breadth of braincase, 8.1 (7.6–8.4), 8.1 (8.0–8.2); length of maxillary toothrow, 5.9 (5.9–6.0), 6.3; palatal length (1 male, 2 females) 6.1, 6.9 (6.5–7.3); length of mandibular toothrow, 6.4 (6.1–6.6), 6.7 (6.7–6.8); greatest length of mandible, 12.0 (11.8–12.3), 12.6 (12.6–12.7).

Remarks. Three of the four females collected in mid-December at Chumbicha were lactating; the fourth had a closed vagina. A fifth individual captured at this site was a male. In a previous year at this same site, seven of the eight bats captured were males.

Histiotus macrotus (Poeppig)

1835. *Nycticeius macrotus* Poeppig, Reise Chile Perú Amaz., 1:451.

Specimens examined (31). Andalgalá, 1 (MACN); Balneario El Caolín, 6 km NW Chumbicha, 2 (OMNH); Chumbicha, 1 km N and W of Balneario by road, 792 m, 8 (3 CML, 2 IADIZA-CM, 3 OMNH); Choya, 13 km NNW Andalgalá, 1,219 m, 1 (OMNH); Cuesta del Clavillo, 5 km al S de La Banderita, 4 (RMB); Dique El Potrero, 12 (9 CMNH, 3 MSB); El Rodeo, 1 (CML); Las Estancias, 1 (CML); Tinogasta, 1 (MACN).

Measurements. External measurements (9 males, 16 females, unless otherwise noted): total length, 109.6 (94–125), 115.8 (106–123); tail, 52.0 (42–57), 55.8 (45–62); hindfoot (8 males, 16 females), 9.7 (9–11.4), 10.1 (9.4–11); ear (7 males, 16 females), 34.3 (33–36.5), 34.0 (31–36.7); weight (7 males, 11 females), 11.2 (9.0–14.4), 1.2 (11.0 -15.0); forearm (5 males, 9 females), 47.1 (45.8–48.0), 48.9 (47.0–51.0). Cranial measurements (6 males , 6 females, unless otherwise noted): greatest length of skull (5 males, 5 females), 17.9 (17.4–18.3), 18.4 (18.0–18.7);

condylobasal length (5 males, 6 females), 17.1 (16.5–17.3), 17.5 (17.0–18.2); least interorbital breadth, 4.2 (4.2–4.3), 4.4 (4.2–4.6); zygomatic breadth (4 males, 6 females), 10.5 (10.3- 10.7), 10.9 (10.5–11.4); breadth of braincase (4 males, 5 females), 8.4 (8.2–8.5), 8.5 (8.2–8.4); length of maxillary toothrow, 6.2 (6.0–6.5), 6.4 (6.1–6.7); palatal length (3 males, 5 females), 7.6 (7.1–8.4), 7.5 (7.1–8.5); length of mandibular toothrow, 6.7 (6.5–7.0), 6.9 (6.6–7.3); greatest length of mandible, 12.4 (11.5–12.9), 12.9 (12.0–13.3).

Remarks. The population of Catamarca has been referred to as *H. m. macrotus* (Barquez, 1987). Of eight individuals captured in mid-December near Chumbicha, six were lactating females. A lactating female was also collected in Choya. A male collected near Chumbicha at the same time period had a testis size of 4 x 2 mm; reproductive data were not recorded for a second male. A male and female were collected at the Chumbicha site in an earlier year. At El Potrero Dike, five of eight individuals captured in late January were females. A female was later captured at this site in March. In Catamarca, we have only collected this species in lower and mid-elevation montane gallery forest along watercourses.

Lasiurus blossevillii (Lesson and Garnot)

1826. *Vespertilio blossevillii* Lesson and Garnot, Bull. Sci. Nat. et Géol., 8:95.

Specimens examined (46). Andalgalá, 2 (1 CMNH, 1 MACN); 13 km N Andalgalá, 20 (CMNH); Río Potrero, 13 km N Andalgalá, 2 (CML); Balneario El Caolín, 6 km NW Chumbicha, 13 (OMNH); Chumbicha, 1 km N and W of Balneario by road, 792 m, 3 (1 CML, 1 IADIZA-CM, 1 OMNH); Dique El Potrero, 5 (2 CML, 2 MSB, 1 CMNH); Río San Pablo, 3 km NW Concepción, 1 (CMNH).

Measurements. External measurements (11 males, 15 females, unless otherwise noted): total length, 98.7 (90–108), 107.9 (99–114); tail, 48.7 (45–54), 52.9 (49–63); hindfoot (10 males, 14 females), 8.3 (7.4–9.4), 8.6 (7.5–10); ear, 11.6 (10–12.9), 11.9 (10–13.4); weight (11 males, 19 females), 8.5 (5.0–10.5), 11.0 (6.0–15.0); forearm (3 males, 4 females), 38.0 (37.9–38.0), 41.1 (39.9–42.4). Cranial measurements (3 males, 5 females, unless otherwise noted): greatest length of skull (2 males, 4 females), 12.4 (12.1–12.7), 13.3 (13.1–13.6); condylobasal length (2 males, 4 females), 11.7 (11.3–12.0), 12.6 (12.5–12.8); least interorbital breadth, 4.2 (4.1–4.3), 4.4 (4.1–4.6); zygomatic breadth (3 males, 4 females), 8.5 (8.2–8.8), 9.3 (9.1–9.5); breadth of braincase, 7.2 (7.0–7.4), 7.4 (7.3–7.7); length of maxillary toothrow, 3.9 (3.8–3.9), 4.1 (3.9–4.3); palatal length (2 males, 5 females), 4.4 (3.8–5.0), 4.3 (3.9–5.0); length of mandibular toothrow, (3 males, 4 females), 4.5 (4.4–4.7), 4.9 (4.8–5.1); greatest length of mandible (3 males, 4 females), 8.7 (8.5–8.8), 9.0 (8.6–9.2).

Remarks. Baker et al. (1988) recommended recognition of western populations of *L. borealis* as a distinct species, *L. blossevillii*. We have chosen, in this case, not to follow Wilson and Reeder (1993). Two of the four specimens collected near Chumbicha in mid-December were molting. One was an adult male and the second was a young male. Of these four specimens, one female was lactating, a young female was in nonbreeding condition, and two males had testes 2.5 x 1.5 and 3 x 2 mm in size. A pregnant female with four embryos was collected at Dique El Potrero near Andalgalá in October; crown-rump lengths ranged from 11.7 to 12.5. Specimens have been collected in January, October, November, and December. This species is generally found in riparian areas or in orchards bordering the drier habitats.

Lasiurus cinereus (Beauvois)

1796. *Vespertilio cinereus* Beauvois, Cat. Raisonne Mus. Peale Philadelphia, p. 18.

Specimens examined (22). 13 km N Andalgalá, 8 (CMNH); 8 km N Chaquiago, along Río Potrero, 2 (OMNH); Balneario El Caolín, 6 km NW Chumbicha, 2 (OMNH); Río San Pablo, 3 km NW Concepción, 5 (1 CML, 4 CMNH); Dique El Potrero, 5 (CMNH).

Measurements. External measurements (16 males, 5 females, unless otherwise noted): total length, 125.7 (119–134), 130.0 (125–

135); tail, 54.6 (51–60), 58.8 (55–61); hindfoot (12 males, 5 females), 10.4 (10–11.2), 11.0 (10.6–11.5); ear, 17.0 (14.9–19.6), 18.0 (16.8–18.6); forearm (5 males), 51.8 (50.2–53.5); weight (15 males, 5 females), 19.1 (13.5–24.0), 21.1 (16.0–26.5). Cranial measurements (7 males, unless otherwise noted): greatest length of skull, 15.3 (14.6–16.6); condylobasal length, 14.8 (14.1–15.8); least interorbital breadth, 5.4 (5.2–5.6); zygomatic breadth, 11.4 (11.2–11.7); breadth of braincase, 8.9 (8.4–9.4); length of maxillary toothrow, 5.4 (5.2–5.6); palatal length (6 males), 5.9 (5.0–6.7); length of mandibular toothrow, 6.1 (5.9–6.3); greatest length of mandible, 11.6 (11.3–12.0).

Remarks. The subspecies in Catamarca is *L. c. villosissimus* (Barquez, 1987). This species is frequently captured with *Lasiurus blossevillii*. Although it is not taken in open desert, it is not uncommon in the mesic peripheries of drier areas. Specimens have been collected in the months of January, March, and November. Almost all of the specimens examined by us were males.

Lasiurus ega (Gervais)

1856. *Nycticeius ega* Gervais, in F. Comte de Castelnau, Exped. Partes Cen. Am. Sud. (Sec. 7), Vol. 1, pt. 2 (Mammifères), p. 73.

Specimens examined (8). Balneario El Caolín, 6 km NW Chumbicha, 8 (OMNH).

Measurements. External measurements (2 males, 5 females): total length, 118.0 (115–121), 125.8 (117–132); tail, 54.0 (53–55), 55.8 (46–60); hindfoot, 10, 10.4 (10–11); ear, 17, 18.0 (17–19); weight, 13.0 (12.0–14.0), 21.4 (18.0–26.0). Cranial measurements: (1 male, 4 females): greatest length of skull, 16.5, 17.6 (17.4–17.7); condylobasal length, 15.7, 16.7 (16.3–17.0); least interorbital breadth, 4.9, 4.9 (4.5–5.2); zygomatic breadth, 11.3, 11.9 (11.8–12.2); breadth of braincase, 8.9, 9.4 (9.2–9.6); length of maxillary toothrow, 5.6, 5.8 (5.7–5.9); palatal length, 5.2, 5.8 (5.6–6.0); length of mandibular toothrow, 6.4, 6.8 (6.7–6.9); greatest length of mandible, 11.4, 12.3 (12.2–12.4).

Remarks. These specimens were collected over water near Chumbicha on two separate nights in November. Two females were pregnant; one had three embryos. These are the first records of this bat for the province. The subspecies in Catamarca is *L. e. argentinus* (Barquez, 1987).

Lasiurus salinae Thomas

1902. *Lasiurus borealis salinae* Thomas, Ann. Mag. Nat. Hist., ser. 7, 9:238.

Specimens examined (7). Balneario El Caolín, 6 km NW Chumbicha, 3 (OMNH); Chumbicha, 1 km N and W of Balneario by road, 792 m, 3 (1 CML, 1 IADIZA-CM, 1 OMNH); Chumbicha, at Balneario, 732 m, 1 (OMNH).

Measurements. External measurements (2 males, 4 females): total length, 100.5 (95–106), 109.3 (105–113); tail, 45.0 (41–50), 52.5 (50–57); hindfoot, 9.0 (8–10), 9.5 (9–10); ear, 10.5 (10–11), 11.3 (11–12); weight, 9.3 (7.5–11.0), 10.0 (7.0 -12.0). Cranial measurements (2 males, 3 females): greatest length of skull, 13.0 (12.7–13.3), 13.3 (12.9 -13.7); condylobasal length, 12.2 (12.0–12.4), 12.7 (12.4–12.9); least interorbital breadth, 4.4 (4.3–4.5), 4.4 (4.2–4.6); zygomatic breadth, 8.9 (8.5–9.2), 9.2 (9.0–9.5); breadth of braincase, 7.4 (7.0–7.7), 7.5 (7.5–7.6); length of maxillary toothrow, 4.1 (4.0–4.2), 4.3 (4.2–4.3); palatal length, 4.2 (4.1–4.3), 4.5 (4.2–4.7); length of mandibular toothrow, 4.7 (4.5–4.8), 4.8 (4.6–4.9); greatest length of mandible, 8.8 (8.5–9.0), 9.1 (8.7–9.5).

Remarks. Our recognition of this species is tentative until additional material can be examined. This is the first citation of the species for the province. The type locality of *salinae* is Cruz del Eje, Córdoba Province (Thomas, 1902). Although described as a subspecies of *borealis*, A. Cabrera (1957) and Shump and Shump (1982) synonomized this taxon with *L. b. varius*. Barquez (1987), after an examination of the type material, suggested that it should be synonomized with *L. b. blossevillii*. Specimens of *L. salinae* are markedly darker and browner than individuals of *L. blossevillii*. Specimens were collected in November and December. One individual collected in December was a male without fully ossified phalanges. This species was captured at Chumbicha along with *L. blossevillii;* both were in the same net at the same time.

Myotis levis (I. Geoffroy)

1824. *Vespertilio levis* I. Geoffroy, Ann. Sci. Nat. Zool., ser. 1, 3:444-445.

Specimens examined (60). Andalgalá, 3 (2 CMNH, 1 MSB); 4 km S Andalgalá, 2 (CMNH); Balneario El Caolín, 6 km NW Chumbicha, 4 (OMNH); Balneario Municipal Capayan, Chumbicha, 1 (OMNH); Chumbicha, 1 km N and W of Balneario by road, 34 (9 CML, 9 IADIZA-CM, 16 OMNH); Cuesta de La Sébila, 2 (CMNH); Cuesta del Clavillo, 5 km al S de La Banderita, 2 (RMB); Dique El Potrero, 1 (CMNH); El Rodeo, 1 (CML); Las Juntas, 2 (RMB); Las Juntas, 50 km NW Catamarca Ciudad, 3 (OMNH); Mollecito, 3 (CML); Río San Pablo, 3 km NW Concepción, 1 (CMNH); Villa El Potrero (N Andalgalá), El Potrero, 1 (CMNH).

Measurements. External measurements (6 males, 41 females, unless otherwise noted): total length, 85.7 (84–87), 85.0 (75–90); tail, 41.5 (37–48), 38.9 (36–43); hindfoot (6 males, 40 females), 7.9 (7–10), 7.1 (6–9); ear (6 males, 40 females), 14.5 (13–15.1), 13.9 (12–17); weight (4 males, 41 females), 5.1 (4.0–6.0), 5.5 (3.5–7.0); forearm (3 males, 35 females), 36.7 (35.0–38.0), 37.3 (35.0–40.0). Cranial measurements (4 males, 17 females, unless otherwise noted): greatest length of skull, 14.3 (13.9–14.6), 14.4 (14.1–14.8); condylobasal length, 13.5 (13.3–13.7), 13.6 (13.2–13.9); least interorbital breadth, 3.6 (3.5–3.7), 3.5 (3.3–3.7); zygomatic breadth (3 males, 11 females), 8.7 (8.6–8.7), 8.6 (8.4–9.0); breadth of braincase, 7.0 (6.9–7.1), 6.9 (6.8–7.1); length of maxillary toothrow, 5.4 (5.3–5.6), 5.4 (5.2–5.7); palatal length (3 males, 16 females), 6.7, 6.9 (6.4–7.7); length of mandibular toothrow (3 males, 17 females), 5.8 (5.7–5.8), 6.0 (5.7–6.2); greatest length of mandible, 10.3 (10.0–10.6), 10.4 (10.0–10.9).

Remarks. This species was captured in the mesic peripheries of desert areas, as well as in drylands well away from water. It has also been collected in forested habitats. It probably occurs throughout the province at lower elevations. Specimens have been collected in Catamarca in January, February, March, July, October, November, and December. Pregnant females were captured in October and November; each had one embryo, the latter with a crown-rump length of 19.5 mm. During three consecutive nights of mist netting (at the Balneario at Chumbicha in December), of 34 individuals captured, only 2 were males, and all but 5 of the 32 females were lactating. Lactating females were also captured in January at Las Juntas and Cuesta del Clavillo. A male collected at Cumbicha in July was found roosting in a building. The subspecies in Catamarca is *M. l. dinellii* (Barquez, 1987).

Family Molossidae

Nyctinomops macrotis (Gray)

1840. *Nyctinomus macrotis* Gray, Ann. Mag. Nat. Hist., 4:5.

Specimens examined (1). Dique El Potrero, Andalgalá, 1 (CMNH).

Measurements. External measurements (1 young male): total length, 136; tail, 60; hindfoot, 11; ear, 29.9; weight, 16.0; forearm, 58.9.

Remarks. Barquez (1987) discusses the differences between *Nyctinomops* and *Tadarida* in Argentina and provides external and cranial measurements of this species from other provinces. This is the first record of this species in Catamarca. The specimen was captured in the winter month of May at the northern edge of the Bolsón de Pipanaco (near Andalgalá in the monte desert) in a moist gallery forest along the foothills of a pre-Andean range; it was a young animal and not reproductively active.

Promops nasutus (Spix)

1823. *Molossus nasutus* Spix, Sim. Vespert. Brasil., p. 59.

Specimens examined (2). Balneario El Caolín, 6 km NW Chumbicha, 1 (OMNH); Dique El Potrero, 1 (CMNH).

Measurements. External measurements (2 females): total length, 122, 126; tail, 52, 58; hindfoot, 9, 10.8; ear, 14, 15.7; weight, 19.0, 14.9. Cranial measurements (1 female):

greatest length of skull, 17.6; condylobasal length, 16.0; zygomatic breadth, 10.7; least interorbital breadth, 4.0; breadth of braincase, 9.4; length of maxillary toothrow, 6.6; palatal length, 6.2; length of mandibular toothrow, 6.8; greatest length of mandible, 11.7.

Remarks. These are the first records for this species in Catamarca. The specimens were both captured in riparian montane gallery forest. One was taken in mesic habitat at an elevation higher than the nearby monte desert (the El Potrero specimen), and the other was captured in mesic habitat above the adjacent chacaon thorn scrub (the Chumbicha specimen). The first specimen was collected in January 1971; the second 16 years later in November 1987.

Tadarida brasiliensis (I. Geoffroy)

1824. *Nyctinomus brasiliensis* I. Geoffroy, Ann. Sci. Nat. Zool., 1:343.

Specimens examined (235). Andalgalá, 2 (1 CMNH, 1 MACN); 8 km N Chaquiago, along Río Potrero, 3 (OMNH); Balneario El Caolín, 6 km NW Chumbicha, 25 (OMNH); Chumbicha, at Balneario, 732 m, 3 (1 CML, 2 OMNH); Chumbicha, 1 km N and W of Balneario by road, 792 m, 3 (OMNH); Cuesta del Clavillo, 5 km al S de La Banderita, 2 (RMB); Dique El Potrero, 31 (CMNH); El Rodeo, 145 (CML); Las Juntas, Estancia de los Figueroa, 50 km NW Catamarca Ciudad, 3 (OMNH); Pomán, 95 km S Andalgalá near Balneario, 1,341 m, 17 (7 CML, 4 IADIZA-CM, 7 OMNH); Río San Pablo, 3 km NW Concepción, 1 (CMNH).

Measurements. External measurements (41 males, 24 females, unless otherwise noted): total length, 96.9 (90–107), 94.7 (89–104); tail (40 males, 23 females), 37.1 (31–46), 35.6 (31–41); hindfoot (40 males, 23 females), 9.0 (7.9–10), 9.1 (7–10.9); ear (40 males, 23 females), 18.2 (14–20.1), 17.1 (14–19.8); forearm (13 males, 18 females), 43.6 (41.8–45.0), 43.8 (37.0–47.0); weight (46 males, 30 females), 11.4 (10.0–13.5), 11.5 (8.0–14.). Cranial measurements (10 males, 7 females, unless otherwise noted): greatest length of skull, 16.7 (15.8–17.5), 16.8 (16.4–17.2); condylobasal length, 15.7 (15.0–16.2), 15.8 (15.5–16.1); least interorbital breadth, 4.0 (4.0–4.1), 4.1 (3.9–4.3); breadth of braincase, 8.5 (8.2–8.8), 8.5 (8.4–8.5); length of maxillary toothrow, 6.1 (5.9–6.3), 6.1 (6.0–6.2); length of mandibular toothrow, 6.7 (6.2–7.1), 6.6 (6.4–6.7); greatest length of mandible, 11.5 (11.0–11.9), 11.8 (11.4–12.2).

Remarks. This species is probably the most common bat in Catamarca, occurring throughout the province at lower elevations and in almost all habitats. Additional measurements and comments may be found in Barquez (1987). Specimens have been captured in January, March, May, October, November, and December. Of the 17 specimens collected at Pomán in December, 14 were females (12 were lactating). The three males had scrotal testes of 3 x 2 mm and one was also molting. A collection of 18 specimens made at Río Potrero in January included 14 males. This site was later sampled in March and only males (12) were collected. The Chumbicha site was sampled several times in November of one year; 18 of 25 bats captured were males. One of the seven females was pregnant. A lactating female and a male with scrotal testes (3 x 2 mm) were captured in December at this site. One night of netting at Dique El Potrero in January yielded 14 males and five females. Later the same year (March), 12 males were captured; no females were collected.

Order Carnivora

Family Canidae

Subfamily Caninae

Pseudalopex culpaeus (Molina)

1782. *Canis culpaeus* Molina, Sagg. Stor. Nat. Chile, p. 293.

Specimens examined (2). Los Nacimientos, 1 (CML); San Francisco, 4,100 m, 1 (MACN).

Additional records. Off Hwy 62, 8 km E jct. Hwy 62 and 1, east of Andalgalá (Mares, 1973).

Measurements. External measurements (1 sex unknown): total length, 935; tail, 375; hindfoot, 110; ear, 75. Cranial measurements (2 sex unknown): greatest length of skull, 162.0, 110.5; condylobasal length, 147.5, 106.2; least interorbital breadth, 25.5, 18.2;

postorbital constriction, 22.5, 22.9; breadth of braincase, 44.4,41.8; length of maxillary toothrow (from the anterior face of canines), 66.6, 49.4; palatal length, 78.9, 53.0; length of mandibular toothrow, —, 60.1; greatest length of mandible, —, 83.9.

Remarks. This is a high mountain species that occasionally descends to the foothills and lowland valleys; it is not common in the lowlands. It is less abundant than other foxes in the province. This and the following two species of foxes were heavily hunted in Argentina, with a total of 101,251 *P. culpaeus* being exported between 1972 and 1979 (Mares and Ojeda, 1984; Ojeda and Mares, 1982, 1984). This species is listed as Appendix II in Argentina (Fuller and Swift, 1984). The subspecies in Catamarca is *P. c. andinus* (A. Cabrera, 1957).

Pseudalopex griseus (Gray)

1837. *Vulpes griseus* Gray, Mag. Nat. Hist. (Charlesworth's), 1:578.

Specimens examined (1). Off Hwy 62 in Río La Cañada, S of Andalgalá, 1 (CMNH).

Additional records. Along Río Amanao, 15 km S jct. Hwy 1 and Río Amanao; Andalgalá Railroad Station, 4 km S Andalgalá; 50 m N, 50 m E jct. Hwy 62 and Amanao Road, approx. 18 km E Andalgalá; off Hwy 62, 8 km E jct. Hwys 62 and 1, E of Andalgalá; 1.5 km S, 150 m W jct. Hwy 1 and railroad track, S of Andalgalá; 6 km N Saujil (Mares, 1973).

Measurements. External measurements (1 adult male): total length, 898; tail, 313; hindfoot, 129.4; ear, 84.2.

Remarks. This species was fairly common in the monte desert of Catamarca in the early 1970s (M.A. Mares, pers. obs.), but it has been hunted heavily and is now uncommon. The subspecies in Catamarca is *P. g. griseus* (A. Cabrera, 1957). This species has an Appendix II listing in Argentina (Fuller and Swift, 1984) and is currently listed as vulnerable (IUCN, 1990).

Pseudalopex gymnocercus (Fischer)

1814. *Procyon gymnocercus* Fischer, Zoognosia, 3:178.

Specimens examined (7). La Chilca, 1 (MACN); Río Charquiadero, 2 (MACN); Río Vallecito, 2,050 m, 1 (MACN); Salar de Pipanaco, 1 (MACN); Singuil, 1,250 m, 2 (MACN).

Measurements. Cranial measurements (2 sex unknown): greatest length of skull, 122.8, 125.8; condylobasal length, 121.3, 121.4; least interorbital breadth, 21.2, 20.5; zygomatic breadth, 63.9, 62.4; postorbital constriction, 23.1, 22.4; breadth of braincase, 43.6, 42.3; length of maxillary toothrow, 56.5, 54.0; palatal length, 64.0, 65.7; length of mandibular toothrow, 63.1, 62.0; greatest length of mandible, 92.0, 91.5.

Remarks. This fox is one of the most abundant foxes in Catamarca (and in Argentina) and one of the most heavily hunted.

Family Felidae

Subfamily Felinae

Herpailurus yaguarondi (Lacépède)

1809. *Felis yaguarondi* Lacépède, in Azara, Voy. Am. Mérid., Atlas, 1809, pl. 10.

Specimens examined. None.

Additional records. Catamarca (Cabrera, 1961a); in Catamarca near the Catamarca-Tucumán border, in the alder forest habitat at 2,000 m (M.A. Mares, pers. obs.).

Measurements. External measurements (1 sex unknown) reported by Cabrera (1961a): total length, 1010; tail, 410; hindfoot, 65; ear, 30.

Remarks. The specimen from Catamarca, described by Cabrera (1961a), was received by the Museo de La Plata but was destroyed. The subspecies in Catamarca is *F. y. ameghinoi* (A. Cabrera, 1957). This species is listed as status indeterminate in Argentina (IUCN, 1990), but is listed as an Appendix II species in Argentina and has protected status (Fuller and Swift, 1984).

Oncifelis colocolo (Molina)

1782. *Felis colocolo* Molina, Sagg. Stor. Nat. Chile, p. 295.

Specimens examined (6). Antofagasta de la Sierra, 1 (MACN); Catamarca, 4 (MACN); La Atravesada, 1800 m, 1 (MACN).

Additional records. Andalgalá (Cabrera, 1961a).

Measurements. Cranial measurements (2 sex unknown): greatest length of skull, 102.0, 92.6; condylobasal length, 94.3, —; least interorbital breadth, 19.7, 15.8; zygomatic breadth, 76.0, 61.9; postorbital constriction, 28.0, 28.0; breadth of braincase, 45.7, —; length of maxillary toothrow, 29.0, 28.0; palatal length, 39.1, 33.7; length of mandibular toothrow, 34.1, 30.9; greatest length of mandible, 67.7, 61.3.

Remarks. Cabrera (1961b) listed the population *F. c. budini* as ranging from northern Argentina to provinces south of Catamarca. The specimen from Andalgalá, listed in Cabrera's (1961a) revision of the Felidae, is a mounted specimen on exhibit in the Museo La Plata. The specimens reported here were collected by Crespo in 1950 (La Atravesada), Olzansky in 1937 (Catamarca), and M. Rumboll (Antofagasta de la Sierra). This is a species of the high (> 3,000 m) grasslands and puna of the Andean and pre-Andean chains (see Mares et al., 1989). This species has an Appendix II listing in Argentina and has protected status (Fuller and Swift, 1984). During the 1970s, 82,195 specimens were exported from Argentina (Mares and Ojeda, 1984; Ojeda and Mares, 1982, 1984).

Oncifelis geoffroyi (d'Orbigny and Gervais)

1844. *Felis geoffroyi* d'Orbigny and Gervais, Bull. Sci. Soc. Philom. Paris, 1844:40.

Specimens examined (1). 14 km S Hualfín, 1 (MACN).

Additional records. Bolsón de Pipanaco (Mares, 1973).

Measurements. None.

Remarks. This is probably the most common wild felid in Catamarca. It occurs in all major habitats. Like most cats, it has been heavily hunted throughout the country, including Catamarca. Between 1972 and 1979, 481,333 were exported from Argentina as furs (Mares and Ojeda, 1984; Ojeda and Mares, 1982, 1984). The subspecies in Catamarca is *O. g. salinarum* (A. Cabrera, 1957). This species has an Appendix II listing in Argentina and has protected status (Fuller and Swift, 1984).

Oreailurus jacobita (Cornalia)

1865. *Felis jacobita* Cornalia, Mem. Soc. Ital. Sci. Nat., 1:3.

Specimens examined (21). Camino a San Francisco, 1 (MACN); Catamarca, 20 (MACN).

Additional records. Laguna Helada, 3,700 m (Cabrera, 1961a).

Measurements. None.

Remarks. Comments on the color variations of the species were reported by Cabrera (1961a). It is known only from the highest grasslands and scrublands above 3,000 m. The specimen from Laguna Helada was part of Shipton's collection, which was destroyed while being acquired by the Fundación Miguel Lillo (Cabrera, 1961a). This species is listed in Appendix I in Argentina and has protected status (Fuller and Swift, 1984) and is listed as Rare by IUCN (1990).

Puma concolor (Linnaeus)

1771. *Felis concolor* Linnaeus, Mantissa Plantarum, 2:522.

Specimens examined (5). El Bolsón, 1 (MACN); La Ciénaga, 1 (MACN); Las Rosas, 1 (MACN); Los Castillos, 2 (MACN).

Additional records. Along Río Amanao, 15 km S jct. Hwy 1 and Río Amanao; off Hwy 1 at km 99, 34 km S Andalgalá (Mares, 1973); Catamarca (Cabrera, 1961a).

Measurements. Cranial measurements (2 males, 1 female): greatest length of skull, 210.3, 181.7, 180.5; condylobasal length, 184.8, 165.7, 173.8; least interorbital breadth, 42.0, 35.6, 39.4; zygomatic breadth, 148.0, 123.3, 133.8; postorbital constriction, 43.6, 44.7, 48.4; breadth of braincase, 74.4, 72.8, 74.0; length of maxillary toothrow, 64.4, 59.5, 56.0; palatal length, 84.4, 76.0, 75.6; length of mandibular toothrow, 72.9, 70.6, 64.3; greatest length of mandible, 137.6, 119.2, 126.4.

Remarks. The specimen from La Ciénaga was assigned to the subspecies *P. c. cabrerae*; the other population that might reach the western mountain ranges of Catamarca is *P. c. puma* Molina (Cabrera, 1961a). According to Cabrera (1961a), the specimen from Las Rosas is from the southern part of the Aconquija Range on the eastern side, near El Alamito. The specimen from Catamarca is in an archeology collection at the Museo La Plata; it was found in a pre-Hispanic Indian tomb (Cabrera, 1961a). Pumas are common in most of the habitats of Catamarca. This species has an Appendix II listing in Argentina (Fuller and Swift, 1984).This species is still heavily hunted in Argentina; between 1976 and 1979, 3,538 specimens were exported from the country (Mares and Ojeda, 1984).

Family Mustelidae
Subfamily Mustelinae
Eira barbara (Linnaeus)

1758. *Mustela barbara* Linnaeus, Syst. Nat., 10th ed., 1:46.

Specimens examined. None.

Additional records. Near Catamarca–Tucumán border along Hwy 65 (M.A. Mares, pers. obs.).

Remarks. The individual sighted by Mares in 1971 was in a densely forested habitat at an elevation above 1,800 m. This may be the southernmost record for this tropical mustelid and is a new record for the province. The subspecies in Catamarca would be *E. b. barbara* (A. Cabrera, 1957).

Lyncodon patagonicus (de Blainville)

1842. *Mustela patagonica* de Blainville, Osteogr. Mamm., pt. 10 (Viverra):1.

Specimens examined (2). Andalgalá, 2 (1 CMNH; 1 CML).

Additional records. Andalgalá (Olrog, 1976); Santa María (Olrog, 1976).

Measurements. External measurements (1 male): total length, 391; tail, 113; hindfoot, 43.3; ear, 15.2; weight, 223.6. Cranial measurements (1 adult male): greatest length of skull, 53.4; condylobasal length, 53.4; least interorbital breadth, 11.2; zygomatic breadth, 31.8; breadth of braincase, 24.1; length of maxillary toothrow (from anterior face of canine), 14.7; palatal length, 22.8; length of mandibular toothrow, 16.4; greatest length of mandible, 30.8.

Remarks. Olrog (1958) reported on cranial measurements and coloration of the specimen from Andalgalá housed at the CML. A second specimen was also cited by Olrog (1958) from Andalgalá, but this specimen has not been located since its listing by Olrog. In 1971, Mares collected an individual that was probably rabid (it was foaming at the mouth, crossing a road during daylight hours, and able to move only with obvious difficulty). Olrog (1976) reported on a specimen from Santa Maria in Catamarca Province, but that specimen could not be located in the Lillo collection. The subspecies in Catamarca is *L. p. thomasi* (A. Cabrera, 1957).

Subfamily Mephitinae
Conepatus chinga (Molina)

1782. *Viverra chinga* Molina, Sagg. Stor. Nat. Chile, p. 288.

Specimens examined (4). Estancia Gracian, Paclin, 1 (CML); Sierra de Humaya, 1 (CML); Hwy 38, 17 km S of La Merced, S of km 613, 640 m, 1 (OMNH); 6 km S of El Rodeo on Hwy 4, 1,371 m, 1 (OMNH).

Additional records. 1.5 km S, 150 m W jct. Hwy 1 and railroad track, S of Andalgalá; 50 m N, 50 m E jct. Hwy 62 and Amanao Road, approximately 18 km E Andalgalá (Mares, 1973); 16 February 1971 at 0500 hrs on the Cuesta de la Chilca between Andalgalá and Las Estancias just before reaching the top of the cuesta (M.A. Mares, pers. obs.); 31 May 1971 at 0530 hrs two young along the road from Las Estancias to S.M. de Tucumán just inside the Catamarca border in subtropical forest (M.A. Mares, pers. obs.); Laguna Blanca, 3200 m (R.A. Ojeda, pers. obs.); Otro Cerro (Thomas, 1919b).

Measurements. External measurements (4 adult males; data for the specimens from El Rodeo and La Merced and from the two specimens from Otro Cerro reported by Thomas, 1919b): total length, 456, 386, 580, 510; tail, 185, 175, 240, 190; hindfoot, 55, 53, 55, 57; ear, 18, 19, 30, 25; weight, 610.0, 1750.0, —, —. Cranial measurements (2 adult males, from Thomas, 1919b): greatest length of skull, 71.5, 69.5; condylobasal length, 66.5, 66.5; least interorbital breadth, 23.0, 21.2; zygomatic breadth, 45.0, 43.0; postorbital constriction, 18.5, 15.6; length of maxillary toothrow, 22.5, 21.5; palatal length, 28.0, 28.0.

Remarks. Type locality of *C. c. budini* Thomas is Otro Cerro. Thomas (1919b) named two new species from Otro Cerro, *C. budini* and *C. calurus*. Both have since been synonymized with *C. chinga*. This skunk has been heavily hunted in Argentina, especially in Buenos Aires Province and in Patagonia. During the 1970s, 1,243,129 animals (includes four species of *Conepatus*) were exported from Argentina (Mares and Ojeda, 1984; Ojeda and Mares, 1982, 1984). The subspecies in the province is *C. c. budini* (A. Cabrera, 1957).

Family Procyonidae
Subfamily Procyoninae
Procyon cancrivorus (Cuvier)

1798. *Ursus cancrivorus* Cuvier, Tabl. Elem. Hist. Nat. Anim., p.113.

Specimens examined (1). Catamarca, 1 (MACN).

Measurements. Cranial measurements (1 sex unknown): greatest length of skull, 129.4; condylobasal length, 123.4; least interorbital breadth, 27.5; postorbital constriction, 26.4; breadth of braincase, 54.3; length of maxillary toothrow, 45.5; palatal length, 74.4.

Remarks. Residents of La Puerta, located 45 km N of the city of Catamarca, indicated in 1987 (M.A. Mares, pers. obs.) that this species is common in the area. The subspecies found in Catamarca is *P. c. nigripes* (A. Cabrera, 1957).

Order Artiodactyla
Family Tayassuidae
Pecari tajacu (Linnaeus)

1758. *Sus tajacu* Linnaeus, Syst. Nat., 10th ed., 1:50.

Specimens examined. None.

Remarks. The southern limits of distribution of the collared peccary are poorly known. The presence of this species in the province of Catamarca and in the neighboring provinces of La Rioja and San Juan has been reported in general distribution accounts (Olrog and Lucero, 1981), but we have been unable to verify their presence with specimens. The subspecies in Catamarca would be *T. t. tajacu* (Cabrera, 1961b). Between 1972 and 1979, 312,000 peccary specimens (includes three species, but *P. tajacu* was probably the most common one) were exported from Argentina (Mares and Ojeda, 1984; Ojeda and Mares, 1982, 1984).

Family Camelidae
Lama guanicoe (Müller)

1776. *Camelus guanicoe* Müller, Linné's Vollstand. Natursyst. Suppl., p. 50.

Specimens examined (1). Pastos Largos, 1 (MACN).

Additional records. Antofagasta de la Sierra; between Amaicha (Province of Tucumán) and Santa María (Province of Catamarca); Cazadero Grande; Cortadera; El Cajón; Laguna Blanca; Tolar (Cajal, 1983a; Holmberg, 1902).

Measurements. Cranial measurements (1 sex unknown): greatest length of skull, 296.0; condylobasal length, 278.0; least interorbital breadth, 103.0; zygomatic breadth, 121.0; breadth of braincase, 72.0; length of maxillary toothrow, 65.5; length of mandibular toothrow, 72.0; greatest length of mandible, 224.0; length of mandibular diastema, 47.5.

Remarks. The status of this species in Catamarca seems delicate. According to reports published early in this century (Holmberg, 1902; see also Roig, 1991), guanacos were present in the puna of Catamarca; in 1971, Mares met a person who had killed a guanaco near the town of Amanao in the 1960s. Research conducted in the 1970s suggested that this species may have been extirpated from the province (Cajal, 1983a; Cajal et al., 1983), but recent field work has located at least a small population of guanacos in the Department of Tinogasta in the western highlands of Catamarca (Cajal, 1991). The subspecies in Catamarca is *L. g. guanicoe* (Cabrera, 1961b). This species has an Appendix II listing in Argentina (Fuller and Swift, 1984). During the 1970s, 444,000 specimens of this species were exported from Argentina (Mares and Ojeda, 1984; Ojeda and Mares, 1982, 1984).

Vicugna vicugna (Molina)

1782. *Camellus* (*sic*) *vicugna* Molina, Sagg. Stor. Nat. Chile, p. 313.

Specimens examined. None.

Additional records. Antofagasta de la Sierra; Cazadero Grande; Cortadera; Laguna

Blanca; Tolar (Cajal, 1983a; J. Cajal, pers. comm.; Holmberg, 1902; R.A. Ojeda, pers. obs.). This species is said to occur in the vicinity of Farallón Negro and Cerro Durazno (personal communication by local inhabitants).

Remarks. Censuses of populations of vicuñas carried out in the Provincial Reserve of Laguna Blanca in 1986 projected a population of 1,192 animals (J. Cajal, pers. comm.). Vicuñas are considered to be competitors of domestic stock (burros, mules, goats, sheep, and llamas, Cajal et al., 1983) and this led to early efforts to control the species. Vicuñas were heavily hunted by the Incas and the species has been persecuted throughout its association with humans in the South American highlands (e.g., Roig, 1991). Vicuñas are now protected in Argentina, and Cajal (1991) estimates a total population of approximately 2,000 animals at four locations that have been censused. This species is listed as vulnerable by the IUCN (1990) and is listed in Appendix I by Argentina and has protected status (Fuller and Swift, 1984).

Family Cervidae

Subfamily Odocoileinae

Hippocamelus antisensis (d'Orbigny)

1834. *Cervus antisensis* d'Orbigny, Ann. Mus. Hist. Nat., 3:91.

Specimens examined (6). Andalgalá, 1 (MACN); Cuesta del Medanito, 3,550 m, 3 (MACN); east of Famatina, 2 (MACN).

Measurements. External measurements (1 male, 1 young male, 1 female): total length, 1420, 1250, 1390; tail, 112, 120, 100; hindfoot, 320, 300, 320; ear, 178, 185, 167. Cranial measurements (1 male, 1 young male, 1 female): greatest length of skull, 248, 204, 239; condylobasal length, 224, 186, 224; least interorbital breadth, 69, 59, 61; zygomatic breadth, 97.4, 90.6, 94.8; breadth of braincase, 69.7, 67.0, 67.4; length of maxillary toothrow, 76.0, 48.4, 77.0; palatal length, 151.6, 118.6, 144.0; length of mandibular toothrow, 75.0, 69.5 (last molars not fully erupted), 84.0; greatest length of mandible, 218.0, 179.0, 207.0; length of mandibular diastema, 58.5, 46.6, 51.0.

Remarks. This species is listed as vulnerable in the northern Andes (IUCN, 1982) and as endangered (Appendix I) in Argentina, where it has protected status (Fuller and Swift, 1984). Comments and measurements regarding the specimens from Cuesta del Medanito were reported by Crespo (1974). Extremely low population numbers have been reported from censuses in the neighboring province of La Ríoja, and factors involved in the reduction in numbers of this species were discussed by Cajal (1983b). The animals are supposedly common (and regularly hunted) in the Sierra de Ambato and Sierra de Manchao in the grasslands and puna above 3,000 m. M.A. Mares has seen photos of several deer taken above Saujil in the Sierra de Manchao/Sierra de Ambato at elevations above 3,000 m.

Mazama gouazoupira (G. Fischer)

1814. *Cervus gouazoupira* Fischer, Zoognosia, 3:465.

Specimens examined (2). Los Varela, 1,250 m, 1 (MACN); Chumbicha, approx. 7 km NW Balneario, 1 (OMNH).

Measurements. External measurements (1 female, from Crespo, 1974): total length, 1080; tail, 80; hindfoot, 235; ear, 105. Cranial measurements (1 male, 1 female): greatest length of skull, —, 182.0; condylobasal length, —, 165.0; least interorbital breadth, 41.6, 42.7; zygomatic breadth, 82.3, 82.0; breadth of braincase, 57.0, 57.0; length of maxillary toothrow, 52.3, 54.4; palatal length, —, 110.6; length of mandibular toothrow, —, 60.0; greatest length of mandible, —, 139.5; length of mandibular diastema, —, 39.0.

Remarks. This is a species of the Chacoan thorn forest. The specimen from Los Varela was collected in 1953 and was reported originally by Crespo (1974). The subspecies in Catamarca is *M. g. gouazoupira* (Cabrera, 1961b).

Ozotoceros bezoarticus (Linnaeus)

1758. *Cervus bezoarticus* Linnaeus, Syst. Nat., 10th ed., 1:67.

Specimens examined (1). Catamarca, 1 (MACN).

Remarks. This species has essentially been extirpated in Argentina. The specimen in the MACN is a partial cranium with an unusual set of antlers, where the tines appear to be reversed from the usual shape, i.e., rather than one lower tine pointing anteriorally located below an upward-pointing fork, there is a forward-pointing fork and a single tine pointing upward. In the absence of other information, we have assumed that the specimen is an *Ozotoceros bezoarticus*.

Order Rodentia

Family Muridae

Subfamily Sigmodontinae

Akodon alterus Thomas

1919. *Akodon alterus* Thomas, Ann. Mag. Nat. Hist., ser. 9, 3:496.

Specimens examined (1). Otro Cerro, 3000 m, 1 (BMNH).

Additional records. Chumbicha (Thomas, 1919a, 1919b); La Merced (Myers et al., 1990); La Puntilla, near Tinogasta (Thomas, 1920a).

Measurements. External measurements (1 adult male): total length, 151; tail, 68; ear, 14. Cranial measurements (1 adult male, from Thomas, 1919b): greatest length of skull, 25.0; condylobasal length, 23.5; least interorbital breadth, 4.6; zygomatic breadth, 12.6; breadth of braincase, 11.0; length of maxillary toothrow, 4.3; palatal length, 10.4.

Remarks. The Otro Cerro site is the type locality for this species. Thomas (1919b) also placed two specimens from Chumbicha in this taxon (one originally identified as *A. arenicola*). We have not examined these specimens but, given the habitat preferred by *alterus*, we question whether they actually are *alterus*. Although Thomas (1919b) described *alterus* as a distinct species, others have questioned the taxonomic status of the species and have since placed it in synonomy with other species of *Akodon*. For example, Cabrera (1961b) synonymized *alterus* with *A. boliviensis tucumanensis*. Recently, Myers et al. (1990) revised the *boliviensis* group of *Akodon*, emphasizing the northern portion of the geographic range of the genus. Myers et al. (1990) maintained the taxonomic arrangement proposed by Cabrera (1961b), but suggested that additional work is needed in order to determine this taxon's distinctiveness. Blaustein et al. (1992) found that *alterus* can be differentiated from *A. tucumanensis* by morphological characters and habitat preferences. Barquez et al. (1991) considered *alterus* to be a distinct species. *Akodon alterus* prefers bunchgrass areas at moderate to high altitude.

Akodon caenosus Thomas

1918. *Akodon caenosus* Thomas, Ann. Mag. Nat. Hist., ser. 9, 1:189.

Specimens examined (15). 6 km SW Hwy 9 on Hwy 18, 1,524 m, 17 (4 CML, 4 IADIZA-CM, 9 OMNH); El Rodeo, 1.5 km NE Hwy 4, 1,371 m, 1 (CML).

Measurements. External measurements (6 males, 9 females): total length, 145.8 (134–155), 139.4 (133–149); tail, 63.0 (56–70), 60.4 (57-65); hindfoot, 20.0 (19-21), 20.0 (19–21); ear, 13.0, 13.1 (13–14); weight, 12.8 (11.0–14.0), 12.2 (11.0–14.0). Cranial measurements (1 male, 2 females, unless noted otherwise): greatest length of skull (2 females), 22.1 (21.7–22.5); condylobasal length, 21.6, 20.1 (19.7–20.5); least interorbital breadth, 4.3, 4.1; zygomatic breadth, 10.8, 11.0 (10.8–11.1); breadth of braincase, 10.0, 10.5 (10.3–10.7); length of maxillary toothrow, 3.9, 3.8 (3.7–3.9); palatal length, 8.4, 8.8 (8.7–8.8); length of mandibular toothrow, 3.9, 3.9; greatest length of mandible, 10.5, 10.8 (10.5–11.0); diastema length, 5.0, 5.2 (5.0–5.4).

Remarks. All specimens were collected in July; all females were nonbreeding and all males had abdominal testes varying in size from 4 x 0.15 to 5 x 1.5 mm. One female was molting on the venter. These animals were collected in grasslands located above the treeline. Although Myers et al. (1990) included this taxon as a subspecies of *A. puer*, we have chosen to recognize it as a distinct species.

Akodon dolores Thomas

1916. *Akodon dolores* Thomas, Ann Mag. Nat. Hist., ser. 8, 18:324.

Specimens examined (3). 1 km E jct. Hwys 60 and 38 along Hwy 60, 2 (OMNH); Chumbicha, 0.5 km E of Hwy 38 along Hwy 60, 457 m, 1 (OMNH).

Additional records. La Carrera (Bianchi and Lizarralde, 1979).

Measurements. External measurements (2 males, 1 female): total length, 198.0 (190–206), 212; tail, 98.0 (97–99), 93; hindfoot, 23.5 (22–25), 23; ear, 18, 18; weight, 34.5 (27–42), 41. Cranial measurements (2 males, 1 female): greatest length of skull, 27.2 (26.8–27.6), 28.2; condylobasal length, 25.9 (25.4–26.4), 27.1; least interorbital breadth, 4.9 (4.7–5.0), 5.3; zygomatic breadth 14.1 (13.9–14.3), 14.8; breadth of braincase, 11.9 (11.8–12.0), 12.5; length of maxillary toothrow, 4.9, 5.4; palatal length, 11.8 (11.6–12.0), 12.5; length of mandibular toothrow, 4.9, 4.9; greatest length of mandible, 13.7 (13.6 13.7), 14.3; diastema length, 6.9 (6.7–7.0), 7.4.

Remarks. The general distribution of this species is central Argentina. Our inclusion of *dolores* for the fauna of Catamarca follows Bianchi and Lizarralde (1979) and Myers (1989). A male, captured in July, had testes that measured 4 x 2 mm. This individual was captured near a ditch in a *Larrea* flat; the low-lying area had chaco and monte plants and was heavily grazed. A male, captured in November, had large scrotal testes, and a female was pregnant with five embryos. These animals are similar to the type specimen that was examined in the British Museum.

Akodon illuteus (Thomas)

1925. *Abrothrix illutea* Thomas, Ann. Mag. Nat. Hist., ser. 9, 15:582.

Specimens examined (6). Cuesta del Clavillo, 5 km al S de La Banderita, 4 (RMB); 5 km S Las Higuerillas on Hwy 9, 1,173 m, 2 (OMNH).

Measurements. External measurements (2 females): total length, 206, 217; tail, 82, 92; hindfoot, 28, 29; ear, 22, 22; weight, 39.0, 41.0. Cranial measurements (1 female): greatest length of skull, 30.5; condylobasal length, 29.0; least interorbital breadth, 5.3; zygomatic breadth, 14.5; breadth of braincase, 13.7; length of maxillary toothrow, 4.6; palatal length, 12.7; length of mandibular toothrow, 5.0; greatest length of mandible, 15.5; diastema length, 8.3.

Remarks. This is the first record of this species for the province. At the time that these specimens were collected, additional specimens were captured just across the border in Tucumán Province. Both individuals were nonbreeding females. One was molting on the head, left forearm, and mid-dorsum.

Akodon simulator Thomas

1916. *Akodon simulator* Thomas, Ann. Mag. Nat. Hist., 8, 18:335.

Specimens examined (67). 13 km N Andalgalá, 1 (CMNH); Río Andalgalá, immed. N Andalgalá, 18 (CMNH); immed. N Andalgalá, 3 (CMNH); 6.5 km (by road) N jct. Hwy 63 to Chaquiago and La Toma Road, 2 (CMNH); Río Andalgalá, 3.5 km N Andalgalá, 2 (CMNH); Choya, 13 km NNW Andalgalá, 1,219 m, 1 (OMNH); 23 km SW Chumbicha, 1 (CMNH); Chumbicha, 600 m, 1 (BMNH); El Rodeo, 1.5 km NE Hwy 4, 1,372 m, 11 (3 CML, 3 IADIZA-CM, 5 OMNH); El Rodeo, 0.75 km S village limits along Hwy 4, 1,372 m, 4 (1 CML, 2 IADIZA-CM, 1 OMNH); El Rodeo, 0.5 km W Hwy 4, 1,372 m, 10 (3 CML, 2 IADIZA-CM, 5 OMNH); 6 km SW Hwy 9 on Hwy 18, 1,524 m, 7 (1 CML, 1 IADIZA-CM, 5 OMNH); 5 km S Las Higuerillas on Hwy 9, 1,091 m, 2 (1 CML, 1 IADIZA-CM); Poman, 95 km S Andalgalá near Balneario, 1,341 m, 4 (1 CML, 1 IADIZA-CM, 2 OMNH).

Additional records. Chumbicha, 600 m (Myers, 1989; Thomas, 1919a); Otro Cerro (Myers, 1989; Thomas, 1919b); 0.5 mi N Andalgalá; 3.5 mi N Andalgalá on Río Andalgalá; El Potrero Dike, 13 km N Andalgalá; 23 km SW Chumbicha; La Toma, 6.5 km N Hwy 63 on La Toma Rd. (Barquez et al., 1980); La Merced; immed. N Andalgalá, 962 m; 3.5 km N Andalgalá, 1,000 m; 13 km N Andalgalá (Myers, 1989).

Measurements. External measurements (24 males, 33 females, unless noted otherwise): total length, 185.1 (168–208), 180.2 (159–210); tail, 79.2 (69–93), 74.6 (55–86); hindfoot, 24.2 (22–26), 23.8 (22.2–26); ear

(22 males, 33 females), 18.5 (17–21.3), 18.2 (14–21.1); weight (22 males, 31 females), 32.6 (18.7–51.0), 33.0 (20.0–62.0). Cranial measurements (7 males, 8 females, unless noted otherwise): greatest length of skull (5 males, 6 females), 28.1 (26.6–29.8), 27.9 (26.6–28.7); condylobasal length, 26.6 (24.8–28.9), 26.4 (24.7–27.4); least interorbital breadth, 5.0 (5.0–5.1), 5.1 (4.8–5.4); zygomatic breadth (6 males, 6 females), 14.2 (13.8–14.7), 13.9 (13.6–14.7); breadth of braincase (7 males, 7 females), 12.5 (12.0–13.0), 12.3 (11.9–12.6); length of maxillary toothrow (6 males, 8 females), 4.9 (4.6–5.1), 4.9 (4.6–5.2); palatal length, 13.0 (112.–15.0), 12. 8 (11.4–14.5); length of mandibular toothrow, 4.9 (4.7–5.1), 5.0 (4.7–5.3); greatest length of mandible, 14.7 (13.9–16.0), 14. 8 (13.9–16.0); diastema length, 7.5 (6.6–8.5), 7.4 (7.0–7.7).

Remarks. The population from Catamarca has been referred to as *A. s. glaucinus* (Myers, 1989) with type locality in Chumbicha (Thomas, 1919a); however, Myers recognized that the specimens "from near Andalgalá are distinctively reddish and perhaps should be considered separately." He tentatively assigned this population to *glaucinus*. *Akodon simulator* is one of the most common species found in the mesic lowland and higher elevation forests of Tucumán, Jujuy, and Salta provinces (e.g., Mares et al., 1989; Barquez et al., 1991). Other than the type series of Thomas, the only specimens known from Catamarca were those taken in riparian habitats above the monte desert. In field work in 1990, however, along the forested portions of eastern Catamarca near El Rodeo and Chumbicha, we found the species to be common. All specimens in this sample possess the distinctive white patch on the chin and/or throat. Specimens have been caught in January, February, April, June, July, and December. Reproductively inactive males and females were found in July (females with closed vaginas; males with abdominal testes ranging in size from 2.5 x 1 to 5 x 2 mm). Males with large scrotal testes were found in February and December (10 x 6 mm). Pregnant females were found in February and December. Number of embryos ranged from five to eight. Molting individuals were collected in July and December.

Akodon spegazzinii Thomas

1897. *Akodon spegazzinii* Thomas, Ann. Mag. Nat. Hist., 6, 20:216.

Specimens examined (17). El Potrero Dike, 13 km N Andalgalá,1 (CMNH); Choya, 13 km NNW Andalgalá, 1,219 m, 14 (4 CML, 3 IADIZA-CM, 7 OMNH); Río Andalgalá, immed. N Andalgalá, 1 (CMNH); Río Andalgalá, 3.5 km N Andalgalá, 1 (CMNH).

Measurements. External measurements (5 males, 9 females): total length, 172.8 (155–187), 171.1 (148–184); tail, 71.0 (64–78), 71.7 (65–81); hindfoot, 22.2 (21–23), 21.3 (20–22); ear, 13.6 (12–15), 13.3 (11–15); weight, 30.7 (21.0–41.5), 31.7 (22.0-46.0). Cranial measurements (2 males, 3 females, unless noted othewise): greatest length of skull, 25.8 (24.8–26.8), 25.3 (24.4–26.2); condylobasal length, 24.0 (22.9–25.1), 23.7 (22.9–24.6); least interorbital breadth, 4.5 (4.4–4.6), 4.5 (4.3–4.7); zygomatic breadth (1 male), 12.6, 12.4 (12.3–12.5); breadth of braincase, 11.6, 11.2 (11.0–11.4); length of maxillary toothrow, 4.5 (4.3–4.6), 4.2 (4.2–4.3); palatal length, 10.7 (10.3–11.1), 10.6 (10.3–11.2); length of mandibular toothrow, 4.4 (4.3–4.5), 4.3 (4.2–4.3); greatest length of mandible, 12.9 (12.4–13.3), 12.9 (12.5–13.1); diastema length, 6.4 (6.1–6.7), 6.3 (6.3–6.4).

Remarks. The specimens from the Andalgalá area were captured in January and February. A pregnant female was collected in February. All of the individuals from Choya were captured in December. Of the nine females, six were pregnant and one was lactating. The number of embryos ranged from five to eight (mean = 6.2), and the crown-rump length (CRL) ranged from 2-22 mm. All of the males had large scrotal testes ranging in size from 8 x 6 mm to 14 x 8 mm. Two of the males were molting. The animals were caught in the town of Choya, which is a rather mesic enclave surrounded by dry monte desert habitat. Specifically, they were captured in an old orchard and in agricultural fields; several were found along a fence row. A number of individuals were caught during daylight hours and in the early evening. This taxon was described by Thomas (1897), the type locality being lower Cachi, Salta Province. Later, it was synonomized with *boliviensis*. Recently, Myers et al. (1990) suggested that the taxon (*boliviensis*) might be divided into *boliviensis* and *spegazzinii* (including *spegazzinii spegazzinii* and *spegazzinii tucumanensis*). Our specimens are similar to the type specimen of *spegazzinii* in measurements and overall coloration. The characters for separating *boliviensis* and *spegazzinii* hold for our sample: greater length of incisive foramina (*spegazzinii* greater than 6.0, *boliviensis* less than 5.1), greater overall skull measurements, and coloration. Barquez et al. (1980) examined *Akodon* from northwestern Argentina, including *caenosus*, *varius* (now *simulator*), and *boliviensis* (including *spegazzinii spegazzinii* and *spegazzinii tucumanensis*). We have also chosen to separate *tucumanensis* from *spegazzinii*. Barquez et al. (1980) may have found differences among their *boliviensis* sample that correspond to *spegazzinii* and *tucumanensis*. We have found the reddish brown coloration of *spegazzinii* to be lighter than that of *tucumanensis*; these animals also appear to prefer non-forested areas.

Akodon tucumanensis J.A. Allen

1901. *Akodon tucumanensis* J.A. Allen, Bull. Amer. Mus. Nat. Hist., 14:410.

Specimens examined (4). 3 km WNW Concepción, 2 (CMNH); 5 km S Las Higuerillas on Hwy 9, 1,173 m, 2 (1 CML, 1 OMNH).

Measurements. External measurements (2 females): total length, 180.0 (179–180); tail, 72.0 (69–75); hindfoot, 22.0; ear, 15.0; weight, 29.0 (27–30). Cranial measurements (1 female): greatest length of skull, 26.7; condylobasal length, 24.6; least interorbital breadth, 4.3; zygomatic breadth, —; breadth of braincase, 11.1; length of maxillary toothrow, 4.3; palatal length, 10.6; length of mandibular toothrow, 4.2; greatest length of mandible, 13.0; diastema length, 6.8.

Remarks. Both females taken at Las Higuerillas were captured in July and were not breeding. The male and female from Concepción were collected in January. These specimens are the first record of this species for the province. The type locality for this species is Tucumán, Tucumán Province. As with the above taxon, we have chosen to recognize *tucumanensis* as a distinct species. Previous authors have placed this taxon in synonomy with *boliviensis*. Our specimens agree with the type material of *tucumanensis* in coloration, measurements, and cranial morphology. In general, *tucumanensis* is much darker than *spegazzinii* and is found in forested areas.

Andalgalomys olrogi Williams and Mares

1978. *Andalgalomys olrogi* Williams and Mares, Ann. Carnegie Mus., 47:197.

Specimens examined (9). Along Río Amanao, about 13 km S, 15 km W Andalgalá, 1 (CMNH); 10 km W Andalgalá (by road on Route 62, at km marker 10), 1 (CMNH); west bank Río Amanao, about 15 km W (on Route 62) Andalgalá, 3 (CMNH); Immed. W Río Amanao, 13 km W Andalgalá, along Hwy 62, 1 (OMNH); along Hwy 62, 8 km W Andalgalá, 2 (1 CMNH, 1 CML); 8.1 km W Andalgalá, 1 (OMNH).

Measurements. External measurements (5 males, 3 females, unless noted otherwise): total length, 223.6 (212-236), 225.3 (218-238); tail, 126.4 (124-130), 128.0 (125-132); hindfoot, 24.3 (23-25), 24.2 (23.8-24.7); ear, 21.1 (20-22), 19.9 (19.6-20.0); weight (3 males, 1 female), 25.0 (23.0-27.0), 30.0. Cranial measurements (3 males, 1 female): greatest length of skull, 28.3 (27.7-28.7), 28.3; condylobasal length, 26.3 (25.9-26.5), 26.6; least interorbital breadth, 4.5 (4.4-4.7), 4.6; zygomatic breadth, 13.2 (13.2-13.2), 13.8; breadth of braincase, 12.7 (12.5-12.9), 13.0; length of maxillary toothrow, 4.7 (4.6-4.7), 4.7; palatal length, 12.4 (12.3-12.5), 12.6; length of mandibular toothrow, 4.5 (4.4-4.6), 4.5; greatest length of mandible, 14.7 (14.6-14.8), 14.7; diastema length, 7.1 (7.0-7.2), 7.1.

Remarks. This genus was discovered by Mares in 1972, but was not described until 1978 (Williams and Mares, 1978), at which time Myers's *Graomys pearsoni* of Paraguay (Myers, 1977) was found to be congeneric with it. Since then, the genus has been reviewed by Olds et al. (1987), who revised the diagnosis and found that this genus is more phenetically similar to *Graomys* than to either *Eligmodontia* or *Calomys*, a view supported by Braun (1993). *Andalgalomys olrogi* was thought by Williams and Mares (1978) to be a monte desert endemic. Indeed, no specimens have ever been taken at other than the type locality. Specimens have been collected in the months of January, April, June, and November. A male in June had inguinal testes. Descriptions, karyology, penal morphology, cranial and dental morphology, and comparisons with other species are given in Williams and Mares (1978), Olds et al. (1987), and Braun (1993).

Andalgalomys roigi

Specimens examined (5). jct. Hwys 60 and 38, immed. S Chumbicha, 1 (OMNH); Chumbicha, 0.5 km E of Hwy 38 along Hwy 60, 1 (CML); Chumbicha, 1.5 km E of Hwy 38 along Hwy 60, near Chumbicha, 3 (1 IADIZA-CM, 2 OMNH).

Measurements. External measurements (3 males, 2 females): total length, 203.7 (195-220), 178.5 (165-192); tail, 114.0 (110-119), 102.0 (89-115); hindfoot, 23.7 (23-24), 22.5 (21-24); ear, 20.0 (20-20), 18.0 (17-19); weight, 22.6 (21.0-25.0), 17.0 (13.0-21.0). Cranial measurements (2 males, 2 females): greatest length of skull, 28.4 (28.1-28.6), 25.9 (24.1-27.7); condylobasal length, 26.4 (26.0-26.7), 23.8 (21.8-25.8); least interorbital breadth, 4.8 (4.8-4.8), 4.2 (4.2-4.2); zygomatic breadth, 13.7 (13.4-13.9), 12.7 (11.8-13.5); breadth of braincase, 12.8 (12.7-12.8), 12.2 (11.8-12.6); length of maxillary toothrow, 4.9 (4.7-5.0), 4.6 (4.4-4.7); palatal length, 12.6 (12.5-12.6), 11.3 (10.58-12.0); length of mandibular toothrow, 4.4 (4.2-4.5), 4.2 (4.2-4.2); greatest length of mandible, 14.4 (13.9-14.8), 13.4 (12.7-14.1); diastema length, 7.1 (7.0-7.1), 6.2 (5.7-6.7).

Remarks. In 1987, Mares collected a specimen of *Andalgalomys* near Chumbicha, in eastern Catamarca, and, in 1990, additional specimens were collected at the same locality. Other individuals have been captured recently in northern San Luis Province, suggesting that this species is a Chacoan species. This species has been taken only in the provinces of Catamarca and San Luis, and recently in La Rioja (R.A. Ojeda, pers. comm.). Specimens have been collected in the months of July and November. Three males collected in July had abdominal testes ranging from 3 x 1.5 mm to 5 x 2 mm. In July, a female was not breeding. A male exhibiting signs of molt was captured in July. Descriptions, karyology, cranial and dental morphology, and comparisons with other species are given in Mares and Braun (1996).

Andinomys edax Thomas

1902. *Andinomys edax* Thomas, Proc. Zool. Soc. London, 1:116.

Specimens examined (2). Otro Cerro, 3,000 m, 1 (MACN); Río Vallecito, 1,500 m, 1 (MACN).

Measurements. External measurements (2 males): total length, 287, 291; tail, 142, 138; hindfoot, 32, 28.6; ear, 23, 24. Cranial measurements (1 male): least interorbital breadth, 4.65; zygomatic breadth, 18.5; length of maxillary toothrow, 8.1; palatal length, 16.6; greatest length of mandible, 20.4; diastema length, 8.9.

Remarks. This population has been referred to as *A. e. lineicaudatus* (Yepes, 1935). Thomas (1920b) noted that confusion was caused when he assigned the locality of Otro Cerro to La Ríoja Province, instead of Catamarca Province (Thomas, 1919b). Additional information regarding variations in color, dentition, general distribution, habitat, and habits are in Hershkovitz (1962). *Andinomys* is primarily a species of high bunchgrass prairie, generally above 2,800 m (e.g., Mares, 1977a; Mares et al., 1989), but it is also found in the mesic yungas forest of Tucumán Province at elevations as low as 700 m (Barquez et al., 1991). In Catamarca, the species may be found to inhabit both the high altitude grasslands that are common on the pre-Andean chains of the eastern parts of the province, as well as the small band of moist forest that extends from Tucumán Province southward into Catamarca.

Bolomys lactens (Thomas)

1918. *Akodon lactens* Thomas, Ann. Mag. Nat. Hist., ser. 9, 1:188.

Specimens examined (1). Otro Cerro, 3,000 m, 1 (BMNH).

Additional records. Laguna Blanca, 3,100 m (Cabrera, 1926).

Measurements. External measurements (1 female): total length, 156; tail, 59; hindfoot, 21; ear, 14. Cranial measurements (1 female): greatest length of skull, 26.3; condylobasal length, 25.1; least interorbital breadth, 4.9; breadth of braincase, 12.0; length of maxillary toothrow, 4.8; palatal length, 13.4; length of mandibular toothrow, 4.8; greatest length of mandible, 14.4; diastema length, 7.8.

Remarks. Reig (1978, 1987) synonymized *Akodon orbus* (Thomas, 1919b) and *Akodon leucolimnaeus* (Cabrera, 1926) with this taxon. The above measurements are of the type specimen of *A. orbus*. In neighboring Tucumán Province, *B. lactens* is a rodent of high (> 1,500 m) habitats, including montane bunchgrass prairie. The only specimen taken in Catamarca was captured in an extensive montane grassland. Anderson and Olds (1989) erroneously placed specimens from Las Pavas, Aconquija, in Catamarca Province; this locality is actually in Tucumán Province.

Calomys musculinus (Thomas)

1913. *Eligmodontia laucha musculina* Thomas, Ann. Mag. Nat. Hist., ser. 8, 11:138.

Specimens examined (20). Balneario El Caolín, 6 km NW Chumbicha, 2 (OMNH); El Rodeo, 0.75 km S of village limits along Hwy 4, 1,372 m, 4 (Arg); Choya, 13 km NNW Andalgalá, 1,219 m, 7 (Arg); Chumbicha, 1 km north and west of Balneario by road, 1 (Arg); 7.5 km W Andalgalá, 1 (CMNH); 23 km SW Chumbicha, 1 (CMNH); immed. N Andalgalá, 2 (CMNH); La Toma Dept., 6.5 km N jct. Hwy 63 (to Chagiago) and La Toma Rd., 1 (CMNH); southern outskirts of Andalgalá, 1 (CMNH).

Additional records. Tinogasta, 1,200 m (Olds, 1988); Chumbicha, 600 m (Hershkovitz, 1962; Olds, 1988; Thomas, 1919a); Puntilla Tinogasta, 1,000 m (Hershkovitz, 1962; Olds, 1988; Thomas, 1920a); Otro Cerro, 3,000 m (Olds, 1988); 10 km (by road on Route 62) W Andalgalá (Williams and Mares, 1978).

Measurements. External measurements (11 males, 7 females, unless otherwise noted): total length, 156.4 (136-195), 167.3 (144-181); tail, 79.2 (66-98), 83.6 (75-90); hindfoot, 18.7 (17-21), 18.1 (16.9-20); ear (10 males, 6 females), 15.1 (14-16.2), 15.6 (15-17); weight (10 males, 7 females), 15.9 (9.0-32.0), 18.9 (10.1-26.5). Cranial measurements (2 males, 3 females): greatest length of skull, 23.5 (22.4-24.6), 22.3 (21.7-23.1); condylobasal length, 22.0 (20.6-23.3), 20.5 (19.7-21.6); least interorbital breadth, 3.7 (3.5-3.8), 3.7 (3.6-3.8); zygomatic breadth, 11.8 (10.9-12.7), 11.6 (10.4-12.7); breadth of braincase, 10.5 (10.0-10.9), 10.3 (10.0-10.7); length of maxillary toothrow, 3.8, 3.7 (3.5-3.9); palatal length, 9.6 (9.0-10.1), 9.6 (9.0-10.5); length of mandibular toothrow, 3.5 (3.4-3.6), 3.5 (3.2-3.8); greatest length of mandible, 12.0 (11.3-12.6), 11.3 (10.5-12.0); diastema length, 5.8 (5.3-6.2), 5.2 (4.6-5.7).

Remarks. *Calomys musculinus* inhabits semiarid areas, disturbed habitats in semidesert, agricultural areas, and thorn scrub. It seldom enters the driest parts of the monte desert, although it has been shown to be remarkably adept at withstanding water deprivation (Mares, 1975a, 1977b). Indeed, it is more tolerant in this respect than any other Argentine murid examined (Mares considered *C. musculinus* as *C. laucha* in the earlier papers). Mares (1977b) hypothesized that *Calomys* would be found to undergo torpor, and Bozinovic and Rosenmann (1988) reported that the species undergoes a daily torpor with greatly reduced metabolism. Olds (1988) recognized *murillus* as a taxon distinct from *musculinus*. We have chosen to follow Wilson and Reeder (1993) by placing *murillus* in synonymy with *musculinus* until further evidence supports taxonomic separation. Thomas (1919a) identified the specimens from Chumbicha as *murillus*; specimens identified as *murillus* and *musculinus* were captured at La Puntilla (Thomas, 1920a). Specimens have been captured in January,

June, July, November, and December. Pregnant females were captured in June (one female with eight embryos), November (one female with six embryos), and December (one female with three embryos, CRL = 8.0 mm; one female with seven embryos, CRL = 10.0 mm). Nonbreeding males were captured in July; males with large scrotal testes were captured in December (7 x 5 mm to 9 x 5 mm).

Calomys venustus (Thomas)

1894. *Oryzomys venustus* Thomas, Ann. Mag. Nat. Hist., ser. 6, 14:359.

Specimens examined. None.

Additional records. La Merced, near abandoned RR tunnels; La Mendicta (Olds, 1988).

Measurements. None.

Remarks. Caviedes–Vidal et al. (1990) found that *C. venustus* from San Luis Province undergoes a daily torpor under laboratory conditions. We have followed Olds (1988) in recognizing *venustus* as occurring in northwestern Argentina. Characters distinguishing this species from other *Calomys* were given by Olds (1988). We were unable to locate La Mendicta in Catamarca Province; there is, however, a La Mendieta in Jujuy Province.

Chroeomys andinus (Philippi)

1858. *Mus andinus* Philippi, Arch. Naturg., 23, 1:77.

Specimens examined (3). Antofagasta de la Sierra, 1 (IADIZA-CM); 7 km N Puesto Laguna Blanca, 1 (IADIZA-CM); Minas Capillitas, 3,200 m, 1 (CMNH).

Measurements. External measurements (1 adult male): total length, 164; tail, 73; hindfoot, 23; ear, 15; weight, 38.0. Cranial measurements (1 adult male): greatest length of skull, 26.6; condylobasal length, 25.6; least interorbital breadth, 4.8; zygomatic breadth, 12.9; breadth of braincase, 11.3; length of maxillary toothrow, 4.3; palatal length, 13.4; length of mandibular toothrow, 4.4; greatest length of mandible, 14.3; diastema length, 7.2.

Remarks. In Catamarca, this species has been found only in bunchgrass prairie and in mesic patches of habitat within the puna and prepuna at elevations above 3,000 m. It has not been commonly trapped anywhere in northwestern Argentina. Mares et al. (1989) reported it occurring as low as 2,000 m in Salta Province (see also Mares et al., 1981). The male captured at Minas Capillitas in January had large scrotal testes.

Eligmodontia marica Thomas

1918. *Eligmodontia marica* Thomas, Ann. Mag. Nat. Hist., ser. 9, 2:483.

Specimens examined (10). Chumbicha, 0.5 km E of Hwy 38 along Hwy 60, 457 m, 5 (Arg); jct. Hwys 60 and 38, immed. S Chumbicha, 1 (OMNH); Chumbicha, 600 m, 4 (BMNH).

Measurements. External measurements (3 males, 7 females, unless otherwise noted): total length, 165.7 (158–175), 157.0 (128–174); tail, 94.7 (92–97), 87.7 (68–99); hindfoot, 21.3 (20–22), 20.6 (19–23); ear, 16.0 (15–17), 15.6 (13–17); weight (1 male, 5 females), 11.5, 11.0 (10.0–12.0). Cranial measurements (3 males, 5 females, unless otherwise noted): greatest length of skull, 22.4 (21.0–23.4), 21.6 (20.4–22.6); condylobasal length (2 males, 5 females), 20.2 (19.0–21.3), 19.3 (16.1–20.4); least interorbital breadth, 3.8 (3.7–4.0), 3.7 (3.6–3.8); zygomatic breadth, 11.7, 11.3 (10.6–11.8); breadth of braincase (1 male, 4 females), 11.0, 10.7 (10.3–11.0); length of maxillary toothrow, 3.5 (3.3–3.7), 3.5 (3.4–3.7); palatal length, 9.7 (8.8–10.2), 9.3 (8.5–9.8); length of mandibular toothrow, 3.5 (3.4–3.6), 3.5 (3.4–3.7); greatest length of mandible (1 males, 4 females), 11.6, 11.1 (10.5–11.4); diastema length, 5.5 (5.3–5.8), 5.1 (4.7–5.4).

Remarks. Thomas (1918) named *E. marica* in honor of his wife; it was his 2,000th type. Cabrera (1961b) included *marica* within *E. typus*, and Hershkovitz (1962) suggested that *Eligmodontia* is a monotypic genus. Mares and Braun (in press) are revising the genus *Eligmodontia*, but initial results show that it is a highly polytypic taxon, containing at least seven species and probably more. *E. marica*, the smallest member of the genus, is slight and blondish, and is known only from at or near the type locality. Until Mares collected a specimen from near Chumbicha in 1987, only the four original specimens in the type series were known. In a subsequent trip to the Chumbicha area, however, Braun also captured five specimens. Thus far, the species is endemic to Catamarca Province in the immediate vicinity of Chumbicha in semiarid disturbed *Larrea* (creosote bush)/ thorn scrub habitat, where it co-occurs with *Thylamys pallidior*, *Akodon dolores*, *Calomys musculinus*, *Graomys griseoflavus*, and *Andalgalomys roigi*.

Eligmodontia moreni Thomas

1896. *Eligmodontia moreni* Thomas, Ann. Mag. Nat. Hist., ser. 6, 18:307.

Specimens examined (17). 1.5 km S El Desmonte, 1 (OMNH); 13 km S of Hwy 62 along Río Amanao, W of Andalgalá, 1 (CMNH); 1.5 km S, 150 m W of jct. of Hwy 1 and RR track, S of Andalgalá, 3 (CMNH); slightly S of Andalgalá, 1 (CMNH); 6 km N Saujil, 1 (CMNH); Laguna Blanca, 3,500 m, 5 (BMNH); Corral Quemado, 1 (FMNH); Minas Capillitas, 3,200 m, 2 (CMNH); San Jose, 4,400 m, 1 (BMNH); Campo de Las Talibus (near Corral Quemado), 1 (IADIZA-CM).

Additional records. Andalgalá Railroad Station (the "Estacion"), 4 km S of Andalgalá; 1.5 km S, 150 m W of jct. Hwy 1 and the railroad track, just S of Andalgalá; 50 m N, 50 m E of jct. of Hwy 62 and Amanao Road, approx. 18 km E of Andalgalá; 6 km N Saujil along Hwy 1; along Río Amanao, 15 km S of jct. Hwy 1 (W of Andalgalá) and the Río Amanao (Mares, 1973).

Measurements. External measurements (7 males, 8 females, unless noted otherwise): total length, 177.7 (146–206), 177.6 (160–195); tail, 98.1 (70–122), 97.0 (79–112); hindfoot, 22.8 (20–26), 22.4 (20.7–25); ear, 17.7 (15–20), 18.1 (15.7–20.7); weight (2 males, 4 females), 18.0 (16.0–20.0), 20.7 (17.5–23.0). Cranial measurements (4 males, 5 females, unless noted otherwise): greatest length of skull (2 males), 23.2 (22.5–23.9), 24.3 (23.0–25.5); condylobasal length (2 males), 20.4 (18.2–22.6), 20.7 (18.2–23.4); least interorbital breadth (3 males), 4.0 (3.9–4.1), 4.0 (3.8–4.2); zygomatic breadth (2 males), 12.0 (11.6–12.3), 12.5 (11.6–13.5); breadth of braincase (1 male, 2 females), 11.8, 11.5; length of maxillary toothrow, 3.7 (3.5–3.9), 3.8 (3.6–4.1); palatal length (3 males), 10.7 (9.5–12.3), 10.7 (9.8–13.0); length of mandibular toothrow, 3.8 (3.7–3.9), 3.9 (3.7–4.1); greatest length of mandible (1 male, 2 females), 12.1, 12.9 (12.7–13.1); diastema length, 5.5 (5.1–6.0), 5.7 (5.3–6.1).

Remarks. This is one the most arid-adapted of the murids of Argentina, extending into the driest parts of the monte desert. Mares (1973, 1975a, 1975b, 1977c, 1988; Mares et al. 1977a) provided detailed information on the biology of this species, including ecology and physiology. Specimens have been collected in February, March, April, June, and November. A pregnant female with three embryos was captured in November. Specimens from Andalgalá and Saujil were listed as *E. typus* by Mares (1973) and Williams and Mares (1978).

Eligmodontia puerulus (Philippi)

1896. *Hesperomys puerulus* Anal. Mus. Nac. Chile, Zool. Ent., 13a:20.

Specimens examined. None.

Additional records. Puesto Laguna Blanca, 3,200 m (J. Cajal, pers. obs.).

Remarks. According to the collectors, four specimens were captured at the locality listed above. We have been unable to locate these specimens.

Eligmodontia sp. 1

Specimens examined (6). 33 km SW El Desmonte, 1 (OMNH); Los Baños Termales, Hualfín, 3 (CML); Agua de Dionisio, Hualfín, 1 (CML); Pozo Jovita, Hualfín, 1 (CML).

Additional records. Pozo Jovita, Hualfín; La Salamanca, Hualfín (Massoia, 1976/77).

Measurements. External measurements (2 males, 1 female, unless noted otherwise): total length, 155.0 (150–159), 136; tail, 77.5 (65–90), 56; hindfoot, 20.0, 20; ear, 16.5 (16-17), 15; weight (1 male), 13.0. Cranial measurements (3 males, 2 females, unless noted otherwise): greatest length of skull (1 male), 23.3; condylobasal length (1 male), 21.8; least interorbital breadth (3 males, 1 female), 3.9 (3.7–4.2), 3.9; zygomatic breadth (2 males), 12.1 (11.5–12.7); breadth of braincase (1 male), 10.9; length of maxillary toothrow, 3.9 (3.7–4.1), 3.8 (3.5–4.0); palatal length (1 male), 10.0; diastema length, 5.7 (5.5–5.8), 5.7 (5.3–6.1).

Remarks. In 1987, Mares collected this small, blond *Eligmodontia* south of El Desmonte. It appeared to be unlike any of the named forms of *Eligmodontia*. Upon comparison with several specimens at the CML, five additional specimens were found that had been collected in the early 1950s. This *Eligmodontia* is the blondest of any of the *Eligmodontia* seen to date; it has hirsute foot pads and a unicolored, nontufted tail. The specimen from south of El Desmonte was collected in November and had large scrotal testes.

Eligmodontia sp. 2

Specimens examined (1). 5 km W, 1 km S Andalgalá, 1 (OMNH).

Measurements. External measurements (1 female): total length, 163; tail, 92; hindfoot, 22; ear, 17; weight, 13.0. Cranial measurements (1 female): greatest length of skull, 22.4; condylobasal length, 20.5; least interorbital breadth, 3.7; zygomatic breadth, 11.6; breadth of braincase, 10.8; length of maxillary toothrow, 3.6; palatal length, 9.3; diastema length, 5.1.

Remarks. In 1987, Mares collected this small *Eligmodontia* near Andalgalá. This individual is unlike any of the other *Eligmodontia* known from the area (*moreni* or *Eligmodontia* sp. 1), but additional specimens may indicate that it is conspecific with *E. marica*.

Graomys edithae Thomas

1919. *Graomys edithae* Thomas, Ann. Mag. Nat. Hist., ser. 9, 3:495.

Specimens examined (8). Otro Cerro, 3,000 m, 1 (BMNH); Chumbicha, 600 m, 7 (BMNH).

Measurements. External measurements (4 males, 4 females): total length, 231.8 (214–263), 223.8 (212–242); tail, 126.5 (115–143), 124.5 (115–132); hindfoot, 25.5 (25–27), 25.0 (23–26); ear, 22.3 (20–24), 22.5 (21–23). Cranial measurements (3 males, 3 females, unless noted otherwise): greatest length of skull, 26.8 (26.3–27.3), 26.8 (26.5–27.2); condylobasal length, 23.3 (22.3–23.8), 23.0 (22.4–23.5); least interorbital breadth, 4.6 (4.4–4.9), 4.6 (4.4–4.9); zygomatic breadth (3 males, 1 female), 14.5 (14.0–15.0), 14.2; breadth of braincase (2 males, 3 females), 13.3 (13.2–13.3), 13.4 (12.9–13.8); length of maxillary toothrow, 5.1 (4.8–5.3), 5.1 (5.0–5.3); palatal length, 13.8 (13.4–14.0), 13.9 (13.8–14.0); diastema length, 7.0 (6.8–7.4), 6.8 (6.6–7.1).

Remarks. Otro Cerro is the type locality for this species described by Thomas (1919b). Some investigators have considered the status of this taxon as uncertain. Comments regarding the status of *G. edithae* may be found in Hershkovitz (1962) and Williams and Mares (1978). Wilson and Reeder (1993) incorrectly place the type locality in La Rioja Province instead of Catamarca Province. Although Thomas (1919b) considered part of a series of specimens from Chumbicha as *G. edithae*, Williams and Mares (1978) suggested that they are *G. griseoflavus*. These specimens are listed above, but we realize that their identification is questionable until more specimens can be examined.

Graomys griseoflavus (Waterhouse)

1837. *Mus griseoflavus* Waterhouse, Proc. Zool. Soc. London, 1837:28.

Specimens examined (138). Km 10, Route 62, 1 (CMNH); 7.5 km W Andalgalá, on Route 62, 1 (CMNH); km 99, S of Andalgalá, 2 (CMNH); along Río Andalgalá, 12 (CMNH); immed. N Andalgalá, 11 (CMNH); Andalgalá waterworks, 4 (CMNH); N waterworks Andalgalá, 12 (CMNH); 8.1 km W Andalgalá, 2 (OMNH); Río Amano, 15 km W Andalgalá, 1 (OMNH); Choya, 13 km NNW Andalgalá, 1,219 m, 14 (3 CML, 3 IADIZA-CM, 8 OMNH); Chumbicha, 0.5 km E of Hwy 38 along Hwy 60, 457 m, 18 (4 CML, 5 IADIZA-CM, 9 OMNH); Chumbicha, 1.5 km E of Hwy 38 along Hwy 60, 457 m, 12 (4 CML, 4 IADIZA-CM, 4 OMNH); Chumbicha, 1 km N and W of Balneario by road, 792 m, 1 (CML); Balneario El Caolín, 6 km NW Chumbicha, 2 (OMNH); jct. Hwy 60 and 38, S Chumbicha, 8 (OMNH); 4 km E Chumbicha, 2 (OMNH); Chumbicha, 1 (BMNH); Cuesta de Zapata, 18 km SW

Londres, 1 (CMNH); Agua Tapada, Hualfín, 1 (CML); Los Nacimientos, Hualfín, 6 (CML); Pié del Tambillo, 25 km W Londres, 1 (CML); Poman, 95 km S of Andalgalá near Balneario, 1,341 m, 8 (1 CML, 1 IADIZA-CM, 6 OMNH); 1 km E jct. Hwys 60 and 38, along Hwy 60, 7 (OMNH); 8 km E jct. Hwys 62 and 1, 4 (CMNH); 6.5 km N jct. Hwy 63 and La Toma Road, 4 (CMNH); along Hwy 40, at km 761, N Tinogasta, 1 (CMNH); Minas Capillitas, 3,200 m, 1 (CMNH).

Additional records. Belén, 800 m; Chumbicha; Corral Quemado, Belén; La Puntilla, Tinogasta, 1,000 m; Otro Cerro, 3,000 m; Tinogasta (Hershkovitz, 1962); La Puntilla, Tinogasta (Thomas, 1920a); Chumbicha (Thomas, 1919a); Otro Cerro (Thomas, 1919b); 35 km SW Catamarca; 165 km WNW Catamarca; 70 km SW Catamarca (Pearson and Patton, 1976); along the Río Andalgalá, approx. 3 km N of Andalgalá; northern Andalgalá water works ("La Toma"), 6.5 km N (by road) of the junction of Hwy 63 to Chaquiago and La Toma Rd.; off Hwy 62, 8 km E of the junction of Hwys 62 and 1, E of Andalgalá; km 761, 18 km N of Tinogasta; off Hwy 1 at km 99, 34 km S of Andalgalá; along the Río Amanao, 15 km S of the junction of Hwy 1 (W of Andalgalá) and the Río Amanao; hills at E end of La Puntilla between Tinogasta and Copacabana (Mares, 1973); Laguna Blanca (Piciucci de Fonolat et al., 1985).

Measurements. External measurements for males and females, as noted: total length (68 males, 50 females), 262.4 (171–322), 259.4 (194–315); tail (68 males, 50 females), 140.1 (56–180), 139.2 (89–180); hindfoot (68 males, 49 females), 28.8 (25.8–32), 28.3 (25–37); ear (66 males, 47 females), 25.6 (21–31.3), 25.4 (18–30); weight (66 males, 49 females), 55.7 (21.0–90.0), 51.6 (23.0–82.0). Cranial measurements for males and females, as noted: greatest length of skull (29 males, 24 females), 32.0 (28.6–36.4), 31.9 (29.0–36.5); condylobasal length (29 males, 24 females), 29.8 (25.8–33.6), 29.7 (26.6–34.5); least interorbital breadth (35 males, 27 females), 5.1 (4.4–6.2), 5.0 (4.5–5.8); zygomatic breadth (31 males, 26 females), 16.4 (14.8–18.2), 16.1 (15.0–17.9); breadth of braincase (32 males, 26 females), 14.1 (13.3–15.0), 13.8 (12.9–14.7); length of maxillary toothrow (36 males, 27 females), 5.4 (4.8–6.0), 5.3 (4.9–5.7); palatal length (35 males, 26 females), 15.2 (12.9–17.4), 14.9 (13.0–17.4); length of mandibular toothrow (35 males, 27 females), 5.6 (4.9–6.1), 5.3 (4.8–5.6); greatest length of mandible (35 males, 27 females), 16.6 (14.0–19.0), 16.6 (14.8–19.1); diastema length (35 males, 27 females), 8.0 (6.7–9.5), 7.9 (7.0–9.1).

Remarks. The specimen from Minas Capillitas cited in the specimens examined is tentatively listed as a *G. griseoflavus*; it is a very young animal. This is one of the most widespread arid-adapted rodents in Argentina. It is found in almost all habitats, from high Andean and pre-Andean hillsides to moist lowland forest, from agricultural areas to mesquite (*Prosopis*) forests in semidesert regions. It can exist for extended periods without free water (Mares, 1977a) and is a strong, aggressive phyllotine that is highly scansorial (Mares et al., 1977b). It does not occur in extensive pure stands of creosote bush in the lowland monte desert. Males and females of this species were captured in January, February, March, April, June, July, November, and December. Males collected in November and December were reproductively active, testes ranging in size from 12 x 7 mm to 16 x 8 mm. Males were in nonbreeding condition the other months, with size of testes ranging from 4 x 2 mm to 9 x 4 mm. Pregnant females were captured in January, November, and December. The number of embryos ranged from four to seven, and the crown-rump length ranged from 5.0–59.0 mm. Nonbreeding females were captured the other months. Individuals that were molting were captured in July and December. This species is considered as injurious wildlife in Argentina (Fuller and Swift, 1984). As noted for several species in the southwestern United States, numerous skeletons of individuals of this species are pink, possibly indicating consumption of cactus parts that are the hosts of the cochineal insect (Braun, pers. observ.).

Neotomys ebriosus Thomas

1894. *Neotomys ebriosus* Thomas, Ann. Mag. Nat. Hist., ser. 6, 14:348.

Specimens examined. None.

Additional records. Mountains of Aconquija, Catamarca Province (Thomas, 1926).

Remarks. The geographic distribution and life history of this species are poorly known. It appears to be restricted to the high puna, along marshes and in grassy areas (Barquez, 1983; Cabrera, 1961b; Olrog and Lucero, 1981; R.A. Ojeda, pers. obs.; Sanborn, 1947). The subspecies in the province is *N. e. vulturnus* (Barquez, 1983; Sanborn, 1947).

Oligoryzomys flavescens (Waterhouse)

1837. *Mus flavescens* Waterhouse, Proc. Zool. Soc. London, 1837:19.

Specimens examined (5). 6 km SW of Hwy 9 on Hwy 18, 1,524 m, 4 (1 CML, 1 IADIZA-CM, 2 OMNH); Cuesta del Clavillo, 5 km al S de La Banderita, 1 (RMB).

Additional records. Chumbicha (Thomas, 1919a).

Measurements. External measurements (3 males, 2 females): total length, 210.3 (201–219), 201.0 (194–208); tail, 123.0 (115–128), 116.5 (115–118); hindfoot, 26.3 (26–27), 24.4 (23.7–25); ear, 14.7 (14–15), 15.8 (15–16.5); weight, 18.7 (16.0–22.0), 17.5 (14.0–21.0). Cranial measurements (3 males, unless noted otherwise): greatest length of skull (2 males), 23.6 (23.2–23.9); condylobasal length, 21.3 (20.5–21.9); least interorbital breadth, 3.7 (3.6–3.7); zygomatic breadth, 12.6 (12.4–12.9); breadth of braincase, 11.1 (11.0–11.3); length of maxillary toothrow, 3.4 (3.4–3.5); palatal length, 9.5 (9.2–9.8); length of mandibular toothrow, 3.4 (3.3–3.4); greatest length of mandible, 11.6 (11.5–11.7); diastema length, 5.5 (5.4–5.7).

Remarks. These specimens represent new records of this species for the province. The specimens collected at Hwys 9 and 18 were captured in a grassland above treeline along a streambed in rocks, shrubs, and overhanging grass. One individual was captured in the same trap with an *Akodon caenosus*. Carleton and Musser (1989) discussed the confusing nature of *Oligoryzomys* species identification. These specimens have the following distinguishing characteristics compared with *Oligoryzomys longicaudatus*: smaller ears, well-haired ears, relatively larger hindfeet, shorter maxillary toothrow. Thomas (1919a) listed a specimen collected at Chumbicha as *O. flavescens*; we have not examined this individual, but we have listed it as an additional record for reference.

Oligoryzomys longicaudatus (Bennett)

1832. *Mus longicaudatus* Bennett, Proc. Zool. Soc. London, 1832:2.

Specimens examined (42). 6 km SW of Hwy 9 on Hwy 18, 1,524 m, 5 (1 CML, 4 OMNH); Choya, 13 km NNW Andalgalá, 1,219 m, 2 (OMNH); El Rodeo, 0.5 km W of Hwy 4, 1,372 m, 18 (5 CML, 6 IADIZA-CM, 7 OMNH); El Rodeo, 1.5 km NE of Hwy 4, 1,372 m, 1 (CML); 35 km N Andalgalá on Río Andalgalá, 1 (CMNH); 6.5 km (by road) N jct. Hwy 63 (to Chaquiago and La Toma Road, La Toma, N Andalgalá Waterworks, 4 (CMNH); along Andalgalá River, immed. N Andalgalá, 2 (CMNH); Río Andalgalá, immed. N Andalgalá, 8 (CMNH); immed. N Andalgalá, 1 (CMNH).

Measurements. External measurements (22 males, 19 females, unless noted otherwise): total length, 235.9 (201–267), 227.3 (187–256); tail, 136.9 (114–152), 131.5 (112–145); hindfoot, 26.1 (24–28), 25.0 (23.2–26); ear (21 males), 19.0 (17–22.7), 18.1 (16.3–20.4); weight (21 males, 16 females), 25.2 (16.0–45.0), 22.6 (14.3–39.7). Cranial measurements (3 males, 4 females, unless noted otherwise): greatest length of skull, 27.1 (26.3–28.6), 27.0 (25.2–28.5); condylobasal length, 24.3 (23.3–26.4), 24.2 (22.4–25.2); least interorbital breadth, 3.8 (3.7–4.0), 3.8 (3.6–4.1); zygomatic breadth (3 females), 14.0 (13.5–14.8), 14.1 (13.2–14.7); breadth of braincase, 12.0 (11.5–12.4), 12.1 (11.5–12.6); length of maxillary toothrow, 4.0 (3.9–4.1), 3.9 (3.9–4.0); palatal length, 12.1 (11.4–12.5), 12.8 (12.0–13.2); length of mandibular toothrow, 4.1 (4.0–4.3), 4.1 (4.0–4.2); greatest length of mandible, 14.1 (13.8–14.4), 13.6 (13.1–13.9); diastema length, 6.4 (6.0–6.9), 6.3 (5.9–6.7).

Remarks. The measurements of these specimens fall within the sizes reported for specimens from the neighboring province of Tucumán by Myers and Carleton (1981). However, the resemblance in cranial characters among the members of the *nigripes–longicaudatus–chacoensis* complex deserves further research at the genetic and ecological levels in order to reach a more conclusive picture regarding the specific status of the forms of *Oligoryzomys* in the province of Catamarca. *O. longicaudatus* is almost always associated with mesic forest situations, whether gallery forest along flowing streams and rivers, or yungas forest. It is a highly scansorial species, although it is easily captured on the ground and inhabits burrows in the ground. Its physiology has been examined by Mares (1977d; Mares et al., 1985): it is not particularly adept at living without free water. This species is considered injurious wildlife in Argentina (Fuller and Swift, 1984). Specimens captured in July were not breeding; young individuals were collected in December. Individuals exhibiting molt patterns were captured in July and December. The individuals captured at El Rodeo were trapped in a dry deciduous scrub forest along a dry arroyo and the neighboring hillsides. The Choya site was an abandoned orchard. The site at Hwys. 9 and 18 was a moist streambed in a high grassland; the individuals were captured beneath the roots of a large shrub and under large clumps of overhanging grass. Thomas (1919b) listed a series of specimens from Otro Cerro that were only identified as *Oryzomys* sp. We have not examined these specimens and thus cannot confirm their identity. The subspecies in the province according to Cabrera (1961b) is *O. l. longicaudatus*.

Oxymycterus akodontinus Thomas

1921. *Oxymycterus akodontinus* Thomas, Ann. Mag. Nat. Hist., ser. 9, 8:615.

Specimens examined (1). Andalgalá, 1 (MACN).

Remarks. This specimen was collected by Crespo in 1950. This species is distributed in northwestern Argentina and it may be conspecific with *O. paramensis* (Wilson and Reeder, 1993).

Phyllotis darwini (Waterhouse)

1837. *Mus darwini* Waterhouse, Proc. Zool. Soc. Lond., 1837:28.

Specimens examined (34). Chumbicha, 1 km N and W of Balneario by road, 792 m, 6 (2 CML, 4 OMNH); El Rodeo, 0.5 km W of Hwy 4, 1,372 m, 2 (1 IADIZA-CM, 1 OMNH); El Rodeo, 1.5 km NE of Hwy 4, 1,372 m., 8 (3 CML, 3 IADIZA-CM, 2 OMNH); Poman, 95 km S Andalgalá, 1,341 m, 2 (OMNH); Minas Capillitas, 3,200 m, 5 (CMNH); along Río Potrero, 10 km N Potrero, 1 (CMNH); hills at end of La Puntilla, between Tinogasta and Copacabana, 2 (CMNH); 13 km N Andalgalá, 2 (CMNH); 6.5 km N jct. Hwy 63 and La Toma Road, 2 (CMNH); Antofagasta de la Sierra, 1 (IADIZA-CM); Laguna Blanca, 1 (CML); 13 km N Puesto Laguna Blanca, 1 (IADIZA-CM); 7 km N Puesto Laguna Blanca, 1 (IADIZA-CM).

Additional records. Chumbicha, 488–610 m (Hershkovitz, 1962; Pearson, 1958; Thomas, 1919a); Otro Cerro, 3,018 m (Hershkovitz, 1962; Pearson, 1958; Thomas, 1919b); Laguna Helada, 3,688 m (Hershkovitz, 1962; Pearson, 1958); Laguna Blanca, 3,505 m (Pearson, 1958); Laguna Blanca, 3,200 m (Yepes, 1933); Pasto Ventura, 3,993 m (Pearson, 1958); Cuesta de Zapata; near Chumbicha; 122 km NNW Tinogasta (Pearson and Patton, 1976); Salar de Antofalla (Hershkovitz, 1962).

Measurements. External measurements (20 males, 10 females, unless noted otherwise): total length, 249.1 (200–288), 237.6 (194–294); tail, 129.7 (95–154), 123.2 (89–166); hindfoot, 27.8 (25.8–30), 28.0 (25–36); ear (19 males, 10 females), 24.1 (21–28), 24.8 (22–29.2); weight, 51.6 (29.0–89.0), 43.4 (20.0–62.0). Cranial measurements (13 males, 6 females): greatest length of skull, 30.9 (27.6–33.0), 30.5 (29.4–31.7); condylobasal length, 28.9 (25.7–30.4), 28.3 (26.8–30.4); least interorbital breadth, 4.4 (4.2–4.7), 4.4 (4.1–4.5); zygomatic breadth, 15.7 (14.3–16.6), 15.5 (14.8–16.4); breadth of braincase, 13.6 (13.2–14.1), 13.4 (12.9–13.6); length of maxillary toothrow, 5.3 (4.8–5.7), 5.3 (5.0–5.7); palatal length, 15.3 (12.2–17.0),

14.7 (13.6–16.2); length of mandibular toothrow, 5.3 (4.8–5.7), 5.2 (4.8–5.7); greatest length of mandible, 16.8 (14.3–18.3), 16.9 (16.1–17.5); diastema length, 8.2 (7.1–8.9), 7.9 (7.5–8.4).

Remarks. Cabrera (1926) described *P. oreigenus* from Laguna Blanca, but Pearson (1958) considered it to be a young individual of *P. darwini*, *P. osilae*, or *P. caprinus*. Pearson (1958) referred to the highland puna populations as *P. d. vaccarum*, whereas he considered the lowland forms to be *P. d. ricardulus*. Hershkovitz (1962) included *P. ricardulus* from Otro Cerro (described by Thomas, 1919b) and *P. oreigenus* (described by Cabrera, 1926) from Laguna Blanca in synonymy with *P. rupestris*. Recent work (Spotorno and Walker, 1983; Walker et al., 1984) has indicated complete genetic isolation of *P. darwini darwini* and *P. darwini vaccarum* in central Chile. Walker et al. (1984) suggested that the two forms are specifically distinct: the oldest available name for the high altitude form (*vaccarum*) and for those from southern Chile would be *P. xanthopygus*. We have chosen to take a conservative approach to systematics of *Phyllotis* in this region until a more detailed study is made using specimens from the area. The above measurements correspond to the lowland populations. Pearson and Patton (1976) commented on the constancy of the chromosomal complements of *P. darwini* throughout its geographic range. A specimen of *Phyllotis* was collected by Budín at La Puntilla, near Tinogasta, but was not identified to species by Thomas (1920a). Reproductively active males (testes medium or large, ranging from 11 x 7 mm to 13 x 8 mm) have been collected in January and December. Nonbreeding males were collected in June and July. Pregnant females were collected in December; number of embryos ranged from three to four, and crown-rump length ranged from 15-21 mm. Nonbreeding females were captured in January and July. Individuals were found to be molting in July and December.

Phyllotis osilae J.A. Allen

1901. *Phyllotis osilae* J.A. Allen, Bull. Am. Mus. Nat. Hist., 14:44.

Specimens examined (12). 6 km SW of Hwy 9 on Hwy 18, 1,524 m, 5 (1 CML, 1 IADIZA-CM, 3 OMNH); Las Cuevas, 6 (MACN); Cerro El Manchao, Los Cajones, 4,280 m, 1 (OMNH).

Additional records. Aconquija, 3,993–4,999 m (Hershkovitz, 1962; Pearson, 1958); Chumbicha, 488 m (Hershkovitz, 1962; Pearson, 1958; Thomas, 1919a); Las Paras, 3,018 m (Hershkovitz, 1962; Pearson, 1958); Otro Cerro, 3,018 m (Hershkovitz, 1962; Pearson, 1958; Thomas, 1919b); Río Vallecito, 2,896 m (Hershkovitz, 1962; Pearson, 1958).

Measurements. External measurements (males and females combined, for individuals from Otro Cerro and Chumbicha reported by Pearson, 1958): head and body length (17), 111.9 (100–134); tail (15), 118.7 (97–145); hindfoot (16), 25.2 (24–27); ear (19), 21.8 (20–24). Cranial measurements (males and females combined, for individuals from Otro Cerro and Chumbicha reported by Pearson, 1958): greatest length of skull (13), 28.9 (27.2–31.8); least interorbital breadth (16), 4.1 (4.0–4.3); length of maxillary toothrow (16), 5.4 (4.7–5.7).

Remarks. The subspecies in Catamarca is *P. o. tucumanus* (Pearson, 1958). A specimen of *Phyllotis* was collected at La Puntilla, near Tinogasta, by Budín; it is a young individual and has not been identified to species (Thomas, 1920a). The Las Paras location is incorrectly assigned to Catamarca Province; this site should be Las Pavas, Tucumán Province. In addition, the specimens from Aconquija may also pertain to Tucumán Province. The specimens in the OMNH collection are nonbreeding individuals that were captured in July. Three of these individuals showed signs of molting. The Hwy 9 and 18 location is a mid-elevational grassland. *P. osilae* is less adept at existing without free water than is *P. darwini* (Mares, 1977a), is a more aggressive species, and is not obligately associated with rocks, as is *darwini*. *P. osilae* is a species of high montane bunchgrass areas. We have not examined the Chumbicha specimens supposedly taken at an elevation of 488 m, but this elevation would place the locality in the desert scrubland of Catamarca. We would be surprised to find *P. osilae* in such a habitat and at such an elevation. Until Budín's field records can be examined, we question whether the specimen listed was actually collected at that elevation.

Reithrodon auritus (Fischer)

1819. *Mus physodes* Fischer, Zoognosia, 3:71.

Specimens examined. None.

Additional records. Otro Cerro (Thomas, 1920b).

Measurements. External measurements (1 male, from Thomas, 1920b): total length, 238; tail, 99; hindfoot, 30.5; ear, 25. Cranial measurements (1 male, from Thomas, 1920b): greatest length of skull, 36.8; zygomatic breadth, 19.7; breadth of braincase, 15.6; maxillary toothrow length, 6.8.

Remarks. This population has been referred to as *R. p. caurinus* (Cabrera, 1961b) with type locality in Otro Cerro (Thomas, 1920b). The biology of this species has not been examined in detail and there are only a few notes on its natural history and taxonomy from neighboring Tucumán Province (Barquez et al., 1991; Dalby and Mares, 1974). Like *Phyllotis osilae*, *Reithrodon* is limited to high elevation bunchgrass prairie, where it does not appear to be common. The species does become quite abundant in the high grasslands of the Andean chains to the south of Catamarca in southern Mendoza Province and in the grasslands of Patagonia (M.A. Mares, pers. obs.).

Family Chinchillidae

Lagidium viscacia (Molina)

1782. *Lepus viscacia* Molina, Sagg. Stor. Nat. Chile, p. 307.

Specimens examined (3). Loma Morada, 2 (CML); Minas Capillitas, 3,200 m, 1 (CMNH).

Additional records. Otro Cerro (Thomas, 1919b); El Pucará (M.A. Mares, pers. obs.).

Measurements. External measurements (3 individuals): total length, 470, 770, 765; tail, 224, 340, 325; hindfoot, 78, 85, 90; ear, 68, 80, 75. Cranial measurements (2 sex unknown): least interorbital breadth, 17.9, 16.3; zygomatic breadth, 43.6 42.5; breadth of braincase, 34.1, 33.4; length of maxillary toothrow, 20.3, 20.6; length of mandibular toothrow, —, 19.6; greatest length of mandible, —, 57.3; diastema length, 24.7, 21.2.

Remarks. This population has been referred to as *L. v. lockwoodi*, with type locality in Otro Cerro, Catamarca Province (Thomas, 1919b). Mountain vizcachas are herbivorous rock specialists of the high Andes and pre-Andean chains of South America (Pearson, 1948; Mares and Lacher, 1987) and are common in boulder areas of Catamarca above 2,500 m, where they occur in extensive colonies. They are diurnal and are frequently hunted for food and sport, although their pelts have little commercial value, a fact which has aided them in escaping depredations such as were visited upon their *Chinchilla* cousins. The individual captured at Minas Capillitas was a nonbreeding female.

Lagostomus maximus (Desmarest)

1817. *Dipus maximus* Desmarest, Nouv. Dict. Hist. Nat., Nouv. ed., 13:117.

Specimens examined (10). Singuil, 1,250 m, 10 (MACN).

Additional records. La Paz, Depto. Icaño; La Paz, Depto. San Antonio; Depto. Paclín; Depto. Ambato; Depto. Santa Rosa (Llanos and Crespo, 1952).

Measurements. Cranial measurements (5 males, 5 females): greatest length of skull, 126.6 (123.6–132.2), 105.1 (102.0–107.0); condylobasal length, 112.7 (108.4–117.8), 95.1 (93.0–96.0); least interorbital breadth, 34.0 (33.0–35.0), 29.2 (27.4–29.9); zygomatic breadth, 76.0 (73.3–78.0), 65.2 (63.8–66.7); breadth of braincase, 38.0 (36.8–40.5), 35.2 (33.1–36.6); length of maxillary toothrow, 28.7 (28.0- 30.3), 25.8 (24.4–27.0); palatal length, 66.3 (65.7–67.5), 53.8 (51.0–56.0); length of mandibular toothrow, 28.3 (27.0–29.4); 25.1 (23.6–26.8); greatest length of mandible, 97.5 (94.3–101.5), 83.3 (79.4–84.9); diastema length, 40.2 (38.6–41.8), 33.2 (33.0–34.0).

Remarks. The national and provincial governments have maintained a strong campaign against this species in an attempt to minimize its competition for forage with cattle, much like the control program utilized against prairie dogs (*Cynomys*) in North America. Like prairie dogs, vizcachas live in large groups in extensive burrow systems. They are largely nocturnal and, in Catamarca, inhabit disturbed areas in Chacoan thorn scrub east of the pre-Andean ranges. The species is now uncommon in the province. The animals are hunted not only in a control program, but also for food (the pickled meat is sold commercially and is considered a delicacy) and leather (the tanned hides, with the fur, are used to make bedspreads, artistic artisan goods, and other products). For example, Llanos and Crespo (1952) noted that in a 1937 control program, 52,460 burrow systems with 135,709 openings distributed over 162,961.5 ha were treated with poison in an attempt to kill the animals. The species has not been studied in Catamarca, but extensive research on its biology has been conducted in the southern grasslands and scrublands of Argentina (Branch, 1993; Branch et al., 1993; Llanos and Crespo, 1952). The subspecies in the province is *L. m. inmollis* (Cabrera, 1961b).

Family Caviidae
Subfamily Caviinae
Galea musteloides Meyen

1832. *Galea musteloides* Meyen, Nouv. Acta Acad. Caes. Leop.-Carol., 16:597.

Specimens examined. None.

Additional records. Near Andalgalá (Mares, 1973); Otro Cerro (Thomas, 1919b).

Remarks. *Galea musteloides* occurs throughout much of Argentina (Cabrera, 1953). It is a ground squirrel-like caviid that primarily inhabits bunchgrass prairie above 2,500 m and mesic (frequently grassy) areas along creeks in areas supporting moist forest, or along roadsides and other disturbed areas in mesic agricultural zones. It is common in such habitats in Tucumán Province (Barquez et al., 1991) and would be expected to occur in similar habitats in Catamarca, especially in the eastern forested and agricultural areas. It can be captured occasionally in open-cage traps, but is seldom caught in any other kinds of traps.

In Mendoza, *Galea* can be found in desert scrublands, coexisting at times with *Microcavia*, although *Galea* is more likely to be captured in grassy portions of the desert scrubland. Pronounced sympatry between *Galea* and *Microcavia* was described by Contreras (1965) and Rood (1972) in Buenos Aires Province.

Cabrera (1953:13-14) also commented on *Galea* and *Microcavia* coexisting near the town of Andalgalá as follows: "In Catamarca, between Andalgalá and the Amanao drainage, I have seen maras (= *Dolichotis*) travel during the early afternoon in January over the burning sand and at midday I have observed *Microcavia* and *Galea* sunning themselves in small groups near the entrances to their burrow systems." Contreras (1965) remarks on this observation to support the idea that these two caviid species coexist with regularity over broad parts of their range. The report of *Galea* in Andalgalá is based on three animals taken by Crespo and Cabrera on a field trip made in the 1940s or 1950s. Mares examined these specimens in 1972 in Buenos Aires (but they were not examined for inclusion in this report). They are listed as being collected from "above" the town of Andalgalá, which could mean from the high grasslands above Andalgalá, as would be expected for *Galea*. Although Mares collected many *Microcavia* from the environs of Andalgalá (especially along the Río Amanao and associated drainage systems), no *Galea* was ever captured. Since these species are difficult to distinguish while they are sunning outside their burrows, it is doubtful that Cabrera saw mixed groups of cavies. More likely, it may have been assumed that the sunning cavies included both species.

The subspecies in the province is *G. m. musteloides* (Cabrera, 1953). This species is considered injurious wildlife in Argentina (Fuller and Swift, 1984).

Microcavia australis
(I. Geoffroy and d'Orbigny)

1833. *Cavia australis* I. Geoffroy and d'Orbigny, Mag. Zool. Paris, p. 3.

Specimens examined (21). Agua de Dionisio, Hualfín, 1 (CML); Agua Tapada, Hualfín,

2,400 m, 3 (CML); Andalgalá, 1 (CMNH); Chumbicha, 2 (BMNH); La Puntilla, 3 (BMNH); Los Médanos, 2 (CML); Recreo (350 m), 7 (BMNH); Río Amanao, 1 (CMNH); San Antonio de La Paz, 1 (MACN).

Additional records. Off Hwy 1, at km 99, 34 km S Andalgalá; Andalgalá Railroad Station, 4 km S Andalgalá; Potrero River dike, approx. 13 km N Andalgalá; along Río Amanao, 15 km S jct. Hwy 1 and Río Amanao (Mares, 1973); Hualfín (Olrog, 1958); La Puntilla, near Tinogasta (Thomas, 1920a); Tinogasta; Chumbicha; Recreo (Thomas, 1921); Recreo; San Antonio de La Paz; Tinogasta; Chumbicha (Cabrera, 1953).

Measurements. External measurements (2 males, 5 females): total length, 229.6 (200–250); hindfoot, 44.4 (42–45.1); ear, 18.2 (15–20). Cranial measurements (4 males, 6 females): greatest length of skull, 51.0 (48.9–54.2); condylobasal length, 46.5 (44.8–50.0); least interorbital breadth, 11.4 (10.6–12.6); zygomatic breadth, 30.0 (28.1–32.0); breadth of braincase, 22.3 (20.7–23.8); length of maxillary toothrow, 12.0 (10.9–12.4); palatal length, 24.6 (23.3–26.1); length of mandibular toothrow, 12.1 (11.5–13.5); greatest length of mandible, 38.6 (36.7–42.1); diastema length, 12.8 (11.9–14.5).

Remarks. Two populations from Catamarca have been referred to as *M. a. salinia* and *M. a. maenas* (Cabrera, 1953). The type locality of *salinia* is Recreo, Catamarca Province (Thomas, 1921), whereas the type locality of *maenas* is Chilecito, in the neighboring province of La Rioja. The above measurements are for the *salinia* population. The *maenas* population occurs in mountain areas, whereas *salinia* is a lowland subspecies (Cabrera, 1953). This guinea pig is ecologically similar to a ground squirrel (e.g., Mares, 1975b) and inhabits most of the major habitats of Catamarca (and of Argentina), especially the drier habitats. The animals are particularly common in mesquite forests that line dry gullies and surround salt flats in the monte desert, as well as in dry thorn scrub, high elevation scrublands, and disturbed areas in mesic habitats at the peripheries of drier regions. The species is strongly herbivorous (e.g., Mares et al., 1977b) and climbs readily into mesquite trees to feed on young leaves. They form dense populations, particularly along forested gullies and in other dry woodlands. The ecology of *M. australis* in provinces further south has been examined by Rood (1970, 1972) and Contreras and Roig (1978), and several comments about the ecology of this species in Catamarca are in Mares (1973, 1976), Mares et al. (1985), and Orians et al. (1977). In Argentina, this species is considered to be injurious wildlife (Fuller and Swift, 1984). A female pregnant with one embryo was captured in August at the Río Amanao.

Microcavia shiptoni (Thomas)

1925. *Nanocavia shiptoni* Thomas, Ann. Mag. Nat. Hist., ser. 9, 15:419.

Specimens examined (8). Laguna Blanca (3400-3500 m), 8 (BMNH).

Measurements. External measurements (2 males, 4 females): total length, 197.8 (180–220); hindfoot, 37.5 (35–40); ear, 17.2 (15–19.9). Cranial measurements (2 males, 4 females, unless otherwise noted): greatest length of skull (1 male, 4 females), 43.8 (42.3–44.6); condylobasal length (1 male, 4 females), 40.8 (38.7–42.2); least interorbital breadth, 9.8 (9.3–10.3); zygomatic breadth, 26.8 (25.7–28.5); breadth of braincase (1 male, 4 females), 19.8 (19.1–20.8); length of maxillary toothrow, 11.0 (10.6–11.5); palatal length, 21.9 (20.4–22.9); length of mandibular toothrow, 11.2 (10.5–11.9); greatest length of mandible, 34.0 (32.8–35.2); diastema length, 11.5 (10.4–12.4).

Remarks. Laguna Blanca is the type locality of the species. Measurements of type and paratype specimens from the BMNH are included. Cabrera (1953) considered *Nanocavia* (described by Thomas, 1925) congeneric with *Microcavia*, despite differences in size, coloration, and habitat.

Subfamily Dolichotinae

Dolichotis patagonum (Zimmermann)

1780. *Cavia patagonum* Zimmermann, Geogr. Gesch. Mensch. Vierf. Thiere, 2:328.

Specimens examined (1). Catamarca, 1 (MACN).

Additional records. El Pajonal (R.A. Ojeda, pers. obs.); 1.5 km S, 150 m W jct. Hwy 1 and railroad track, just S Andalgalá; 50 m N, 50 m E jct. Hwy 62 and Amanao Road, approx. 18 km E Andalgalá; 6 km N Saujil, along Hwy 1 (Mares, 1973); 6 March 1971 in the Río Amanao, 8 km S of the jct. of the river and Hwy 62, west of Andalgalá (M.A. Mares, pers. obs.); Cuesta de Zapata; Tinogasta (Catalog MACN).

Measurements. Cranial measurements (1 sex unknown): greatest length of skull, 130.1; condylobasal length, 122.5; least interorbital breadth, 40.3; zygomatic breadth, 64.4; postorbital constriction, 47.9; breadth of braincase, 50.3; length of maxillary toothrow, 28.7; palatal length, 63.2; length of mandibular toothrow, 29.0; greatest length of mandible, 102.7; diastema length, 44.0.

Remarks. Cabrera (1953) reported this specimen as belonging to the population described as *D. p. centricola*. Patagonian "hares," or maras, are likely the fastest rodent in the world; Mares clocked several that were running up the dry Río Amanao just west of Andalgalá at 40-50 km/hr. They are browsers, foraging on most green vegetation, including *Prosopis, Larrea*, and various cacti. The species has been studied by Taber and Macdonald (1984) in Patagonia. Scattered information on its ecology in Catamarca is available in Mares (1975b), Mares and Hulse (1977), and Mares et al. (1977a, 1977b). The species is heavily hunted for meat and for its skin, which is used in various artisan goods. Unlike *Lagostomus*, the meat of *Dolichotis* is not commercialized, possibly because the species generally occurs in low abundance (usually in small family groups of 3-4 animals) and is difficult to collect due to its speed and large home range.

Pediolagus salinicola (Burmeister)

1875 (1876). *Dolichotis salinicola* Burmeister, Proc. Zool. Soc. London, 1875 (1876):634.

Specimens examined. None.

Additional records. SE of Catamarca, between Totoralejos and Recreo (Cabrera, 1953).

Remarks. Southeast of Catamarca, between Totoralejos and Recreo, is the type locality of the species (Cabrera, 1953). The species is uncommon in Catamarca and we have never observed animals in the expected habitat of the Chacoan thorn forest of the eastern parts of the province. However, it is heavily hunted throughout its range and its numbers in Catamarca may have been greatly reduced.

Family Ctenomyidae

Ctenomys coludo Thomas

1920. *Ctenomys coludo* Thomas, Ann. Mag. Nat. Hist., Ser. 9, 6:119.

Specimens examined. None.

Additional records. La Puntilla (Thomas, 1920a).

Measurements. External measurement (1 adult male, from Thomas, 1920a): total length, 302; tail, 97; hindfoot, 36.5. Cranial measurements (1 adult male, from Thomas, 1920a): greatest length of skull, 47.0; condylobasal length, 47.5; least interorbital breadth, 9.3; zygomatic breadth, 31.0; braincase breadth, 18.0; length of maxillary toothrow, 8.8; palatal length, 21.7.

Remarks. La Puntilla is the type locality for this species. This taxon was later treated as a subspecies of *C. fulvus* (Cabrera, 1961b). *C. fulvus* is a Chilean species from the high deserts, and it is doubtful that the Catamarca specimens are conspecific with the Chilean ones.

Ctenomys fochi Thomas

1919. *Ctenomys fochi* Thomas, Ann. Mag. Nat. Hist., ser. 9, 3:117.

Specimens examined (1). 1.5 km E of Hwy 38 along Hwy 60, Chumbicha, 457 m, 1 (OMNH).

Additional records. Chumbicha (Thomas, 1919a).

Measurements. External measurement (1 adult female): total length, 230; tail, 72; hindfoot, 32; ear, 6; weight, 119.0. Cranial measurements (1 adult female): greatest length of skull, 37.4; condylobasal length, 36.6; least interorbital breadth, 7.6; zygomatic breadth, 21.9; braincase breadth, 15.2; length of maxillary toothrow, 8.2; palatal length, 15.7; length of mandibular toothrow, 8.1; greatest length of mandible, 28.2; diastema length, 9.2

Remarks. This female, collected in July, was molting but not breeding. Described by Thomas (1919a), this species has a distinctive dark area on the top of the head and muzzle. This taxon was synonomized with *mendocinus* by Cabrera (1961b), and this arrangement is generally followed by other investigators (Redford and Eisenberg, 1992). We have opted to recognize it as distinct from the other species of *Ctenomys* until further work is done on this group in northwestern Argentina. The measurements of this individual match those of the type specimen. Although numerous active burrows were observed and calls heard on successive nights in July 1990, when this site was surveyed again in December of the same year, they were not seen (J.K. Braun, pers. obs.; R.L. Humphrey, pers. obs.). Mares has observed that *Ctenomys* often become inactive for long periods during the hottest parts of the summer.

Ctenomys knighti Thomas

1919. *Ctenomys knighti* Thomas, Ann. Mag. Nat. Hist., ser. 9, 3:498.

Specimens examined. None.

Additional records. Otro Cerro (Thomas, 1919b).

Measurements. External measurement (1 adult male, 1 adult female, from Thomas, 1919b): total length, 285, 252; tail, 82, 74; hindfoot, 36, 30; ear, 7, 7. Cranial measurements (1 adult male, 1 adult female, from Thomas, 1919b): greatest length of skull, 49.5, 44.0; condylobasal length, 50.5, 43.0; least interorbital breadth, 12.0, 9.8; zygomatic breadth, 30.2, 27.0; braincase breadth, 19.8, 17.8; length of maxillary toothrow, 11.1, 10.0; palatal length, 23.7, 20.0.

Remarks. The type locality for this species is Otro Cerro. Thomas (1919b) reported that twelve animals, including the type, were collected by Budín at this locality. Barquez et al. (1991) comment on the ecology of this species in neighboring Tucumán Province.

Ctenomys opimus Wagner

1848. *Ctenomys opimus* Wagner, Arch. Naturgesch., 1:75.

Specimens examined (1). Vicinity of Antofagasta de la Sierra, 1 (IADIZA-CM).

Measurements. External measurements (1 female): total length, 290; tail, 100; hindfoot, 41.2; ear, 10.8; weight, 315.

Remarks. The biology of this species in Perú was examined in detail by Pearson (1959). Additional comments on *C. opimus* in neighboring Salta Province are in Mares et al. (1989).

Ctenomys sp. 1

Specimens examined (1). 7 km N Laguna Blanca, 3,200 m, 1 (IADIZA-CM).

Measurements. External measurements (1 female): total length, 255; tail, 67; hindfoot, 31.4; ear, 6.8; weight, 141.0.

Remarks. This animal, captured in sandy soils, is similar to *C. opimus* in its yellowish general coloration, but is much smaller in size.

Ctenomys sp. 2

Specimens examined (18). 2 km N jct. Hwy 62 and Amanao Road (W Andalgalá), 8 (CMNH); 200 m N jct. Hwy 62 and Amanao Road, 4 (CMNH); 50 m N, 50 m E jct. Hwy 62 and Amanao Road, 1 (CMNH); 50 m S jct. Hwy 62 and Amanao Road, 3 (CMNH); 1 km E jct. Hwy 62 and Amanao Road, 1 (CMNH); 8.1 km W Andalgalá, 1 (OMNH).

Measurements. External measurements (9 males, 8 females): total length, 246.7 (222–267), 219.6 (186–266); tail, 84.3 (75–94), 74.0 (61–85); hindfoot, 35.2 (33.9–36.4), 31.3 (28.6–33.9); ear, 8.1 (6.9–9.4), 6.8 (5.7–7.7); weight, 152.6 (108.9–219.6), 107.2 (58.7–180.0). Cranial measurements (4 males, 3 females): greatest length of skull, 41.4 (39.8–42.8), 38.8 (35.6–41.8); condylobasal length, 39.9 (38.0–42.0), 37.8 (34.6–40.7); least interorbital breadth, 8.7 (8.2–9.3), 8.5 (7.6–9.2); zygomatic breadth, 24.9 (23.9–26.4), 22.9 (21.4–24.0); breadth of braincase, 17.3 (16.8–18.0), 16.6 (16.0–17.0); length of maxillary toothrow, 8.7 (8.2–9.1), 8.6 (7.6–9.2); palatal length, 19.9 (18.5–21.0), 17.8 (14.7–19.6); length of mandibular toothrow, 9.2 (8.8–9.5), 8.8 (8.3–9.1); greatest length of mandible, 32.7 (31.8–34.6), 30.0 (27.7–32.4); diastema length, 11.1 (10.2–12.0), 10.1 (8.8–10.8).

Remarks. In several papers (Mares, 1975b; Mares and Hulse, 1977; Mares et al., 1977b, 1985; and others), animals from these monte localities listed above were referred to as *C. fulvus*. The animals are much smaller than the type specimen of *fulvus* from Tinogasta. Olrog and Lucero (1981) also listed *C. fulvus* as present in the monte of Catamarca Province. Mares and Ojeda (1982), however, limited *C. fulvus* to the Andes of the Chilean and Argentine border (but with no definite records in Catamarca). At present, the taxonomic status of the *Ctenomys* from the Bolsón de Pipanaco is uncertain. Nevertheless, there is a good deal of information on the ecology of the species because of research carried out near Andalgalá by Mares (e.g., 1973, 1975b), Mares and Hulse (1977), Mares et al. (1977b), and Mares (unpubl.). Animals have been collected in February, April, May, and August. Males had scrotal testes in February; males with abdominal or inguinal testes were sampled in May and August. Juvenile males were found in February. Nonbreeding females were collected in all months. A pregnant female with an open vagina and large mammae was collected in August. The two embryos had crown–rump lengths of 24.6 and 23.8 mm.

Family Octodontidae

Octomys mimax Thomas

1920. *Octomys mimax* Thomas, Ann. Mag. Nat. Hist., ser. 9, 6:118.

Specimens examined (5). La Puntilla, 1000 m, 5 (BMNH).

Measurements. External measurements (1 male and 2 females): total length, 315.0, 331.5 (319–344); tail, 161.0, 174.5 (170–179); hindfoot, 32.3, 33.9 (33.2–34.6); ear, 21.5, 23.4 (21.4–25.3). Cranial measurements (1 male and 2 females, unless noted otherwise): greatest length of skull, 42.5, 43.8 (43.5–44.0); least interorbital breadth, 9.5, 9.3 (9.1–9.4); zygomatic breadth (1 male, 1 female), 21.1, 20.8; length of maxillary toothrow, 8.1, 8.1 (7.8–8.3); palatal length, 16.2, 16.1 (15.8–16.4); length of mandibular toothrow, 8.0, 8.0 (7.5–8.5); diastema length, 10.0, 9.8 (9.6–10.0).

Remarks. The measurements provided above include the type, paratype, and topotypes from La Puntilla, collected by Budín in 1920 (Thomas, 1920a). Although Mares looked for this species near the type locality in the early 1970s, no individuals were captured. It has not been well studied anywhere, but there is some information available on its ecology in San Juan Province (Mares, 1973, 1975b, 1980; Mares et al., 1977b). Thomas (1920a) stated that Budín noted that the specimens were collected among some large rocks that had fallen into a ravine; little water was present. He further noted that he was able to find only two "habitations" that were in close proximity and was unable to locate any others. This animal was not known by the local people.

Family Abrocomidae

Abrocoma cinerea Thomas

1919. *Abrocoma cinerea* Thomas, Ann. Mag. Nat. Hist., ser. 9, 4:132.

Specimens examined. None.

Additional records. Otro Cerro (Thomas, 1920b).

Measurements. External measurements (1 adult male, from Thomas, 1920b): total length, 345; tail, 144; hindfoot, 31; ear, 24.2. Cranial measurements (1 adult male, from Thomas, 1920b): greatest length of skull, 50.0; least interorbital breadth, 6.4; zygomatic breadth, 24.5; breadth of braincase, 20.0; length of maxillary toothrow, 10.2.

Remarks. This population has been referred to as *A. c. budini* (Cabrera, 1961b), with type locality at Otro Cerro (Thomas, 1920b). Thomas (1920b) noted that four individuals were collected at this site.

Order Lagomorpha

Family Leporidae

Lepus europaeus Pallas

1778. *Lepus europaeus* Pallas, Nova. Spec. Quad. Glir. Ord., p. 30.

Specimens examined (2). 1 km N, 1 km W Andalgalá, 1 (CMNH); 37 km S La Merced along Hwy 38, 1 (OMNH).

Additional records. Mares observed two animals along the road at the base of the Cuesta de la Chilca, near Andalgalá in 15 July 1971 at 0530 hrs.

Measurements. External measurements (1 adult male, 1 adult female): total length, 745, 360; tail, 97, 66; hindfoot, 150, 96; ear, 110, 79.6.

Remarks. This introduced species has become common throughout Argentina in most major habitats and is considered as injurious wildlife (Fuller and Swift, 1984).

DISCUSSION

Catamarca Province supports about 9% of South America's approximately 900 species of mammals. This might seem surprising for a province that is largely desert or semidesert, but the drylands of the continent have been shown to be especially important repositories of species and higher taxa (Mares, 1992). Of the 81 species of mammals, 51 are considered as uncommon or rare, but for most of these (e.g., *Desmodus rotundus*, *Thylamys pallidior*, *Sturnira lilium*, *Nyctinomops macrotus*, *Akodon alterus*, *Bolomys lactens*, *Galea musteloides*, *Ctenomys knightii*, *Andalgalomys olrogi*, *Andalgalomys roigi*, *Lyncodon patagonicus*), their low abundance is probably not related to human activities. Some (e.g., *Sturnira lilium*, *Dolichotis patagonum*, *Herpailurus yaguarondi*) occur in Catamarca at the limits of their geographic distribution and often are inhabitants of marginal habitats or of habitats that are themselves at the limits of their distribution. Others (e.g., *Bolomys lactens*, *Andalgalomys olrogi*, *Lyncodon patagonicus*) display that general rarity that characterizes so many species in nature.

For several species, however, their rarity is clearly the result of either hunting pressures or habitat modification. Most of these are animals of the high mountains and all have some economic importance as food, leather, wool, or as competitors of domestic stock. All native artiodactyls in the province, for example, are either rare or uncommon. Several carnivores (e.g., the canids, *Pseudalopex culpaeus*, *P. griseus*, *P. gymnocercus*), two rodents (*Chinchilla brevicaudata* and *Lagostomus maximus*), and at least two armadillos (*Tolypeutes matacus* and *Euphractus sexcinctus*) have been heavily hunted and are now uncommon throughout most of the province (although we listed *E. sexcinctus* as probable, in the past, it was likely common in easternmost Catamarca). Some carnivores (e.g., *Oncifelis colocolo*, *Oreailurus jacobita*) seem to be generally uncommon by nature, but have also been hunted extensively.

Even though several species of mammals have clearly been adversely affected by humans, the province, with its low human population, continues to support a rich fauna. Indeed, Catamarca's mammal fauna might be considered to be roughly similar to the fauna that was found in the region when the Spaniards first arrived more than four centuries ago. However, the situation is of greater complexity than is apparent at first glance. There is evidence that many species that today are restricted to extreme northern Argentina once occurred well to the south of Catamarca at the time of the Conquest. Roig (1991), for example, has examined letters written by Jesuits and other travelers in Argentina in the 17th century and found that they made comments about species that no longer occur in the areas where the observations were made. Such species as the capybara (*Hydrochaeris hydrochaeris*), the maned wolf (*Chrysocyon brachyurus*), the giant armadillo (*Priodontes maximus*), the southern three-banded armadillo (*Tolypeutes matacus*), and the jaguar (*Panthera onca*) very likely included Catamarca within their geographic distribution at one time. We also expect that the giant anteater (*Myrmecophaga tridactyla*), the Chacoan peccary (*Catagonus wagneri*), and the tapir (*Tapirus terrestris*) may have extended into eastern Catamarca in the Chacoan thorn scrub. Desertification, changing patterns of water courses, construction of dams, agricultural expansion, and deforestation probably led to the extirpation of these species from Catamarca.

Viewed from this standpoint, there have been great changes in the faunal composition of the province and no major habitat type has been spared. The Chacoan thorn forest may have lost up to a dozen species, all of them mid- to large-size mammals, including anteaters, deer, peccaries, armadillos, and rodents. The high montane habitats have seen the near extermination of the huemul (*Hippocamelus antisensis*), the vicuña (*Vicugna vicugna*), the guanaco (*Lama guanicoe*), the chinchilla (*Chinchilla brevicaudata*), and the Andean red fox (*Pseudalopex culpaeus*). Even the relatively species-poor monte desert has seen serious declines in populations of the gray fox (*Pseudalopex griseus*) and the spotted cat (*Oncifelis geoffroyi*). Thus, even though Catamarca has never supported a dense human population except for the most mesic eastern parts of the province, it has not been spared from significant effects of humans on the mammal populations of the province. Why should this be so?

In considering the changes in faunal composition experienced by the mammals of Catamarca over the last several centuries, and particularly over the last 100 years, we can speculate as to the causes. Originally, the province supported fairly sophisticated populations of Native Americans; undoubtedly mammals were utilized extensively for food and other products. We know, for example, that the Incas expended great efforts in hunting vicuñas and guanacos (e.g., Roig, 1991). However, great habitat changes were not produced by indigenous populations. Extensive deforestation, for example, is directly related to the needs of Europeans for housing, fuel, rails, and other construction needs (Roig, 1991). Similarly, advances in technology made it possible to redirect river flow, control run-off, and construct dams for hydroelectric use. All of these activities occurred in Catamarca over the last century. Their influence was greatest in the areas that had significant rivers and that could support agriculture and cities, i.e., the eastern mesic and semiarid lowlands. Clearly the extirpation of species that has occurred in the Chacoan woodlands and forested lowlands of eastern Catamarca has been extensive: most of the species of mammals of the province that have been eliminated or reduced in abundance were originally found in this region.

Nevertheless, there have been significant faunal changes in habitats other than those that are more mesic, especially in the driest portions of the high puna desert and in the monte. These effects are related to a different set of human activities, one not necessarily associated with population density or technological development. In the high Andes and in the lowland monte, human population levels are quite low. In these regions, one finds subsistence agriculture based on irrigation, mining, and grazing of domestic stock, especially goats, and, in the high grasslands, sheep. A large herd of goats can be maintained at minimal cost. A small family may have a herd of a hundred or more goats. The animals require some protection from predators (usually pumas in Catamarca) and daily fresh water, often supplied by an artesian well or small pond. Throughout the entire year, the herd grazes across the driest parts of the monte, the most unproductive puna habitats, and the most inaccessible mountainsides, foraging on almost any type of green vegetation. In the mesquite forests of the monte, for example, young trees are eaten as seedlings, well before they can become established as mature trees. The forests are heavily utilized for fuel and other wood needs, and the incessant grazing by goats, coupled with human needs for wood, are leading to the elimination of the once extensive monte woodlands.

Goats also disturb the shrublands in the monte and puna, feeding on almost any plant, including creosote bush, and trampling most of the vegetation. Goat paths, visible even from airplanes, extend throughout the arid portions of Catamarca, whether up steep mountain sides, across xeric scrublands, or along gullies. Yet this extensive habitat destruction is done for the benefit of very few people. Although goats supply meat, milk, and leather, relatively few people are actually dependent upon goats for their welfare. When one considers the small number of people living outside of towns and cities in Catamarca, there are many more goats in the remaining habitats than there are people. Goats contribute little to the monetary economy of the province. We believe that goats and other domestic stock have a strong deleterious affect on the xeric ecosystems of Catamarca, primarily through habitat conversion and desertification. Moreover, the goat economy permits those few people involved with goat herds to supplement their income with money obtained through hunting species of commercial importance, such as foxes, cats, and other fur bearers. There have been recent attempts to limit commercial utilization of wildlife throughout Argentina, including Catamarca, but, to date, the results of these conservation efforts have not been assessed. There have been no attempts to limit grazing by domestic livestock.

Catamarca's mammal fauna is less threatened than that of many other parts of South America, or even of Argentina. The several extirpations that may have occurred over the last two centuries are unfortunate, but the low human population density in the province, the vast xeric habitats that will never support dense human populations, and the inaccessibility of most of the province should ensure that further extirpation of species will occur at reduced rates. Indeed, the recent moves to control commercial hunting will ultimately have a significant positive effect on species having commercial value. Most of the mammals of the province, however, will be spared from extirpation by their good fortune in inhabiting one of the truly undeveloped parts of Argentina, a province that never has, and likely never will, yield readily to human domination.

ACKNOWLEDGMENTS

We would like to thank the many individuals who have aided in field and laboratory work since 1970. Field work during 1990-1993 was supported by a grant from the National Science Foundation (BSR-8906665) to M.A. Mares. Field work in 1993 was made possible by a grant to M.A. Mares from the National Geographic Society (4820-92). Field and laboratory research, manuscript preparation, and computer facilities were provided by the Oklahoma Museum of Natural History. C. Kacmarcik aided in figure preparation. We thank T.L. Yates, S. Anderson, and H.H. Genoways for their valuable comments on the manuscript.

LITERATURE CITED

Anderson, S., and N. Olds. 1989. Notes on Bolivian mammals. 5. Taxonomy and distribution of *Bolomys* (Muridae, Rodentia). American Museum Novitates, 2935:1-22.

Anonymous. 1981. Atlas físico de la República Argentina. Vol. 1. Centro Editor de América Latina, Buenos Aires, Argentina, 192 pp.

Anonymous. 1982. Atlas político de la República Argentina. Centro Editor de América Latina, Buenos Aires, Argentina, 287 pp.

Anonymous. 1988. Atlas de la República Argentina. Ejercito Argentino, Instituto Geográfico Militar, Buenos Aires, Argentina, 81 pp.

Baker, R.J., J.C. Patton, H.H. Genoways, and J.W. Bickham. 1988. Genic studies of *Lasiurus* (Chiroptera: Vespertilionidae). Occasional Papers, The Museum, Texas Tech University, 117:1-15.

Barquez, R.M. 1983. La distribución de *Neotomys ebriosus* Thomas en la Argentina y su presencia en la Provincia de San Juan (Mammalia, Rodentia, Cricetidae). Historia Natural, 3:189-191.

Barquez, R.M. 1987. Los murciélagos de Argentina. Tesis Doctoral, Facultad de Ciencias Naturales, Universidad Nacional de Tucumán, San Miguel de Tucumán, Argentina, 525 pp.

Barquez, R.M., N. Giannini, and M.A. Mares. 1993. Guide to the bats of Argentina (Guía de los murciélagos de Argentina). Special Publication, Oklahoma Museum of Natural History, Norman, Oklahoma, 119 pp.

Barquez, R.M., M.A. Mares, and R.A. Ojeda. 1991. The mammals of Tucumán (Mamíferos de Tucumán). Special Publication, Oklahoma Museum of Natural History, Norman, Oklahoma, 282 pp.

Barquez, R.M., D.F. Williams, M.A. Mares, and H.H. Genoways. 1980. Karyology and morphometrics of three species of *Akodon* (Mammalia: Muridae) from northwestern Argentina. Annals of Carnegie Museum, 49:379-403.

Bianchi, N.O., and M. Lizarralde. 1979. Genetics of the South American akodont rodents (Cricetidae). VI. Polymorphism in *Akodon dolores* (Thomas). Genetica, 50:99-104.

Blaustein, S.A., R.C. Liascovich, L.I. Apfelbaum, L. Daleffe, R.M. Barquez, and O.A. Reig. 1992. Correlates of systematic differentiation between two closely related allopatric populations of the *Akodon boliviensis* group from NW Argentina (Rodentia: Cricetidae). Zeitschrift für Säugetierkunde, 57:1-13.

Bozinovic, F., and M. Rosenmann. 1988. Daily torpor in *Calomys musculinus*, a South American rodent. Journal of Mammalogy, 69:150-152.

Branch, L.C. 1993. Social organization and mating system of the plains vizcacha (*Lagostomus maximus*). Journal of Zoology (London), 229:473-491.

Branch, L.C., D. Villarreal, and G.S. Fowler. 1993. Recruitment, dispersal, and group fusion in a declining population of the plains vizcacha (*Lagostomus maximus*; Chinchillidae). Journal of Mammalogy, 74:9-20.

Braun, J.K. 1993. Systematic relationships of the tribe Phyllotini (Muridae: Sigmodontinae) of South America. Special Publication, Oklahoma Museum of Natural History, Norman, Oklahoma, 50 pp.

Cabrera, A. 1926. Dos roedores nuevos de las montañas de Catamarca. Revista Chilena de Historia Natural, 30:319-321.

Cabrera, A. 1953. Los roedores argentinos de la familia Caviidae. Publicaciones de la Escuela de Veterinaria, Buenos Aires, 6:1-93.

Cabrera, A. 1957. Catálogo de los mamíferos de América del Sur. I (Metatheria–Unguiculata–Carnivora). Revista del Museo Argentino de Ciencias Naturales "Bernardino Rivadavia," Ciencias Zoológicas, 4:1-307.

Cabrera, A. 1961a. Los félidos vivientes de la República Argentina. Revista del Museo Argentino de Ciencias Naturales "Bernardino Rivadavia," Ciencias Zoológicas, 4:161-247.

Cabrera, A. 1961b. Catálogo de los mamíferos de América del Sur. II (Sirenia–Perissodactyla–Artiodactyla–Lagomorpha–Rodentia–Cetacea). Revista del Museo Argentino de Ciencias Naturales "Bernardino Rivadavia," Ciencias Zoológicas, 4:308-732.

Cabrera, A.L. 1957. La vegetación de la puna, Argentina. Revista de Investigaciones Agrícolas, Buenos Aires, Argentina, 370 pp.

Cabrera, A.L. 1976. Regiones fotogeográficas Argentinas. Enciclopedia Argentino de Agricultura y Jardinería, T. 1, fasc. 1:1-85.

Cajal, J. 1981. Estudios preliminares sobre el área de acción en marsupiales (Mammalia, Marsupialia). Physis, 40:27-37.

Cajal, J. 1983a. Situación del guanaco en la República Argentina. Pp. 1-21, *in* Situación del guanaco y estrategia de la conservación de los camélidos en la Argentina. Subsecretaría de Ciencia y Tecnología, Buenos Aires.

Cajal, J. 1983b. La situación del taruca en la provincia de La Rioja, República Argentina. Subsecretaría de Ciencia y Tecnología, Buenos Aires, 15 pp.

Cajal, J. 1991. An integrated approach to the management of wild camelids in Argentina. Pp. 305-321, *in* Latin American mammalogy: history, biodiversity, and conservation (M.A. Mares and D.J. Schmidly, eds.), University of Oklahoma Press, Norman, Oklahoma, 468 pp.

Cajal, J., A.R. Reca, and J.C. Pujalte. 1983. Situación de los camélidos en las reservas de Laguna Blanca, Laguna Brava y San Guillermo. Comunicado en la XI Reunión Argentina de Ecología Villa Giardino, Córdoba. 17-23 Abril.

Carleton, M.D., and G.G. Musser. 1989. Systematic studies of oryzomyine rodents (Muridae, Sigmodontinae): A synopsis of *Microryzomys*. Bulletin of the American Museum of Natural History, 191:1-83.

Caviedes–Vidal, E., E.C. Codelia, V. Roig, and R. Doña. 1990. Facultative torpor in the South American rodent *Calomys venustus* (Rodentia: Cricetidae). Journal of Mammalogy, 71:72-75.

Contreras, J.R. 1965. Un caso de simpatría entre dos géneros de roedores de la subfamilia Caviinae. Neotropica, 11:81-83.

Contreras, J.R., and V.G. Roig. 1978. Observaciones sobre la organización social, la ecología y la estructura de los habitáculos de *Microcavia australis australis* en Ñacuñan, Provincia de Mendoza. Ecosur, 5:191-199.

Crespo, J. 1974. Comentarios sobre nuevas localidades para mamíferos de Argentina y de Bolivia. Revista del Museo Argentino de Ciencias Naturales "Bernardino Rivadavia," Ciencias Zoológicas, 11:1-31.

Dalby, P., and M.A. Mares. 1974. Notes on the distribution of the coney rat, *Reithrodon auritus*, in northwestern Argentina. American Midland Naturalist, 92:205-206.

Digilio, A.P.L., and P.R. Legname. 1966. Los árboles indígenas de la Provincia de Tucumán. Opera Liiloana XV, Instituto Miguel Lillo, Tucumán, Argentina.

Fuller, K.S., and B. Swift. 1984. Latin American wildlife trade laws. World Wildlife Fund-U.S., 354 pp.

Greegor, D.H. 1975. Renal capabilities of an Argentine desert armadillo. Journal of Mammalogy, 56:626-632.

Greegor, D.H. 1980a. Diet of the little hairy armadillo, *Chaetophractus vellerosus*, of northwestern Argentina. Journal of Mammalogy, 61:331-334.

Greegor, D.H. 1980b. Preliminary study of movements and home range of the armadillo, *Chaetophractus vellerosus*. Journal of Mammalogy, 61:334-335.

Greegor, D.H. 1985. Ecology of the little hairy armadillo *Chaetophractus vellerosus*. Pp. 397-405, *in* The evolution and ecology of armadillos, sloths, and vermilinguas (G.G. Montgomery, ed.). Smithsonian Institution Press, 451 pp.

Hershkovitz, P. 1962. Evolution of neotropical cricetine rodents (Muridae) with special reference to the phyllotine group. Fieldiana, Zoology, 46:1-524.

Hoffman, M.S. (ed.). 1992. The world almanac and book of facts. 1993. World Almanac, New York, 960 pp.

Holmberg, E.A. 1902. El guanaco. Boletín del Ministerio de Agricultura de la Nación Argentina, 2:1157-1167.

IUCN. 1982. The IUCN mammal red data book. Part 1. The threatened mammalian taxa of the Americas and the Australasian zoogeographic region (excluding Cetacea). IUCN, Gland, Switzerland, 516 pp.

IUCN. 1990. 1990 ICUN Red list of threatened animals. IUCN, Gland, Switzerland, 228 pp.

Llanos, A.C., and J.A. Crespo. 1952. Ecología de la vizcacha (*Lagostomus maximus maximus* Blainv.) en el nordeste de la provincia de Entre Rios. Revista de Investigaciones Agrícolas, 6:289-378.

Mabry, T.J., J.H. Hunziker, and D.R. DiFeo, Jr. (eds.). 1977. Creosote bush: Biology and chemistry of *Larrea* in New World deserts. Dowden, Hutchinson & Ross, Inc., Stroudsburg, Pennsylvania, 284 pp.

Mares, M.A. 1973. Climates, mammalian communities, and desert rodent adaptations: An investigation into evolutionary convergence. Unpubl. Ph.D. Thesis, University of Texas, Austin, 345 pp.

Mares, M.A. 1975a. Observations of Argentine desert rodent ecology, with emphasis on water relations of *Eligmodontia typus*. Pp. 155-175, *in* Rodents in desert environments (I. Prakash and P.K. Ghosh, eds.), B.V. Junk, The Hague, The Netherlands, 624 pp.

Mares, M.A. 1975b. South American mammal zoogeography: Evidence from convergent evolution of desert rodents. Proceedings of the National Academy of Sciences, USA, 72:1702-1706.

Mares, M.A. 1976. Convergent evolution of desert rodents: Multivariate analysis and zoogeographic implications. Paleobiology, 2:39-64.

Mares, M.A. 1977a. The comparative water relations and habitat requirements of three South American phyllotine rodents. Journal of Mammalogy, 58:514-520.

Mares, M.A. 1977b. Water independence in a South American non-desert rodent. Journal of Mammalogy, 58:653-656.

Mares, M.A. 1977c. Water economy and salt balance in a South American desert rodent, *Eligmodontia typus*. Comparative Biochemical Physiology, 56A:325-332.

Mares, M.A. 1977d. Aspects of the water balance of *Oryzomys longicaudatus* from Northwest Argentina. Comparative Biochemistry and Physiology, 57A:237-238.

Mares, M.A. 1980. Convergent evolution among desert rodents: A global perspective. Bulletin of Carnegie Museum of Natural History, 16:1-51.

Mares, M.A. 1988. Reproduction, growth, and development in Argentine gerbil mice, *Eligmodontia typus*. Journal of Mammalogy, 69:852-854.

Mares, M.A. 1992. Neotropical mammals and the myth of Amazonian biodiversity. Science, 255:976-979.

Mares, M.A., and J.K. Braun. 1996. A new species of phyllotine rodent, genus *Andalgalomys* (Muridae: Sigmodontinae) from Argentina. Journal of Mammalogy, 77:928-941.

Mares, M.A., and A.C. Hulse. 1977. Patterns of some vertebrate communities in creosote bush deserts. Pp. 209-226, *in* Creosote bush: Biology and chemistry of *Larrea* in New World deserts (T.J. Mabry, J.H. Hunziker, and D.R. DiFeo, Jr., eds.). Dowden, Hutchinson & Ross, Inc., Stroudsburg, Pennsylvania, 284 pp.

Mares, M.A., and T.E. Lacher, Jr. 1987. Ecological, morphological, and behavioral convergence in rock-dwelling mammals. Current Mammalogy, 1:307-348.

Mares, M.A., and R.A. Ojeda. 1982. Patterns of diversity and adaptation in South American hystricognath rodents. Pp. 393-432, *in* Mammalian biology in South America (M.A. Mares and H.H. Genoways, eds.). Pymatuning Laboratory of Ecology, Special Publication No. 6, Linesville, Pennsylvania, 539 pp.

Mares, M.A., and R.A. Ojeda. 1984. Faunal commercialization and conservation in South America. BioScience, 34:580-584.

Mares, M.A., W.F. Blair, F.A. Enders, D.H. Greegor, J. Hunt, A.C. Hulse, D. Otte, R. Sage, and C. Tomoff. 1977a. The strategies and community patterns of desert animals. Chap. 5, pp. 107-163, *in* Convergent evolution in warm deserts (G.H. Orians and O.T. Solbrig, eds.). Dowden, Hutchison & Ross, Inc., Stroudsburg, Pennsylvania, 333 pp.

Mares, M.A., F. Enders, J. Kingsolver, J. Neff, and B.B. Simpson. 1977b. *Prosopis* as a niche component. Pp. 123-149, *in* Mesquite: Its biology in two desert ecosystems (B.B. Simpson, ed.). Dowden, Hutchison & Ross, Inc., Stroudsburg, Pennsylvania, 250 pp.

Mares, M.A., J. Morello, and G. Goldstein. 1985. The Monte Desert and other subtropical semiarid biomes of Argentina, with comments on their relation to North American arid areas. Pp. 203-237, *in* Ecosystems of the world (M. Evanari and I. Noy-Meir, eds.). Elsevier, The Netherlands, 365 pp.

Mares, M.A., R.A. Ojeda, and R.M. Barquez. 1989. Guide to the mammals of Salta Province, Argentina (Guía de los mamíferos de la Provincia de Salta, Argentina). University of Oklahoma Press, Norman, Oklahoma, 303 pp.

Mares, M.A., R.A. Ojeda, and M.P. Kosco. 1981. Observations on the distribution and ecology of the mammals of Salta Province, Argentina. Annals of Carnegie Museum, 50:151-206.

Massoia, E. 1976/77. Sobre la identidad del holotip de *Graomys hypogaeus* Cabrera, 1934 (Mammalia-Rodentia-Cricetidae). Revista de Investigaciones Agropecuarias, Instituto Nacional de Tecnología Agropecuaria, Buenos Aires, Serie 5, Patoloía Vegetal, 13:15-20.

Minoprio, J.D.L. 1945. Sobre el *Chlamyphorus truncatus* Harlan. Acta Zoológica Lilloana, 3:5-58.

Morello, J. 1958. La provincia fitogeográfica del Monte. Opera Liiloana 2, Instituto Miguel Lillo, Tucumán, Argentina, 155 pp.

Morello, J., and J. Adámoli. 1968. Las grandes unidades de vegetación y ambiente del Chaco Argentino. Primera parte: Objetivos y metodología. INTA. La Vegetación de la República Argentina, 10:1-125.

Morello, J., and C.S. Toledo. 1959a. El bosque chaqueño. I. Paisaje primitivo, paisaje natural y paisaje cultural en el oriente de Salta. Revista de Agronomía de Noroeste Argentino, 3:5-81.

Morello, J., and C.S. Toledo. 1959b. El bosque chaqueño. II. La ganadería y el bosque en el oriente de Salta. Revista de Agronomía de Noroeste Argentino, 3:209-258.

Myers, P. 1977. A new phyllotine rodent (Genus *Graomys*) from Paraguay. Occasional Papers Museum of Zoology, University of Michigan, 676:1-7.

Myers, P. 1989. A preliminary revision of the *varius* group of *Akodon* (*A. dayi*, *dolores*, *molinae*, *neocenus*, *simulator*, *toba*, and *varius*). Pp. 5-54, *in* Advances in Neotropical mammalogy (K.H. Redford and J.F. Eisenberg, eds.). The Sandhill Crane Press, Inc., Gainesville, Florida, 614 pp.

Myers, P., and M.D. Carleton. 1981. The species of *Oryzomys* (*Oligoryzomys*) in Paraguay and the identity of Azara's "Rat sixieme ou Rat a Tarse Noir." Miscellaneous Publications, Museum of Zoology, University of Michigan, 161:1-41.

Myers, P., J.L. Patton, and M.F. Smith. 1990. A review of the *boliviensis* group of *Akodon* (Muridae: Sigmodontinae), with emphasis on Perú and Bolivia. Miscellaneous Publications, Museum of Zoology, University of Michigan, 177:1-104.

Ojeda, R.A., and M.A. Mares. 1982. Conservation of South American mammals: Argentina as a paradigm. Pp. 505-521, *in* Mammalian biology in South America (M.A. Mares and H.H. Genoways, eds.). Pymatuning Laboratory of Ecology, Special Publication No. 6, Linesville, Pennsylvania, 539 pp.

Ojeda, R.A., and M.A. Mares. 1984. La degradación de los recursos naturales y la fauna silvestre en la Argentina. Interciencia, 9:21-26.

Ojeda, R.A., and M.A. Mares. 1989. A biogeographic analysis of the mammals of Salta Province, Argentina: Patterns of species assemblage in the Neotropics. Special Publications, The Museum, Texas Tech University, 27:1-66.

Olds, N. 1988. A revision of the genus *Calomys* (Rodentia: Muridae). Unpubl. Ph.D. dissertation, City University of New York, 372 pp.

Olds, N., S. Anderson, and T.L. Yates. 1987. Notes on Bolivian mammals 3: A revised diagnosis of *Andalgalomys* (Rodentia, Muridae) and the description of a new subspecies. American Museum Novitates, 2890:1-7.

Olrog, C.C. 1958. Notas mastozoológicas sobre la colección del Instituto Miguel Lillo (Tucumán). Acta Zoológica Lilloana, 16:91-95.

Olrog, C.C. 1976. Sobre mamíferos del noroeste argentino. Acta Zoológica Lilloana, 32:5-12.

Olrog, C.C., and M. Lucero. 1981. Guía de los mamíferos argentinos. Ministerio de Cultura y Educación, Fundación Miguel Lillo, Tucumán, Argentina, 151 pp.

Orians, G.H., and O.T. Solbrig (eds.). 1977. Convergent evolution in warm deserts. Dowden, Hutchison and Ross, Inc., Stroudsburg, Pennsylvania, 333 pp.

Orians, G.H., R.G. Cates, M.A. Mares, A. Moldenke, J. Neff, D.F. Rhoades, M.L. Rosenzweig, B.B. Simpson, J.C. Schultz, and C.S. Tomoff. 1977. Resource utilization systems. Pp. 164-224, *in* Convergent evolution in warm deserts (G.H. Orians and O.T. Solbrig, eds.). Dowden, Hutchison & Ross, Inc., Stroudsburg, Pennsylvania, 333 pp.

Papadakis, J. 1974. Ecología: Posibilidades agropecuarias de las provincias Argentinas. Enciclopedia Argentino de Agricultura y Jardinería, T. II, fasc. 3:1-86.

Pearson, O.P. 1948. Life history of the mountain viscacha in Perú. Journal of Mammalogy, 29:345-374.

Pearson, O.P. 1958. A taxonomic revision of the rodent genus *Phyllotis*. University of California Publications in Zoology, 56:391-496.

Pearson, O.P. 1959. Biology of the subterranean rodents, *Ctenomys*, in Perú. Memorias del Museo de Historia Natural "Javier Prado," 9:1-56.

Pearson, O.P., and J.L. Patton. 1976. Relationships among South American phyllotine rodents based on chromosomal analysis. Journal of Mammalogy, 57:339-350.

Piciucci de Fonalat, A.M., G. Scrocchi, M.R. Aguirre, and J.L. Orgeira. 1985. Aportes a la morfometría de *Graomys griseoflavus* (Rodentia: Cricetidae). Historia Natural, 5:225-230.

Redford, K.H., and J.F. Eisenberg. 1992. Mammals of the Neotropics. The southern cone. Vol. 2. Chile, Argentina, Uruguay, Paraguay. The University of Chicago Press, Chicago, 430 pp.

Reig, O.A. 1978. Roedores cricetidos del Plioceno Superior de la Provincia de Buenos Aires (Argentina). Publicaciones del Museo Municipal de Ciencias Naturales de Mar del Plata, "Lorenzo Scaglia" 2:164-190.

Reig, O.A. 1987. An assessment of the systematics and evolution of the Akodontini, with the description of new fossil species of *Akodon* (Cricetinae: Sigmodontinae). Pp. 347-399, *in* Studies in Neotropical mammalogy. Essays in honor of Philip Hershkovitz (B.D. Patterson and R.M. Timm, eds.). Field Museum of Natural History, 506 pp.

Roig, V.G. 1991. Desertification and distribution of mammals in the southern cone of South America. Pp. 239-279, *in* Latin American mammalogy: History, biodiversity, and conservation (M.A. Mares and D.J. Schmidly, eds.). University of Oklahoma Press, Norman, Oklahoma, 468 pp.

Rood, J.P. 1970. Ecology and social behavior of the desert cavy (*Microcavia australis*). The American Midland Naturalist, 83:415-454.

Rood, J.P. 1972. Ecological and behavioural comparisons of three genera of Argentine cavies. Animal Behaviour Monographs, 5:1-83.

Sanborn, C.C. 1947. The South American rodents of the genus *Neotomys*. Fieldiana: Zoology, 31:51-57.

Schaller, G.B. 1983. Mammals and their biomass on a Brazilian ranch. Arquivos de Zoologia, 31:1-36.

Shump, K.A., and A.U. Shump. 1982. *Lasiurus borealis*. Mammalian Species, 183:1-6.

Simpson, B.B. (ed.). 1977. Mesquite: Its biology in two desert ecosystems. Dowden, Hutchison and Ross, Inc., Stroudsburg, Pennsylvania, 250 pp.

Spotorno, A.E., and L.I. Walker. 1983. Análisis electroforético y biométrico de dos especies de *Phyllotis* en Chile Central y sus híbridos experimentales. Revista Chilena de Historia Natural, 56:51-59.

Taber, A., and D.W. Macdonald. 1984. Scent dispensing papillae and associated behaviour of the mara, *Dolichotis patagonum* (Rodentia: Caviomorpha). Journal of Zoology, 203:298-

301.

Tate, G.H.H. 1933. A systematic revision of the marsupial genus *Marmosa*. Bulletin of the American Museum of Natural History, 66:1-250.

Thomas, O. 1897. On some small mammals from Salta, N. Argentina. Annals and Magazine of Natural History, (6), 20:214-218.

Thomas, O. 1902. On mammals collected at Cruz del Eje, central Córdoba. Annals and Magazine of Natural History, (7), 9:237-245.

Thomas, O. 1918. A new species of *Eligmodontia* from Catamarca. Annals and Magazine of Natural History, (9), 2:482-484.

Thomas, O. 1919a. On some small mammals from Catamarca. Annals and Magazine of Natural History, (9), 3:115-118.

Thomas, O. 1919b. On small mammals from "Otro Cerro," northeastern Rioja, collected by Sr. L. Budin. Annals and Magazine of Natural History, (9), 3:489-500.

Thomas, O. 1920a. On mammals from near Tinogasta, Catamarca, collected by Sr. Budin. Annals and Magazine of Natural History, (9), 6:117-120.

Thomas, O. 1920b. New species of *Reithrodon*, *Abrocoma*, and *Scapteromys* from Argentina. Annals and Magazine of Natural History, (9), 5:473-478.

Thomas, O. 1921. On cavies of the genus *Caviella*. Annals and Magazine of Natural History, (9), 7:445-448.

Thomas, O. 1925. A new genus of cavy from Catamarca. Annals and Magazine of Natural History, (9), 15:418-420.

Thomas, O. 1926. The Spedan-Lewis South American Exploration. II. On mammals collected in the Tarija Department, southern Bolivia. Annals and Magazine of Natural History, (9), 17:318-328.

Vervoorst, F. 1951. Resultados de un viaje a la cuenca de la Laguna Verde (Tinogasta, Catamarca). III. Pp. 61-67, *in* Observaciones sobre la vegetación entre Tinogasta y la cuenca. Actas de la XV Semana de Geografía.

Vervoorst, F. 1954. El bosque de algarrobos de Pipanaco (Catamarca). Ph.D. thesis, Universidad de Buenos Aires.

Vervoorst, F. 1966. Guía para la excursión de la Sociedad Argentina de Botánica por las provincias de Tucumán y Catamarca, conmemorando el sesquicentenario de la Declaración de Independencia. Unpubl. ms, 32 pp.

Vervoorst, F. 1967. La vegetación de la República Argentina. VII. Las comunidades vegetales de la depresión del salado (Provincia de Buenos Aires). INTA, Instituto de Botánica Agrícola, Serie Fitogeográfica, 7:1-257.

Vervoorst, F. 1982. Noroeste. Pp. 9-24, *in* Conservación de la vegetación natural en la República Argentina. Simposio XVIII Jornadas de Botánica, Serie Conservación de la Naturaleza, Fundación Miguel Lillo, Tucumán, Argentina, 127 pp.

Walker, L.I., A.E. Spotorno, and J. Arrau. 1984. Cytogenetic and reproductive studies of two nominal subspecies of *Phyllotis darwini* and their experimental hybrids. Journal of Mammalogy, 65:220-230.

Wetzel, R. 1982. Systematics, distribution, ecology, and conservation of South American edentates. Pp. 345-375, *in* Mammalian biology in South America (M.A. Mares and H.H. Genoways, eds.). Pymatuning Laboratory of Ecology, Special Publication No. 6, Linesville, Pennsylvania, 539 pp.

Wetzel, R. 1985. Taxonomy and distribution of armadillos, Dasypodidae. Pp. 23-46, *in* The evolution and ecology of armadillos, sloths, and vermilinguas (G. Montgomery, ed.). Smithsonian Institution Press, Washington, DC, 451 pp.

Williams, D.F., and M.A. Mares. 1978. A new genus and species of phyllotine rodent (Mammalia: Muridae) from northwestern Argentina. Annals of Carnegie Museum, 47:192-221.

Wilson, D.E., and D.M. Reeder (eds.). 1993. Mammal species of the world. A taxonomic and geographic reference. Second Edition. Smithsonian Institution Press, Washington, DC, 1206 pp.

Yepes, J. 1929a. Notas sobre la distribución geográfica del "pichi ciego menor" (*Chlamyphorus truncatus*) y piche llorón (*Chaetophractus vellerosus*) Physis, 9:439-446.

Yepes, J. 1929b. Notas sobre la distribución geográfica del "pichi ciego menor" (*Chlamyphorus truncatus*). Physis, 9:2-4.

Yepes, J. 1932. Las formas geográficas del "pichi ciego menor" (*Chlamyphorus truncatus*

Harl.). Physis, 11:9-18.

Yepes, J. 1933. Nuevos roedores para la fauna Argentina. Revista Chilena de Historia Natural, 37:46-49.

Yepes, J. 1935. Consideraciones sobre el género *Andinomys* (Cricetinae) y descripción de una forma nueva. Anales del Museo Argentino de Ciencias Naturales "Bernardino Rivadavia," 38:333-348.

Gazetteer of Localities

1 Aconquija–27° 13', 66° 08'
2 Agua de Dionisio–27° 17', 66° 43'
3 Amaicha–26° 36', 65° 55'
4 Ambato–28° 30', 66° 10'
5 Andalgalá–27° 36', 66° 20'
6 Antofagasta de la Sierra–26° 04', 67° 25'
Balneario El Caolín (see Chumbicha)
7 Belén–27° 39', 67° 02'
Bolsón de Pipanaco (see Salar de Pipanaco)
8 Campo de las Talibas–27° 12', 66° 57'
9 Catamarca–28° 28', 65° 47'
10 Cazadero Grande–27° 25', 68° 11'
11 Cerro Durazno–27° 16', 66° 32'
12 Cerro El Manchao–28° 16', 66° 02'
13 Chaquiago–27° 32', 66° 21'
14 Choya–27° 32', 66° 24'
15 Chumbicha–28° 52', 66° 14'
16 Concepción–28° 41', 66° 04'
17 Copacabana–28° 12', 67° 29'
18 Corral Quemado–27° 05', 66° 54'
19 Cortadera–27° 33', 68° 09'
Cuesta de La Sébila (see La Sébila)
20 Cuesta de Zapata–27° 55', 67° 23'
21 Cuesta del Clavillo–27° 13', 66° 06'
22 Cuesta del Medanito–27° 22', 66° 08'
Dique El Potrero (see El Potrero)
23 El Alamito–27° 28', 66° 02'
24 El Bolsón–27° 55', 65° 53'
25 El Cajón–26° 26', 66° 16'
26 El Desmonte–26° 54', 66° 06'
27 El Pajonal–28° 22', 66° 18'
28 El Potrero–27° 32', 66° 20'
29 El Pucará–27° 43', 66° 01'
30 El Rodeo–28° 13', 65° 52'
31 Estancia Gracian–28° 08', 65° 40'
32 Famatina–28° 55', 67° 31'
33 Farallón Negro–27° 19', 66° 44'
34 Hualfín–27° 14', 66° 50'
35 Hwy 9 on Hwy 18–27° 48', 65° 48'
Hwy 60 and 38 (see Chumbicha)
Hwy 62 and Amanao Road (see Andalgalá)
36 Icaño–28° 54', 65° 19'
37 La Atravesada–27° 21', 65° 59'
38 La Carrera–28° 22', 65° 43'
39 La Chilca–27° 38', 66° 12'
40 La Ciénaga–27° 28', 66° 59'
41 La Cienaguita–28° 10', 67° 40'
42 La Guardia–29° 33', 65° 27'
43 La Merced–28° 10', 65° 41'
44 La Puerta–28° 10', 65° 48'
45 La Puntilla–28° 08', 67° 30'
46 La Sébila–28° 43', 66° 22'
47 Las Cuevas–27° 02', 66° 39'
48 Las Estancias–27° 28', 66° 02'
49 Las Higuerillas–27° 47', 65° 51'
50 Las Juntas, 2100 m–28° 08', 65° 54'
51 Laguna Blanca–26° 37', 66° 57'
52 Laguna Helada–27° 05', 67° 09'
La Mendicta (could not be located)
53 La Salamanca–27° 13', 66° 20'
Las Rosas (could not be located)
Loma Morada (could not be located; possibly Cerro Morado)
54 Londres–27° 43', 67° 07'
55 Los Castillos–27° 58', 65° 48'
Los Médanos (could not be located)
56 Los Nacimientos–27° 11', 66° 44'
57 Los Varela–27° 57', 65° 52'
58 Minas Capillitas–27° 20', 66° 25'
59 Mollecito–27° 47', 66° 12'
60 Nevados del Aconquija–27° 13', 66° 08'
61 Otro Cerro–28° 44', 66° 17'
62 Paclín–28° 07', 65° 40'
63 Pasto Ventura–26° 43', 67° 13'
64 Pastos Largos–27° 39', 68° 09'
65 Poman–28° 24', 66° 13'
66 Potrero–27° 32', 66° 20'
Potrero River Dike (see El Potrero)
Puesto Laguna Blanca (see Laguna Blanca)
Puntilla Tinogasta (see La Puntilla)
67 Quirós–28° 47', 65° 07'
68 Recreo–29° 16', 65° 04'
69 Río Andalgalá–27° 48', 66° 17'
70 Río Amanao–27° 53', 66° 22'
71 Río Charquiadero–27° 25', 66° 05'
Río Cañada (could not be located)
72 Río Potrero–27° 23', 65° 58'
73 Río San Pablo–28° 42', 66° 04'
74 Río Vallecito–27° 20', 66° 00'
75 Salar de Antofalla–25° 44', 67° 45'
76 Salar de Pipanaco–28° 07', 66° 25'

77 San Antonio de la Falda–28° 24', 65° 44'
78 San Antonio de la Paz–28° 56', 65° 06'
79 San Francisco–26° 36', 66° 11' (it is possible that one or both of these localities could be Cerro San Francisco at 26° 53', 68° 15')
San José (could not be located)
80 Santa María–26° 41', 66° 02'
81 Santa Rosa–28° 05', 65° 15'
82 Saujil–28° 11', 66° 14'
83 Sierra de Humaya–27° 50', 65° 59'
84 Singuil–27° 49', 65° 52'
85 Tinogasta–28° 04', 67° 34'
86 Tolar–28° 02', 68° 15'
87 Totoralejos–29° 38', 64° 51'

Michael A. Mares, Oklahoma Museum of Natural History and Department of Zoology, University of Oklahoma, Norman, Oklahoma 73019; Ricardo A. Ojeda, IADIZA, Centro Regional de Investigaciones Científicas y Técnicas (CRICYT), Zoología y Ecología Animal, C.C. 507, 5500 Mendoza, Argentina; Janet K. Braun, Oklahoma Museum of Natural History, University of Oklahoma, Norman, Oklahoma 73019; and Ruben M. Barquez, Instituto Miguel Lillo, Universidad Nacional de Tucumán, 4400 Tucumán, Argentina

Similarity Coefficients and Relationships of Wisconsin-Age Faunas New Mexico and Trans-Pecos Texas

Arthur H. Harris

Abstract

Presence–absence coefficients of similarity were calculated between 23 fossil vertebrate faunas (predominantly mammals) from New Mexico and the Guadalupe Mountains of Trans-Pecos Texas, and also between these faunas and six modern regional faunas thought to possibly represent approximate modern analogues. These quantitative results are compared to past qualitative assessments with variable results. The coefficients of similarity often agree with qualitative analyses but lack the sensitivity of qualitative methods; they are potentially important adjuncts to qualitative analysis.

INTRODUCTION

The vertebrate faunas of present-day New Mexico and the adjacent Guadalupe Mountains of Trans-Pecos Texas display significant geographic differences among themselves, reflecting varying origins, different histories, and the region's varied climate and topography. Without doubt, past faunas also varied from place to place and, within a single area, through time. Interpretation of past patterns of variation, however, is particularly difficult due to spotty distribution of sites in space and time, in addition to the normal difficulties of working with fossil faunas: biases in deposition, preservation, and recovery; the difficulty of sure identification based on fragmentary material; and mixing of faunas of different ages.

The most common approach to interpretation of Pleistocene fossil faunas has been qualitative. That is, using a knowledge of present-day ecologic distributions and environmental limiting factors of extant members of the fauna and applying these data to individual taxa within the fossil faunas. The results are taken to characterize the environmental conditions at the time of faunal deposition if all treated taxa are consistent in their requirements; inconsistencies not explainable by accepted theory require further investigation or are taken as indications that the fauna is not a natural unit (e.g., is a mixed fauna or encompasses an extended time span that includes significant environmental change).

Qualitative evaluation of similarities between fossil faunas and between fossil and modern faunas has long been applied to Late Pleistocene cave faunas of the region. For example, Stearns (1942) suggested a major lowering of life zones, allowing marmots to descend from the high mountains of northern New Mexico to occupy New La Bajada Hill southwest of Santa Fe; later, Murray (1957) implied a relationship between southeastern New Mexico and the highlands of northern New Mexico based on the fauna of Burnet Cave. Similar efforts directed toward other faunas (discussed further below) have compared various Pleistocene faunas to the modern faunas of northeastern New Mexico, southeastern Idaho, or to major mixtures of more than one area of similarity (Harris, 1985, 1987, 1989; Slaughter, 1975). In terms of chronology, Harris (1977) attempted to place a series of New Mexican sites into segments of time based on faunal similarities.

Life Among the Muses: Papers in Honor of James S. Findley. Edited by T.L. Yates, W.L. Gannon & D.E. Wilson.

Although such qualitative efforts may be valuable, there is danger that excessive subjectivity may undermine their usefulness. There also are dangers associated with blind acceptance of quantitative methodologies, but a combination of different approaches may clarify problems not apparent when only one pathway is taken. In this study, several quantitative measures of faunal similarity were applied to a group consisting of major faunas from the terminal age of the Late Pleistocene (the Wisconsinan age), plus modern faunas of areas deemed possible approximate analogues of Wisconsinan faunas; the results are compared with published qualitative evaluations.

MATERIALS AND METHODS

Measurements of Similarity

A number of objective measures of similarity based on presence–absence data for modern and fossil faunas have been proposed—and criticized (e.g., Cheetham and Hazel, 1969; Hazel, 1970; Raup and Crick, 1979). Three measures of similarity widely used in paleontology are the coefficients of Jaccard, Dice, and Simpson (Fallaw, 1979). Where C equals the number of taxa in common and N_1 and N_2 the number of taxa in the smaller and larger faunas, respectively, the Jaccard Coefficient is $C/(N_1+N_2-C)$, the Dice Coefficient is $2C/(N_1+N_2)$, and the Simpson Coefficient is C/N_1. Of the three, the Simpson Coefficient is claimed to best compare faunas of unequal size (Fallaw, 1979), although Raup and Crick (1979) point out serious potential shortcomings to this coefficient. All three coefficients vary from 0 (no taxa in common) to 1 (complete similarity); in general, when based on the same faunas, the Simpson Coefficient is larger than the other two, and the Dice Coefficient is larger than the Jaccard.

There are, however, several different classes of data obtainable from such coefficients. One thing is the degree of similarity between the various fossil faunas. A high degree of similarity between two faunas indicates similar climatic and historic factors, while low similarity indicates differences requiring explanation. The Simpson Coefficient seems best for evaluating similarity between faunas that vary widely in size due to sampling error, as commonly is the case with fossil faunas.

A second important datum is the proportion of a fossil fauna present in a given modern fauna. The higher the proportion, the more likely that critical environmental features of the fossil and modern faunas were similar. Again, the Simpson Coefficient seems best since the smaller fauna (usually the fossil fauna) governs the number of potential matches.

How well a fossil fauna matches an entire modern fauna also is pertinent. Once the number of taxa in common is calculated, this becomes, in large part, a question of faunal sizes: a small modern fauna with a given number of taxa in common with an ancient fauna is more similar to the latter than is a large modern fauna with the same number of shared taxa (conversely, a small fossil fauna with a given number of taxa in common with a modern fauna is more similar to that fauna than is a large fossil fauna with the same number of shared taxa). The Simpson Coefficient is inappropriate for this measure since it is governed by the size of the smaller fauna. Either the Jaccard or Dice Coefficient is suitable, though the effect of unequal sample sizes adversely affects the results. In the material considered here, there is little difference in the results of the two coefficients other than of scale, and the Dice Coefficient has been used.

Two presence/absence matrices were prepared (data available from the author). In one, 124 taxa were utilized for examination with the Dice Coefficient; these taxa included those mammals expected to be identifiable with reasonable certainty, plus taxa meeting criteria 2 and 4, below. The other matrix, for investigating the first two questions, included a more restricted set of taxa (93) meeting the following guidelines: 1) taxa common to all faunas were omitted; 2) large extinct forms that probably were ubiquitous (e.g., horses, camels, mammoths) were not used; 3) only taxa occurring in one or more fossil faunas were included, thus omitting modern-faunal elements that would not show up in the Simpson Coefficient; and 4) several non-mammalian taxa (e.g., *Gopherus*) were used that could be expected to be identifiable with fair certainty and were notably out of their range in some or all of the fossil faunas.

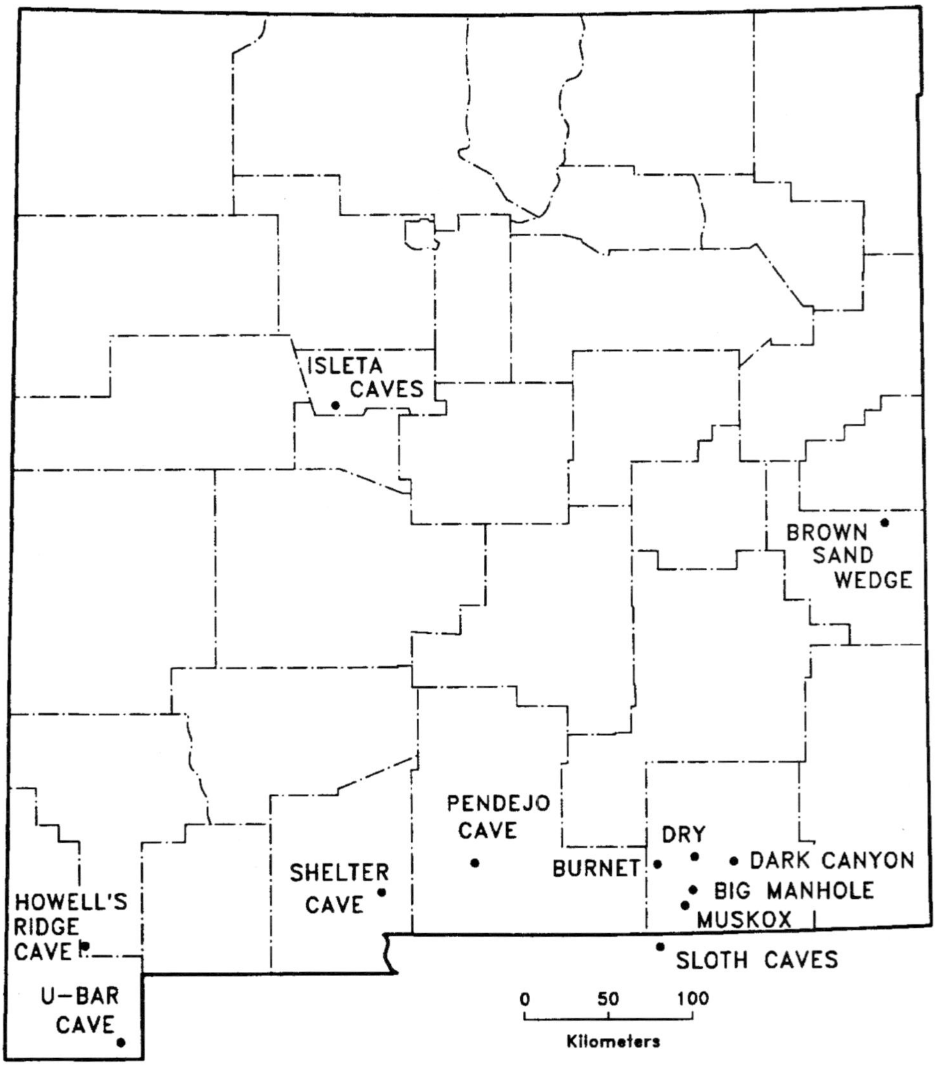

Fig. 1. Locations of Wisconsinan fossil faunas.

Similarity coefficients were set to zero between modern faunas so that these faunas were compared only to the fossil faunas, not to each other. Clustering and dendrogram preparation, utilizing NTSYS-PC (Rohlf, 1990), was by the sequential agglomerative hierarchical non-overlapping method (SAHN) using unweighted pair-group arithmetic averages (UPGMA).

Faunas

Locations of fossil sites are shown in Fig. 1. Modern areas for comparison include regions suggested or reasonable as approximate modern analogues for Wisconsinan faunas. Each modern area is large enough to include a variety of habitats including, in the West, a large elevational range. Thus, modern faunal comparisons are regional comparisons, not comparisons with biotic communities. This strategy allows comparison with fossil sites where deposition of prey animals from some distance away from the site may have occurred and encompasses the effects of minor range fluctuations due to short-term climatic trends. Pertinent details of each fauna follow, including entry to the literature for the fossil sites (abbreviated names as used in Table 1; N is the number of taxa used in this study):

E KANSAS: Modern fauna of the eastern four tiers of counties of Kansas (N = 46).

WYOMING: Modern fauna of southeastern Wyoming: Carbon, Laramie, Goshen, Platte, and Albany counties (N = 76).

NE NM: Modern fauna of northeastern New Mexico, including those parts of Union, Harding, and Colfax counties east of the Canadian River (N = 52).

SE NM: Modern fauna of southeastern New Mexico south of the pass between the Gallinas and Jicarilla mountains; includes most of Chaves, Lea, Lincoln, Otero, and Eddy counties and the southern half of Roosevelt County (N = 71).

NO NM: Modern fauna of north–central New Mexico, including Rio Arriba, Taos, and Colfax counties; northern portions of Sandoval and Santa Fe counties; and the western halves of Mora and San Miguel counties (N = 85).

SE ID: Modern fauna of southeastern Idaho (N = 72). Based on Caribou, Bear Lake, Bannock, Franklin, and Oneida counties and the southeastern two-thirds of Power County.

FULL: Full-glacial fauna (N = 35) from Animal Fair, Grids A and B, Levels 1-4, Dry Cave, Eddy Co., NM, 1,280 m. Harris (1989, 1993b).

C PARLOR: Charlies Parlor, Dry Cave. This is a small fauna (N = 17) biased toward larger forms. A radiocarbon date of 15,030 ± 210 B.P. (before present) is from near the base of the deposits; the date, based on bone collagen, may be somewhat too recent.

HARRIS P: Harris' Pocket (N = 30). This is a Dry Cave fauna radiocarbon dated on rodent feces at 14,470 ± 250 B.P. Harris (1970, 1993b); (Holman, 1970).

BISON CH: Bison Chamber, Dry Cave (N = 23). By interpolation with dated strata, this fauna is believed to be < 14,470 and > 10,730 ^{14}C-years B.P. Harris (1970, 1993b); Holman (1970).

LESS 13: Several pre-Holocene Dry Cave sites dating to < 13,000 B.P. (N = 43); includes TTII, Human Corridor, Camel Room, Boulder Room, and Stalag 17. Harris (1989, 1993b).

HOWELLS: Howell's Ridge Cave, Grant Co., NM, 1,675 m. A mixed, late Wisconsinan–Holocene fauna (N = 28) encompassing a span from > 13,600 radiocarbon years B.P. to the present. Harris (1985; 1993b); Howard (1962); Van Devender and Worthington (1977).

UB18 20: U-Bar Cave full glacial, approximately 18-20 ka (thousands of years ago), Hidalgo Co., NM, 1,570 m (N = 26). All Pleistocene U-Bar Cave dates are based on radiocarbon dating of bulk samples of matrix; the organic portion of the matrix is believed to consist largely of decomposed bat guano and other organic debris. Harris (1989).

UB15 18: U-Bar Cave, approximately 15-18 ka (N = 32). Harris (1989).

UB14 15: U-Bar Cave, approximately 14-15 ka (N = 45). Harris (1989).

UB13 14: U-Bar Cave, approximately 13-14 ka (N = 23). Harris (1989).

UBAR MID: A number of radiocarbon dates are available, ranging from 26,150 ± 1450 to 35,890 ± 2640 B.P. Harris (1987). N = 33.

DRY EARL: Dry Cave faunas (N = 33) from University of Texas at El Paso (UTEP) localities 1 and 17 (Lost Valley), 2 and 5 (Sabertooth Camel Maze), and 26 and 27 (Rm of the Vanishing Floor). Radiocarbon dates on bone carbonates (which are considered unreliable for dating) run from 25,160 ± 1730 (Locs. 2, 5) to 29,290 ± 1060 (Locs. 1, 17) and 33,590 ± 1500 (Locs. 26, 27); until recently, the faunas were considered to be later mid-Wisconsinan in age, in keeping with the radiocarbon dates and comparable to the mid-Wisconsinan U-Bar Cave fauna in age (Harris, 1987). These sites are now thought to have been deposited earlier, possibly even in the early Wisconsinan. N = 33.

PEND MID: Middle Wisconsinan faunas from Pendejo Cave, Otero Co., NM, 1,470 m (N = 43). Radiocarbon dates range from about 26 to > 55 ka. The fauna is currently under study and further refinements are expected. Harris (1991).

ISLETA 1: Isleta Cave No. 1, Bernalillo Co., NM, 1,716 m. This is a mixed late Wisconsinan–Holocene fauna (N = 27). Two ^{14}C-dates on bone collagen are 17,240 ± 260 (AA 1208) and 16,430 ± 205 (AA 1209). Harris (1993b); Harris and Findley (1964).

Table 1. Simpson Coefficient similarity matrix comparing all fossil faunas with each other.

		A	B	C	D	E	F	G	H	I	J	K	L	M	N	O	P	Q	R	S	T	U	V	W
FULL	A	1.00																						
C PARLOR	B	0.82	1.00																					
HARRIS P	C	0.83	0.59	1.00																				
BISON CH	D	0.74	0.59	0.74	1.00																			
LESS 13	E	0.77	0.94	0.77	0.96	1.00																		
HOWELLS	F	0.39	0.47	0.39	0.48	0.61	1.00																	
UB18 20	G	0.35	0.35	0.35	0.44	0.58	0.46	1.00																
UB15 18	H	0.44	0.59	0.40	0.44	0.66	0.50	0.89	1.00															
UB14 15	I	0.60	0.65	0.57	0.52	0.61	0.75	0.89	0.94	1.00														
UB13 14	J	0.48	0.35	0.39	0.35	0.61	0.48	0.74	0.78	0.91	1.00													
UBAR MID	K	0.55	0.53	0.37	0.52	0.58	0.61	0.73	0.72	0.91	0.70	1.00												
DRY EARL	L	0.39	0.41	0.30	0.35	0.49	0.36	0.46	0.56	0.70	0.65	0.52	1.00											
PEND MID	M	0.57	0.82	0.57	0.52	0.56	0.61	0.69	0.78	0.79	0.65	0.61	0.70	1.00										
ISLETA 1	N	0.48	0.59	0.37	0.48	0.70	0.52	0.39	0.41	0.56	0.44	0.48	0.48	0.67	1.00									
ISLETA 2	O	0.52	0.47	0.48	0.35	0.56	0.44	0.32	0.48	0.56	0.35	0.44	0.40	0.68	0.72	1.00								
SHELTER	P	0.36	0.36	0.21	0.14	0.57	0.29	0.43	0.64	0.71	0.50	0.36	0.36	0.71	0.29	0.29	1.00							
BIGMAN C	Q	0.60	0.59	0.56	0.48	0.68	0.40	0.40	0.60	0.72	0.35	0.44	0.56	0.76	0.48	0.52	0.50	1.00						
BURNET C	R	0.48	0.65	0.37	0.35	0.52	0.37	0.31	0.41	0.56	0.30	0.33	0.44	0.56	0.44	0.44	0.36	0.48	1.00					
BROWN	S	0.33	0.12	0.22	0.22	0.44	0.39	0.11	0.17	0.33	0.06	0.22	0.33	0.33	0.28	0.17	0.07	0.17	0.44	1.00				
MUSKOX	T	0.54	0.53	0.57	0.70	0.51	0.50	0.50	0.47	0.62	0.48	0.49	0.61	0.54	0.56	0.36	0.36	0.60	0.52	0.39	1.00			
UP SLOTH	U	0.55	0.24	0.60	0.05	0.50	0.40	0.50	0.55	0.80	0.50	0.40	0.60	0.55	0.25	0.25	0.14	0.45	0.45	0.11	0.90	1.00		
LO SLOTH	V	0.48	0.35	0.48	0.44	0.52	0.36	0.50	0.52	0.72	0.39	0.35	0.55	0.52	0.30	0.28	0.36	0.52	0.33	0.17	0.86	0.85	1.00	
DARK CAN	W	0.72	0.59	0.72	0.61	0.72	0.48	0.32	0.48	0.72	0.35	0.40	0.52	0.80	0.48	0.52	0.29	0.52	0.48	0.28	0.56	0.40	0.36	1.00

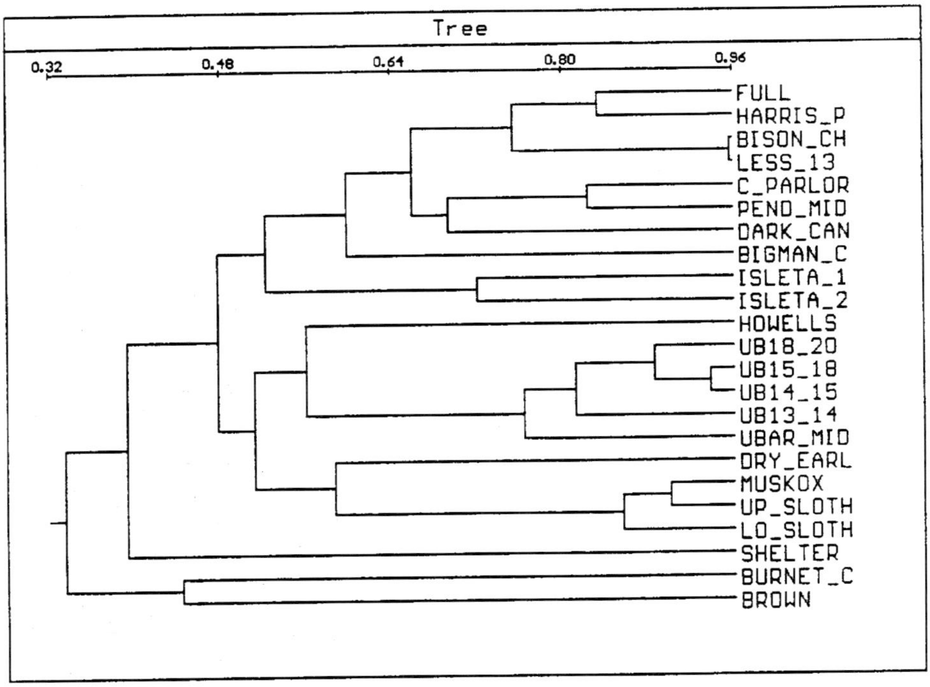

Fig. 2. Dendrogram showing similarities among Wisconsinan fossil faunas. Abbreviations as in text.

ISLETA 2: Isleta Cave No. 2, Bernalillo Co., NM, 1,716 m. Adjacent to Isleta Cave No. 1, this cave likewise has a mixed Wisconsinan-Holocene fauna (N = 25). The sole date available is 11,230 ± 140 on bone collagen (AA 1207). Harris (1993b); Harris and Findley (1964).

SHELTER: Shelter Cave, Dona Ana Co., NM, 1,435 m (N = 14). This cave was excavated in the late 1920s and the material never properly worked up. Packrat midden dates (Thompson et al., 1980) indicate that deposition likely started during the mid-Wisconsinan (31,250 ± 2200, A-2140); other dates and archaeological material indicate deposition into the Holocene and probably to the present.

BIGMAN C: Big Manhole Cave lies at 1,388 m just north of the boundary of Carlsbad Cavern National Park, Eddy Co., NM. Most material is late Wisconsinan (< 20,000 B.P.); some may be several thousands of years older. This fauna (N = 25) is currently under study. Harris (1993b).

BURNET C: Burnet Cave is a mixed late Wisconsinan-Holocene fauna west of Dry Cave in Eddy Co., NM, 1,435 m (N = 27). Harris (1985); Murray (1957); Schultz and Howard (1935).

BROWN: Brown Sand Wedge, Blackwater Loc. No. 1, Roosevelt Co., NM, ca. 1,280 m. A date of 11,170 ± 360 B.P. appears to apply to this late Wisconsinan fauna (N = 18). This is the only non-cave site considered here. Paleodrainage was to the east. Slaughter (1975).

MUSKOX: Muskox Cave, Eddy Co., NM, 1,600 m (N = 39). Dates range from 15,500 ± 1100 to 18,140 ± 200 for much of the deposit (Logan, 1981). Unfortunately, a full report with stratigraphic data has never been published.

UP SLOTH: Upper Sloth Cave, Hudspeth Co., TX, 2,000 m (N = 20). Located in the southern Guadalupe Mountains, this cave and the nearby Lower Sloth Cave were excavated by Logan. A small portion of the fauna may be Holocene. Logan and Black (1979).

LO SLOTH: Lower Sloth Cave, Hudspeth Co., TX, 2,000 m (N = 29). Some Holocene material may be present. Logan (1983).

DARK CAN: Dark Canyon Cave, Eddy Co., NM, 1,100 m (N = 25). Probably (on subjective faunal grounds) pre-pleniglacial late Wisconsinan and/or late mid Wisconsinan. Harris (1985, 1993b); Howard (1971).

RESULTS AND DISCUSSION

The Simpson Coefficient similarity matrix is shown in Table 1 and a dendrogram based on it is shown in Fig. 2; Table 2 includes the portion of the Dice Coefficient similarity matrix relating the fossil faunas to modern faunas. Pertinent points are discussed below.

Full Glacial

Southeastern Idaho, northeastern New Mexico, northern New Mexico, and Wyoming form a compact group of modern faunas varying in similarity (in that order) by the Dice Coefficient with the full-glacial fauna from 0.52 to 0.48. On qualitative criteria, Harris (1989:216) placed the area of most similarity as

Table 2. Similarity matrix (Dice Coefficient) based on 129 taxa, comparing modern faunas with fossil faunas.

	E KANSAS	WYOMING	NE NM	SE NM	NO NM	SE ID
FULL	0.365	0.482	0.500	0.425	0.495	0.518
C PARLOR	0.187	0.319	0.428	0.288	0.291	0.355
HARRIS P	0.311	0.429	0.409	0.388	0.448	0.427
BISON CH	0.257	0.340	0.394	0.375	0.366	0.333
LESS 13	0.488	0.533	0.604	0.568	0.573	0.500
HOWELLS	0.266	0.323	0.469	0.415	0.438	0.316
UB18 20	0.191	0.233	0.455	0.404	0.392	0.202
UB15 18	0.329	0.330	0.588	0.495	0.457	0.304
UB14 15	0.413	0.426	0.673	0.559	0.610	0.406
UB13 14	0.257	0.260	0.421	0.375	0.348	0.250
UBAR MID	0.350	0.327	0.581	0.452	0.487	0.377
DRY EARL	0.425	0.363	0.558	0.528	0.436	0.320
PEND MID	0.444	0.483	0.666	0.551	0.558	0.465
ISLETA 1	0.324	0.403	0.475	0.460	0.442	0.360
ISLETA 2	0.309	0.356	0.363	0.350	0.400	0.350
SHELTER	0.200	0.266	0.242	0.209	0.242	0.232
BIGMAN C	0.333	0.392	0.461	0.428	0.432	0.367
BURNET C	0.324	0.365	0.425	0.440	0.407	0.380
BROWN	0.461	0.273	0.309	0.285	0.269	0.241
MUSKOX	0.372	0.448	0.500	0.553	0.560	0.446
UP SLOTH	0.208	0.185	0.301	0.365	0.339	0.215
LO SLOTH	0.263	0.301	0.463	0.490	0.469	0.294
DARK CAN	0.277	0.372	0.461	0.367	0.414	0.346

the southeastern Idaho region "with a secondary area of sympatry in northeastern New Mexico." Thus, the two methods agree in this case (however, similarities by the Simpson Coefficient do not). Within the fossil suite, the most similar faunas are the subsequent faunas from Dry Cave and the fauna from nearby Dark Canyon Cave (0.83–0.72). Most different are the Brown Sand Wedge fauna (0.33) and the full-glacial fauna from U-Bar Cave (0.35).

Charlies Parlor

This fauna shows the most similarity among the modern faunas to that of northeastern New Mexico (0.43). Among the late Wisconsinan faunas, affinity is to the late (< 13,000 B.P.) Dry Cave fossil faunas (0.94), although there is also a rather strong similarity (0.82) to the mid-Wisconsinan Pendejo Cave fauna. The small size and bias toward larger animals, however, likely distorts relationships.

Harris' Pocket

This fauna is very similar to the full-glacial fauna of Dry Cave (0.83) and other glacial Dry Cave faunas; as might be expected, relationships to modern faunas are similar to those of the Dry Cave full-glacial faunas.

Bison Chamber

This fauna follows the general pattern of the late-Wisconsinan Dry Cave faunas in showing high relationships to northern modern faunas; however, these relationships, in general, are somewhat weaker with the greatest similarity to the northeastern New Mexico fauna (0.39). The probable approach in time (or even chronological overlap) with the < 13,000 B.P. faunas shows up in the near identity with that group (0.96).

Less Than 13,000 B.P.

These faunas show a notable change from the faunas thus far considered in affinity to modern faunas, not in decreased similarities to northern faunas (which remain high), but in increased likeness to eastern faunas (eastern Kansas, 0.49; southeastern New Mexico, 0.57). Features of the Dry Cave faunas similar to those displayed here were noted by Harris (1989:213): "Most of the extant fauna earlier characterized as cognate with that of southeastern Idaho was maintained from the pleniglacial to the end of the Pleistocene . . . . During the post-pleniglacial, southern and eastern species were added." Affinity with Dry Cave and southeastern New Mexico late-Wisconsinan faunas is high.

Howell's Ridge Cave

Similarity with the three New Mexican modern faunas is relatively high (0.42-0.47), while the northern faunas form a distinctly lower-similarity cluster (0.32). Highest value with the fossil faunas is with the geographically nearby 14-15 ka U-Bar Cave fauna (0.75). Also, there are fairly high values with the mid-Wisconsinan faunas of U-Bar and Pendejo caves; intermixture of some modern taxa into the Howell's Ridge Cave fauna undoubtedly influences these values to some degree.

U-Bar Cave, Late Wisconsinan

The northeastern New Mexico fauna is the modern fauna most similar to all of the U-Bar Cave Wisconsinan faunas; similarities of the fossil faunas to other modern faunas are low except that the U-Bar 14-15 ka fauna is rather similar to both northern (0.61) and southeastern (0.56) New Mexico faunas.

The four late Wisconsinan U-Bar Cave samples are similar among themselves (0.89–0.94) and, secondarily, to the U-Bar and Pendejo Cave mid-Wisconsinan faunas (0.65–0.91). The 14-15 ka fauna consistently shows higher similarity to all but the terminal Dry Cave faunas than do the other three and also to most other faunas in the Guadalupe Mountains area. In part, this may be due to the relatively large size of the fauna.

Harris (1989) interpreted the late-Wisconsinan faunas of U-Bar Cave as a mixture of taxa showing no strong affinity with any one modern area. Northeastern New Mexico is the only geographic area represented in this study by a modern fauna that Harris (1989:Fig 1) showed as a source area.

U-Bar Cave, Mid-Wisconsinan

The pattern of affinity with modern faunas is similar to that of other U-Bar Cave faunas. Among fossil faunas, the U-Bar Cave late-

Wisconsinan sites are most similar (topping out at 0.91 at 14-15 ka); moderate similarity with the Pendejo Cave fauna (0.70) also is displayed. The more subjective assessment of Harris (1987) was that, with the possible exception of winters being some colder, there was no major change from mid-Wisconsinan through late-Wisconsinan times at U-Bar Cave. The findings appear concordant.

Pendejo Cave, Mid-Wisconsinan

The modern relationships are similar to those of the U-Bar Cave middle-Wisconsinan fauna except that similarities to all five other modern faunas are somewhat greater. Among the larger fossil faunas, the similarities to all of the U-Bar Cave faunas are moderate; the Isleta Caves, Shelter Cave, and Big Manhole Cave also show moderate affinities, though the Holocene components of the Isleta Caves may be skewing the figures. Dark Canyon Cave, which may be late mid-Wisconsinan, has the highest similarity value (0.80).

Dry Cave, Early Fauna

Greatest similarity with modern faunas is to northeastern New Mexico (0.56) and southeastern New Mexico (0.53). This fauna differs strongly from all late-Wisconsinan faunas except the terminal U-Bar Cave faunas (0.65–0.70); there also is some affinity with the Pendejo Cave mid-Wisconsinan (0.61). Harris (1987) hypothesized greatest similarity with northeastern New Mexico.

Isleta Caves

No one modern fauna stands out as notably similar to these two caves; northeastern New Mexico is most similar to the Isleta Cave No. 1 fauna (0.48) and northern New Mexico to the second cave (0.40). Presence of Holocene taxa undoubtedly biases the similarity values.

The two faunas most closely resemble each other (0.72), though Isleta Cave No. 1 also is quite similar to the latest Dry Cave faunas (0.70). The only other nearly comparable similarity values are with the Pendejo Cave fauna.

Harris and Findley (1964) suggested that the fauna was most similar to that of present-day southeastern Wyoming and adjacent Colorado; the Dice Coefficients do not agree with this assessment.

Shelter Cave

This very small fauna (14 taxa considered) shows only weak affinity with any modern fauna; Wyoming has a slight edge on the others. The only relatively large similarity values with fossil faunas are with the 14-15 and 15-18 ka U-Bar Cave faunas and with the Pendejo Cave mid-Wisconsinan. Small faunas consisting largely of widespread taxa (taxa most likely to be absent from other faunas solely because of sampling error) tend to have those few faunal members that do have a high informational content overwhelmed by the common members. This appears likely to be the case here.

Big Manhole Cave

This fauna shows a rather diffuse pattern of affinities among the modern; northeastern New Mexico is most similar at 0.46. The fauna shows moderately high similarities to the latest Wisconsinan Dry Cave faunas and to two of the U-Bar Cave faunas, but most closely resembles the middle Wisconsinan of Pendejo Cave (0.76). A relative paucity of small-sized faunal elements may well be biasing the fauna.

Burnet Cave

Although perhaps the most famous of the New Mexican sites, affinities with other Pleistocene sites are relatively low, except with the very small Charlies Parlor fauna at 0.65 (otherwise, the highest similarity value is 0.56 with the 14-15 ka U-Bar Cave fauna and with the middle Wisconsinan Pendejo Cave fauna). Modern affinities are strongest with southeastern New Mexico (0.44) and northeastern New Mexico (0.43). The mixing of Holocene materials with those of Pleistocene origin introduces bias.

Brown Sand Wedge

This is the most distinctive of the fossil faunas studied, grouping with the eastern Kansas modern fauna (0.46) and, next, with northeastern New Mexico (0.31). Affinity with fossil faunas is low, reaching a high of 0.44 with Burnet Cave and with the < 13 ka Dry Cave fauna. Slaughter (1975) appeared most impressed by the combination of taxa now found only to the north with those now found only to the south. Harris (1985) noted a distinctly eastern riparian element.

Muskox Cave

This cave has several apparently anomalous taxa for the time period involved (Harris, 1985), leading to the suspicion that it is temporally mixed. Strongest modern affinities are to northern New Mexico (0.56) and southeastern New Mexico (0.55); presence of the out-of-place taxa may be reflected in the affinity to the latter. Similarity with most fossil faunas is moderate (e.g., 0.70 with the Bison Chamber fauna). However, the other two sites with faunas identified by Logan (both relatively high-elevation sites for the region) are high at 0.90 and 0.86. Logan (1981) interpreted the habitat as spruce-fir forest and meadows; Harris (1985) suggested open coniferous forest with a boreal aspect, grasslands, and possibly woodland elements.

Sloth Caves

Both of these sites were excavated by Logan, who also did the faunal identifications (Logan, 1983; Logan and Black, 1979). There is some suggestion that the faunas may be mixed, in part, with Holocene elements (Harris, 1985). The most similar modern faunas are those of southeastern New Mexico for both (0.37, 0.49). The faunas are similar to each other (0.85), to Muskox Cave (0.90, 0.86), and to U-Bar Cave 14-15 ka (0.80, 0.72), but otherwise are fairly distinctive. Logan and Black (1979) compared the climatic conditions represented by the Upper Sloth Cave fauna with those of the modern Black Hills of South Dakota.

Dark Canyon Cave

Modern faunas most like that of Dark Canyon Cave are those of northeastern New Mexico (0.46) and northern New Mexico (0.41). The fauna is moderately similar to the Dry Cave late-Wisconsinan faunas (0.72 for most) and to the U-Bar Cave 14-15 ka fauna (also 0.72), but shows most in common with the Pendejo Cave mid-Wisconsinan (0.80).

CONCLUSIONS

The quantitative methods, as used here, tend to agree with qualitative assessments to a considerable degree, but with notable exceptions. For example, sensitivity seems to be notably less: there are very important things that purely quantitative analysis is not showing. Sensitive faunal indicators of various environmental aspects have no more weight than insensitive taxa in the quantitative analyses. For example, both *Lemmiscus curtatus* and *Brachylagus idahoensis* are far out of range at Isleta Cave No. 2, generally being considered typically Great Basin forms—yet, no hint of this appears in the quantitative treatment.

The coefficients of similarity do appear to be a useful adjunct to qualitative interpretation in guiding attention to features otherwise submerged in an extensive database. As a case in point, the relatively high similarity values of the 14-15 ka U-Bar Cave fauna with southeastern New Mexico glacial faunas compared to the lower values for other U-Bar Cave faunas indicate further interpretation is necessary: Are the U-Bar Cave dates skewed? Did climatic change to conditions more typical of the southeastern New Mexico Wisconsinan occur at that time? Or is this an artifact caused by the somewhat larger sample size of this fauna?

Although in theory information-laden taxa could be weighted in quantitative studies, both the subjectivity involved in determining weights and the labor involved in adjusting these weights for geographic and ecological positions would seem to make such schemes impractical. A melding of relatively simple quantitative methods, such as those utilized here, with qualitative data should result in inferences more useful than either alone. On the other hand, quantitative methods involving relative frequencies of taxa within faunas joined with interpretation of qualitative data may give results superior to those produced by these simpler methods (Harris, 1993a), but their labor intensive nature may prevent widespread use.

LITERATURE CITED

Cheetham, A.H., and J.E. Hazel. 1969. Binary (presence–absence) similarity coefficients.

Journal of Paleontology, 43:1130-1136.

Fallaw, W.C. 1979. A test of the Simpson Coefficient and other binary coefficients of faunal similarity. Journal of Paleontology, 53:1029-1034.

Harris, A.H. 1970. The Dry Cave mammalian fauna and late pluvial conditions in southeastern New Mexico. Texas Journal of Science, 22:3-27.

Harris, A.H. 1977. Wisconsin age environments in the northern Chihuahuan Desert: Evidence from the higher vertebrates. Pp. 23-52, *in* Transactions of the symposium on the biological resources of the Chihuahuan Desert region, United States and Mexico (R.H. Wauer and D.H. Riskind, eds.), National Park Service Transactions and Proceedings Series, 3:1-658.

Harris, A.H. 1985. Late Pleistocene vertebrate paleoecology of the West. University of Texas Press, Austin, 293 pp.

Harris, A.H. 1987. Reconstruction of Mid-Wisconsin environments in southern New Mexico. National Geographic Research, 3:142-151.

Harris, A.H. 1989. The New Mexican late Wisconsin—East versus west. National Geographic Research, 5:205-217.

Harris, A.H. 1991. Preliminary report on the fauna of Pendejo Cave, Otero County, New Mexico. Current Research in the Pleistocene, 8:92-93.

Harris, A.H. 1993a. Wisconsinan pre-pleniglacial biotic change in southeastern New Mexico. Quaternary Research 40:127-133.

Harris, A.H. 1993b. Quaternary vertebrates of New Mexico. Pp. 179-197, *in* Vertebrate Paleontology in New Mexico, New Mexico Museum of Natural History Bulletin, 2:i-vii, 1-338.

Harris, A.H., and J.S. Findley. 1964. Pleistocene-Recent fauna of the Isleta Caves, Bernalillo County, New Mexico. American Journal of Science, 262:114-120.

Hazel, J.E. 1970. Binary coefficients and clustering in biostratigraphy. Geological Society of America Bulletin, 81:3237-3252.

Holman, J.A. 1970. A Pleistocene herpetofauna from Eddy County, New Mexico. Texas Journal of Science, 22:29-39.

Howard, H. 1962. Bird remains from a prehistoric cave deposit in Grant County, New Mexico. Condor, 64:241-242.

Howard, H. 1971. Quaternary avian remains from Dark Canyon Cave, New Mexico. Condor, 73:237-240.

Logan, L.E. 1981. The mammalian fossils of Muskox Cave, Eddy County, New Mexico. Pp. 159-160, *in* Proceedings of the Eighth International Congress of Speleology, 1:159-160.

Logan, L.E. 1983. Paleoecological implications of the mammalian fauna of Lower Sloth Cave, Guadalupe Mountains, Texas. National Speleological Society Bulletin, 45:3-11.

Logan, L.E., and C.C. Black. 1979. The Quaternary vertebrate fauna of Upper Sloth Cave, Guadalupe Mountains National Park, Texas. Pp. 141-158, *in* Biological investigations in the Guadalupe Mountains National Park (H.H. Genoways and R.J. Baker, eds.), National Park Service Proceedings and Transactions Series, 4:1-442.

Murray, K.F. 1957. Pleistocene climate and the fauna of Burnet Cave, New Mexico. Ecology, 38:129-132.

Raup, D.M., and R.E. Crick. 1979. Measurement of faunal similarity in paleontology. Journal of Paleontology, 53:1213-1227.

Rohlf, F.J. 1990. NTSYS-PC. Numerical taxonomy and multivariate analysis system. Exeter Software, Setauket, New York.

Schultz, C.B., and E.B. Howard. 1935. The fauna of Burnet Cave, Guadalupe Mountains, New Mexico. Proceedings of the Academy of Natural Sciences of Philadelphia, 87:273-298.

Slaughter, B.H. 1975. Ecological interpretation of the Brown Sand Wedge local fauna. Pp. 179-192, *in* Late Pleistocene environments of the Southern High Plains (F. Wendorf and J.J. Hester, eds.). Fort Burgwin Research Center, Publication, 9:1-290.

Stearns, C.E. 1942. A fossil marmot from New Mexico and its climatic significance. American Journal of Science, 240:867-878.

Thompson, R.S., T.R. Van Devender, P.S. Martin, T. Foppe, and A. Long. 1980. Shasta ground sloth (*Nothrotheriops shastense* Hoffstetter) at Shelter Cave, New Mexico: Environment, diet, and extinction. Quaternary Research, 14:360-376.

Van Devender, T.R., and R.D. Worthington. 1977. The herpetofauna of Howell's Ridge Cave and the paleoecology of the northwestern Chihuahuan Desert. Pp. 85-106, *in* Transactions of the symposium on the biological resources of the Chihuahuan Desert region, United States and Mexico (R.H. Wauer and D.H. Riskind, eds.), National Park Service Transactions and Proceedings Series, 3:1-658.

Arthur H. Harris, Faculty Curator, Vertebrate Paleobiology, Centennial Museum's Laboratory for Environmental Biology, University of Texas at El Paso, El Paso, Texas 79968-0915

Historical Implications and Characteristics of Assemblages of Small Mammals in West-Central Kansas

E.D. Fleharty and Rob Channell

Abstract

Small mammals were trapped on four habitats: cropland, roadside ditch, pasture, and remnant grassland. The small mammal community of each of these habitats was compared on the basis of richness, evenness, diversity, and similarity. The cropland was found to most resemble the pasture. When richness, evenness, and diversity were comparison characters, the ditch most resembled the remnant grassland; however, when similarity was the comparison character, the ditch most closely resembled the pasture.

Population dynamics, habitat preference, probable historical occurrence, and impact of human activities on individual species were also examined. The only species that was permanent in all four habitats was the deer mouse, *Peromyscus maniculatus*. The cotton rat, *Sigmodon hispidus,* was a permanent species in the ditch. *Reithrodontomys megalotis* was a transient on all of the habitats but the remnant grassland, where it was a permanent species.

INTRODUCTION

There is probably no native prairie ecosystem remaining in western Kansas that is reflective of the biotic communities that were extant prior to the arrival of the European settlers. At that time, this area was inhabited primarily by Arapaho and Cheyenne Indians with occasional incursions by the Kaw, Pawnee, Sioux, Kiowa, and Comanche tribes (Choate and Fleharty, 1975) as they exploited the vast herds of bison that roamed these prairies (Darton, 1916).

The discovery of gold in Colorado led to the establishment of the Smoky Hill Trail in 1859 (Choate and Fleharty, 1975). Military outposts, like Fort Hays (founded in 1865), were established to protect the ever increasing numbers of white settlers as they moved westward on wagon trains, stage lines, and eventually the railroad. The Kansas–Pacific railroad reached Hays on 16 October 1867, and Hays City was formally founded on 23 November 1868. According to records kept at Fort Hays, the bison had disappeared by 1877.

Agricultural experimentation began at Hays City in 1867 or 1868, when W.E. Webb enclosed 10 ha with a board fence and, in an attempt to turn over 2 ha of native grassland, ruined a number of plows (Hays Sentinel, 18 January 1878). By 1870, Hays City had a population of 320, as did the adjacent fort (Walker, 1872). In 1873 and 1874, Martin Allen planted a variety of trees but grasshoppers and drought killed them all. Allen wrote about the hard surface of the ground, stating that it "would shed rain, nearly equal to a shingle roof" (Hays Sentinel, 8 August 1879). Richard Smith Elliott sowed wheat, rye, and barley at Ellis on 20 October 1870 and wrote, "the work of redeeming the domain of the buffalo was begun" (Elliott, 1883). In April 1871, 48.5 ha of grassland were turned over near Ellis by Dr. Louis Watson. Forty and a half ha of this were planted to corn and sorghum; the remainder were sowed to wheat, barley, and rye in September. This experiment by Dr. Watson failed in the winter of 1872 (Watson, 1872).

Louis Agassiz from Harvard came to Hays on the Kansas–Pacific railroad and found the

Life Among the Muses: Papers in Honor of James S. Findley. Edited by T.L. Yates, W.L. Gannon & D.E. Wilson.

area "eminently adapted to the culture of wheat." He predicted that this area soon would be one of the leading wheat-producing areas of the world (Hays Sentinel, 8 March 1876).

These feeble attempts at agriculture were only harbingers of things to follow. There was no large-scale agriculture in Ellis County until 1876, when German–Russian immigrants arrived. By 1877, 535 ha were planted to wheat and 530 to corn; this increased the following year to 1,634 ha of wheat and 1,306 ha of corn (Hays Sentinel, 9 May 1879).

As prairies were plowed under to serve as croplands, roads and borrow ditches constructed, and small villages founded, native vegetation was impacted and small mammal assemblages were altered. A few remnants of the native prairie were "saved," but by closing them to grazing and "protecting" them from fire, these, too, were altered. The remaining prairie was fenced and subjected to various intensities of grazing by domestic livestock.

Investigations in western Kansas have examined various aspects of small mammal ecology as they relate to these altered ecosystems. These have included studies on the use of fencerows, croplands, and grazed lands (Fleharty and Navo, 1983; Navo and Fleharty, 1983; Kaufman and Kaufman, 1989; Kaufman and Kaufman, 1990), remnant prairies (Fleharty, 1972; Fleharty and Mares, 1973; Hansen and Fleharty, 1974; Choate and Fleharty, 1975), and riparian communities (Fleharty and Stadel, 1968; Frydendall, 1969) as small mammal habitat.

Although these studies have provided significant information on small mammal ecology, many were of relatively short duration or had relatively few sampling periods and were, of course, not conducted under similar climatic regimes. Additionally, cropland ecosystems have not been studied systematically during a complete rotational cycle as often is practiced in dryland farming of western Kansas.

The purpose of this study was to sample small mammal populations from four ecosystems that have been established under the influence of human activities to better understand the effects that these changes have had on the assemblages of small prairie mammals.

DESCRIPTION OF STUDY AREAS

The study was conducted in west–central Kansas in Ellis County from December 1980 through October 1983. In this area, habitat types in Ellis County can be grouped into six categories: 1) remnant grassland; 2) range or pasture land; 3) cropland; 4) riparian habitats; 5) "waste areas" such as field borders, roadside ditches, fencerows, shelterbelts, and rights-of-way; and 6) urban areas. In 1981, croplands accounted for 117,378 ha of the 230,675 ha in the county, and rangeland or pastures, 109,111 ha. Of the cropland under cultivation, 49% was planted in winter wheat and 9-10% in grain sorghum (Virgil Quint, Agricultural Stabilization and Conservation Service, pers. comm.).

Four sites representing four habitat types were selected for study. Cover in these habitats was estimated using a modified step-point technique as described by Evans and Love (1957). The first study site (T14S, R19W, SE 1/4 SEC. 1, Ellis Co.) was situated on a 14 ha remnant mixed grassland owned by Fort Hays State University that has not been farmed, grazed, or burned since 1910. The remnant grassland, bisected by a drainage draw, has been the subject of extensive investigation and has been described in detail (Albertson and Tomanek, 1965; Hulett et al., 1972). The remnant grassland was dominated by big bluestem (*Andropogon gerardi*) and little bluestem (*Andropogon scoparius*). Big bluestem accounted for 54.7% of the total cover, little bluestem 20.1%, sideoats grama (*Bouteloua curtinpendula*) 9.7%, kochia (*Kochia scoparius*) 8.0%, and common sunflower (*Helianthus annuus*) 0.8% (Hulett et al., 1972). The draw was dominated by kochia and sunflower and differed from the rest of the grassland in its "weedy" structure (Hulett et al., 1972). The remnant grassland maintained a cover estimate of 99% throughout the study.

A moderately grazed pasture on the Fort Hays State University farm (T14S, R19W, SE 1/4 SEC 1, Ellis Co.) was selected as the second study site. This area, representative of the shortgrass prairie, was dominated by sideoats grama, blue grama (*Bouteloua gracilis*), buffalo grass (*Buchloe dactyloides*), and western ragweed (*Ambrosia psilostachya*). Cover was maintained throughout the year at more than 90%.

The third study site (T14S, R18W, SE 1/4 SEC 7, Ellis Co.), a 70-ha cropland on the Fort Hays State University farm, was bordered on the south and east by roadside ditches approximately 6 m wide and on the north and west by fencerows and grazed pastures. The dry-land farming regimen practiced in west–central Kansas consists of a three-year rotation. Trapping began in winter wheat stubble (December 1980–June 1981). On 31 March 1981, the cropland was subtilled for weed control and anhydrous ammonia (44.5 kg/ha) was applied. An herbicide for additional weed control was applied on 26 May 1981. The cropland was disked and planted in grain sorghum (49,000–74,000 seeds/ha) and "Milo-guard" herbicide was applied during 6-12 June 1981. Grain sorghum was harvested during 28-30 October 1981. The grain sorghum stubble remained from harvest until the field was disked for weed control on 13 July 1982. These weeds included kochia, field bindweed (*Convolvulus arvensis*), and green foxtail (*Setaria virdis*). On 19 August 1982, the cropland was prepared for planting wheat by subtilling and by applying anhydrous ammonia. Winter wheat was planted on 21 September 1982 and harvested on 7 July 1983. Harvest returned the cropland to wheat stubble, in which it remained until the end of the study.

Cover estimates in the cropland were zero for only three to five days after the planting of the two crops and ranged between 36% and 75%, with an average of 61% at other times. The highest cover readings were recorded in August 1981 (72%), when 52% of the cover was sorghum that was 50.8 cm to 91.4 cm tall and 20% was weedy vegetation; in June 1982 (72%), when the crop phase was sorghum stubble and 15% of the cover was provided by the stubble and 57% by weedy vegetation; in May 1983 (73%) and June 1983 (74%), when the wheat was rapidly growing; in July 1983 (67%), September 1983 (75%), and October 1983 (65%), when the field was in wheat stubble and weedy vegetation comprised about 40% of the cover (until October, when weedy cover was reduced to 18%).

The fourth study site was a roadside ditch that bordered the south edge of the cropland and paralleled a county road. Vegetative samples of ditch material were taken in May, August, and September of 1982. Samples were taken from the slope bordering the field, the bottom of the ditch, and the slope bordering the road to check for variation of plant community composition within the ditch.

The species composition in the ditch consisted of 28% western wheatgrass (*Agropyron smithii*), 27% Kentucky bluegrass (*Poa pratensis*), 16% kochia, 9% smooth brome (*Bromus inermis*), and 6% Japanese brome (*Bromus japonicus*). Western wheatgrass was evenly distributed over the sample periods but was found more often on the slopes than in the bottom. Kentucky bluegrass and kochia were most prevalent in May and declined in numbers to October. Kochia was most abundant on the slope bordering the field and bluegrass predominanted on the slope bordering the road. Smooth brome favored the field slope, whereas Japanese brome showed no preference within the ditch. In late October 1982, living plants contributed only 3% of the cover, and no species made up more then 1% of the cover. Dead plant litter made up 89% of the cover, and 8% of the ditch was exposed soil. Precipitation and temperature data were obtained from the Fort Hays Experiment Station. The data are summarized in Figs. 1 and 2.

METHODS AND MATERIALS

Within each study area, a trapline of 20 stations at 15 m intervals was established. The trapline in the remnant grassland was oriented in an east–west direction in the center of the area and crossed the drainage, whereas the trapline in the pasture was located near the center and ran from north to south. The trapline in the cropland was oriented in a north–south direction with the traps 200 m from the southern border of the field. The trapline was centered within the field and was 320 m from the east and west boundaries. The trapline in the ditch was oriented in an east–west direction.

Three Sherman live traps (7.6 by 8.9 by 22.9 cm) were placed within 1 m of each station in the remnant grassland, pasture, and cropland. At each station within the ditch, one trap was placed on the slope adjacent to the road, one in the bottom, and a third under the fence

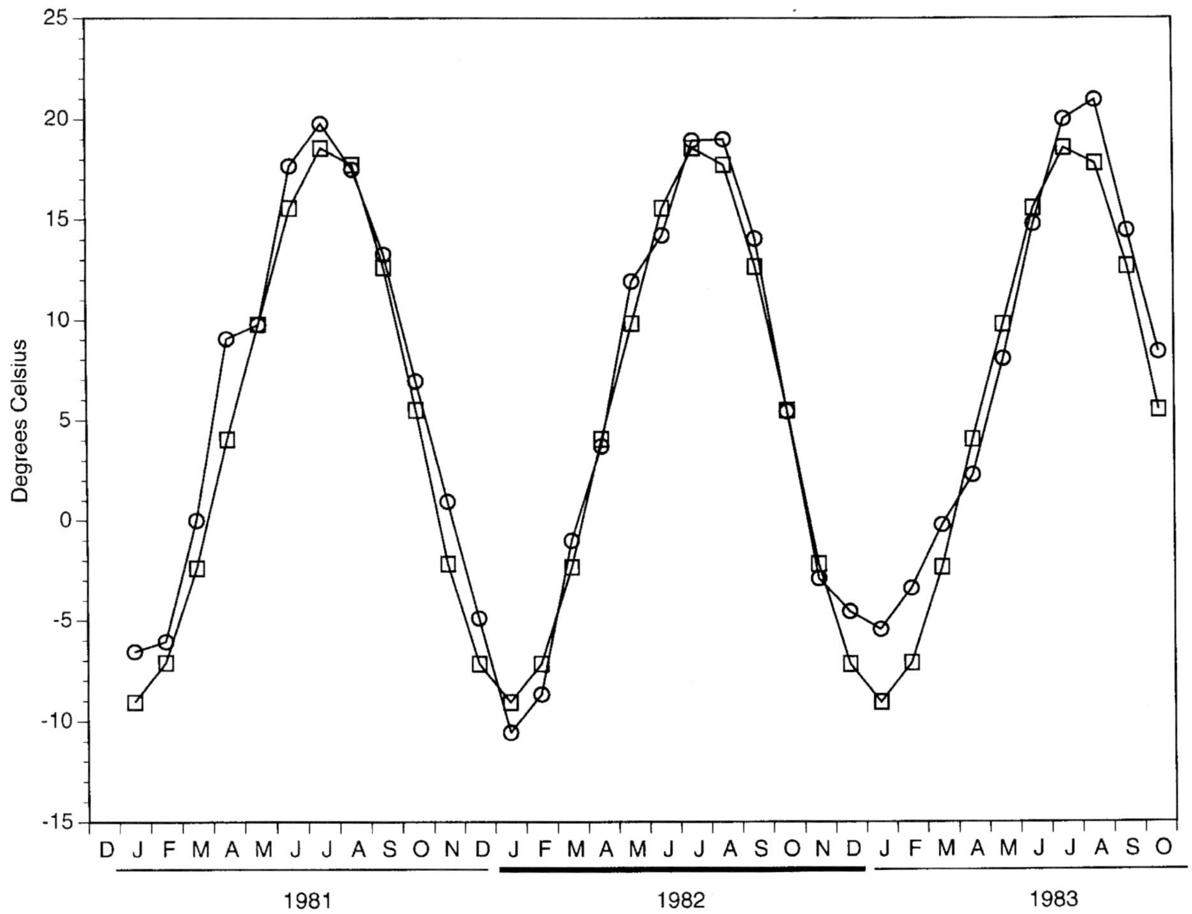

Fig. 1. The average monthly temperature as recorded at the Fort Hays Experiment Station (squares = long term mean, circles = monthly mean for that month of the study).

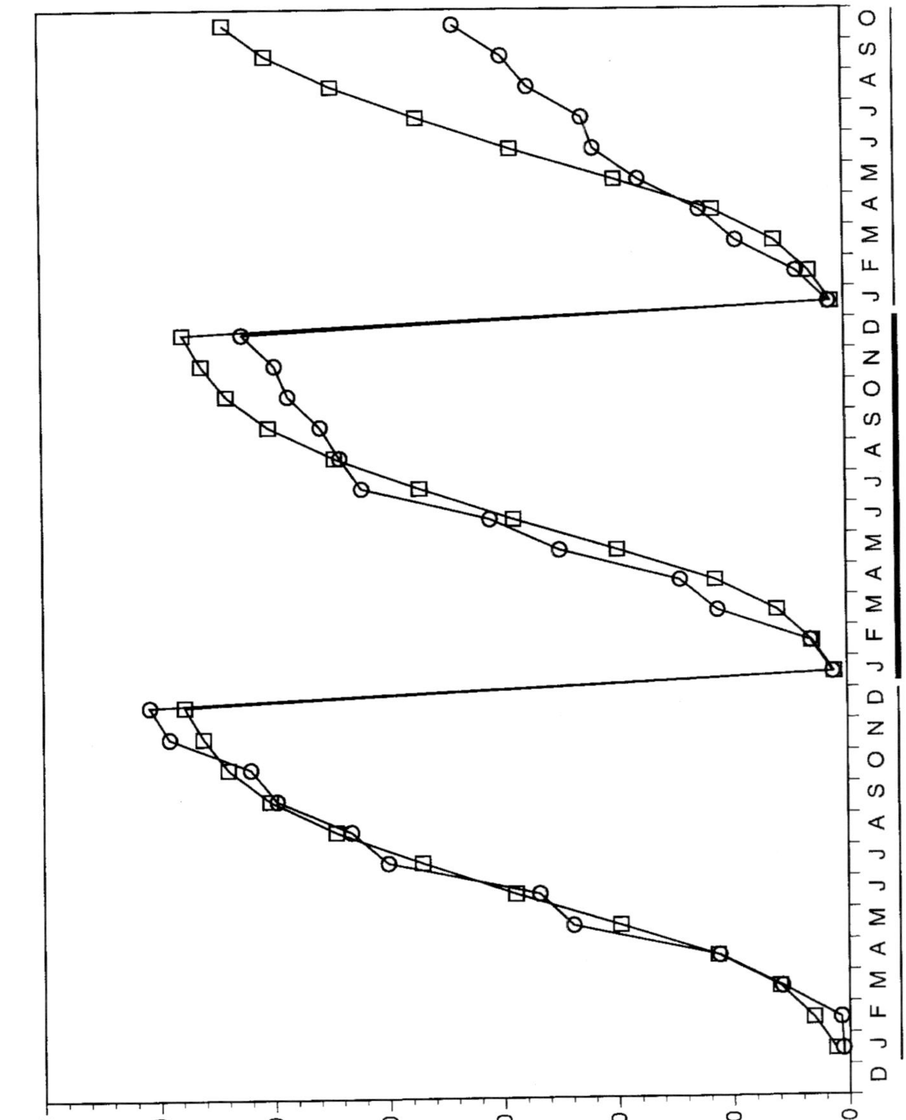

Fig. 2. The accumulated annual precipitation as recorded at the Fort Hays Experiment Station (square = the long term mean, circle = the monthly mean for that month of the study).

next to the cropland. Closed traps were placed at each station one night before trapping was initiated. Traps were set the following afternoon and baited with a mixture of peanut butter, wild bird seed, and raisins as described by Stout and Sonenshine (1973). During cold weather, cotton was placed in each trap for nesting material. Traps were checked morning and evening for three consecutive days. During summer months, traps were closed after the morning check and re-opened in late afternoon. Animals removed from traps were identified to species, weighed, sexed, examined for reproductive condition, marked by toe clipping, and released at the capture site. Reproductive condition was described by position of testes or development of mammae and condition of pubic symphysis.

Trapping was initiated in the cropland and remnant grassland in December 1980, and traplines in the ditch and pasture were added in October of 1981. With few exceptions, trapping was conducted once a month within a week of the new phase of the moon. In the cropland, additional trapping periods were conducted both before and after major farming perturbations.

Population size was determined for each species in each community for each trapping period using the minimum number known to be alive (MNA) and, as such, included both resident and recruited individuals (Hilborn et al., 1976). Individuals were described as resident if they had been captured in any two trapping periods. Recruited individuals were those that were captured for the first time and could, of course, become part of the resident population if captured in subsequent periods.

Species were described as being permanent, semi-permanent, or transient on the basis of capture records. Permanent species were those that always were present in the habitat in question; at least one individual that was captured in a previous trapping period had to be caught during each trapping period for a species to be a permanent species. Semi-permanent species were those in which a resident population was established at some point in the study but not throughout the entire study. Species that never established resident populations were described as transient.

Trappability (number actually caught/MNA) of all permanent and semi-permanent species exceeded 80%, indicating low bias (Hilborn et al., 1976). From the population data, community characteristics of richness (number of species), evenness (J'), and diversity (H') (Shannon, 1949) were calculated for all four habitats for each trapping period. Hutcheson's (1970) modified t-test was used to compare trapping period diversities between communities (inter-community) and within communities (intra-community) for consecutive trapping periods. Richness, evenness, and diversity of small mammal assemblages in each habitat were compared for the entire study with a non-parametric Tukey's multiple range test and on a more sensitive pair-wise basis using a paired sample Wilcoxon's test (Zar, 1984). An index of community similarity based on species composition and abundance was used to determine inter-community and intra-community similarity (Horn, 1966) and their means.

Species habitat breadth was calculated (Whittaker and Levin, 1975) as a measure of the number of habitats that a species occupied during the trapping period and the evenness in which they occupied them. A mean species habitat breadth was calculated for each species for the entire study.

Spearman correlations were run on each species in each community to look for interspecific interactions and to examine species between communities (intra-community) for resident and recruited individuals (Zar, 1984). Statistical significance was set at 0.05 for all tests.

RESULTS

The accumulated precipitation and monthly mean temperature in 1981 and most of 1982 closely approximated the long-term means (Figs. 1 and 2). However, a drought began in August 1982 and lasted through the end of the study. December 1982 through March 1983 had appreciably milder temperatures than the long-term mean, whereas July and August of 1983 were two of the warmest months on record for Ellis County.

Populations of all species, broken down by habitat, are shown in Fig. 3. The total individuals in the cropland and remnant grassland were relatively stable throughout the study, with slight

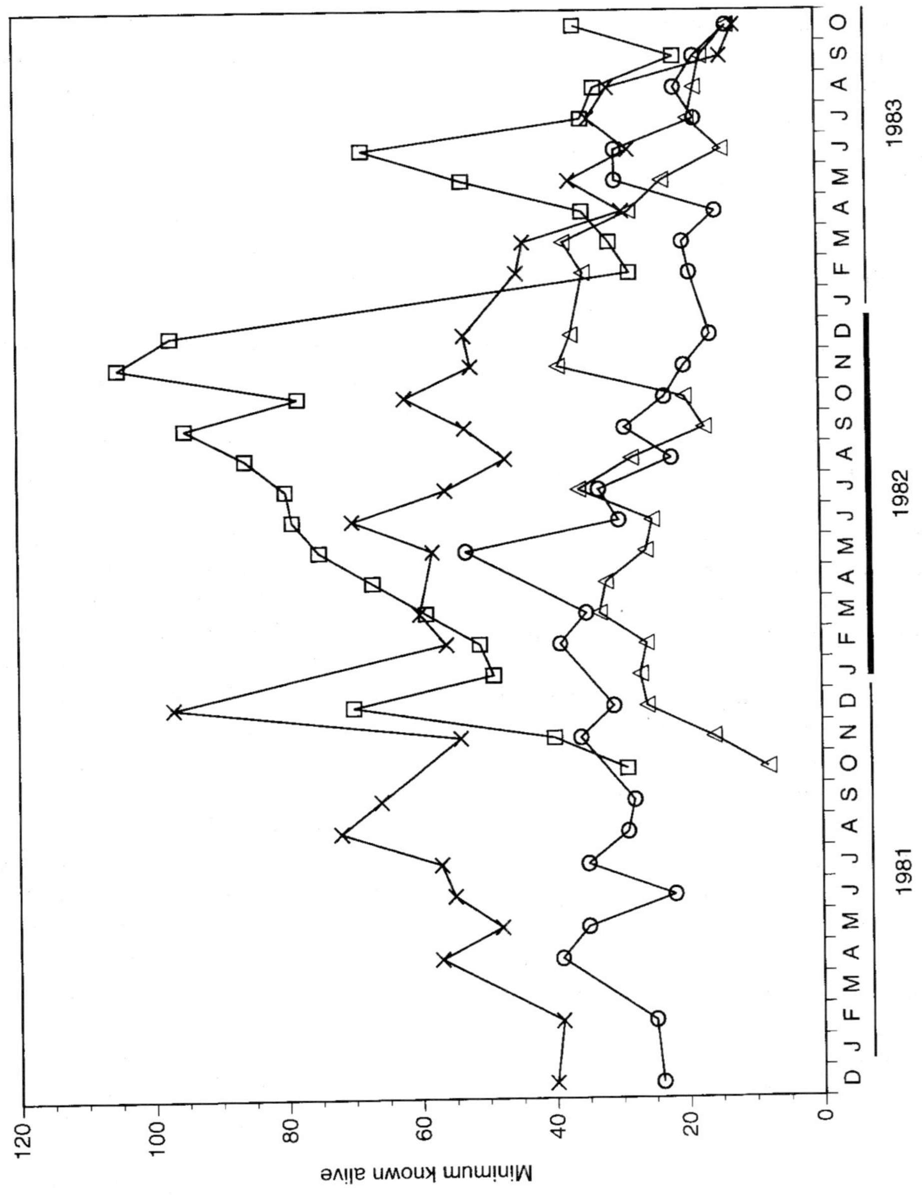

Fig. 3. Total individuals of all species known to be alive in each habitat (circle = cropland, square = ditch, X = remnant grassland, triangle = pasture).

declines during the drought. In the pasture, total numbers were relatively stable but increased slightly during the mild winter of 1982–83. The increased total numbers persisted until June of 1983. Total number of individuals in the ditch rose steadily from the beginning of trapping in October 1981 through 1982, but early in 1983 the number of small mammals in the ditch declined sharply and became comparable to the numbers of the other habitats.

The greatest mean richness was demonstrated in the ditch (5.45 species/trapping period). The greatest absolute richness of nine species in July and August of 1982 was also found in the ditch. The remnant grassland had the next highest mean richness value of 4.76. The mean species richness in the cropland was 1.97, whereas that in the pasture was 2.42 (Table 1).

The largest mean evenness (0.80) was found in the remnant grassland. Next was the ditch (0.69), pasture (0.37), and cropland (0.27) (Table 1). Monthly evenness values in the remnant grassland were greater than those in all other habitats except in October 1982, April–-June 1983, August–September 1983, when values in the ditch were greater, and in December 1982 when the value in the cropland was the greatest (Fig. 4). Evenness in the pasture and ditch increased during the drought of 1983, whereas that in the remnant grassland remained unchanged and that in the cropland remained characteristically erratic.

The remnant grassland had the greatest mean diversity (0.52) of small mammals, followed closely by that in the ditch (0.50) and more distantly by those in the pasture (0.15) and cropland (0.11) (Table 1). The single highest monthly diversity (0.74) was recorded in the remnant grassland in July 1982. Monthly diversities in the remnant grassland were relatively stable and were only exceeded in April and May–October 1983 by those in the ditch (Fig. 5). Diversity of small mammals in the ditch reached peaks in late summer of 1982 and 1983 and were lowest in late winter and early spring of those years. Although no pattern appears in the diversity of small mammal species in the pasture, it (like the ditch) exhibited increased diversity in the dry summer of 1983.

Tukey's multiple range test performed on each of the community characteristics (richness, evenness, diversity) showed that the cropland and pasture were not different from each other and that the remnant grassland did not differ from the ditch. All other comparisons of community characteristics among the habitats (cropland/remnant grassland, cropland/ditch, pasture/remnant grassland, pasture/ditch) were significantly different. The Wilcoxon's test for paired samples supported the results from Tukey's test.

Hutcheson's (1970) t-test for diversities (Table 2) demonstrates that, when the cropland and pasture and remnant grassland and ditch were compared, only 4 of 21 diversity comparisons for each pair of habitats exhibited significant differences. However, when other comparisons (cropland/remnant grassland, cropland/ditch, pasture/remnant grassland, pasture/ditch) were made, more deviations were found (e.g., 18 of 20 were significantly different for the pasture/remnant grassland comparison). Therefore, from the perspective of diversity, cropland and pasture were similar, as were ditch and remnant grassland.

Horn's (1966) similarity index provided a slightly different perspective. Although the mean similarity of cropland to pasture was highest (0.90), the mean similarity of pasture to ditch (0.72) and of cropland to ditch (0.71) were higher than that of the remnant grassland to the ditch (0.64) (Table 2). This was a deviation from the pattern of remnant grassland/ditch resemblance established by the use of community characters mentioned above. In the eleven months (November 1981–September 1982) before the drought that began in October of 1982, mean similarity values in the ditch to the other habitats were approximately equal (0.72 to the remnant grassland, 0.74 to the cropland, and 0.74 to the pasture). After October 1982, during the drought, the mean similarity values of the ditch (November 1982–October 1983) to the remnant grassland (0.54) fell; its similarity to the cropland (0.72) and pasture (0.72) stayed roughly the same as before the drought.

Diversities within all four communities did not differ from one month to the next except for 5 of 26 comparisons that were significantly different for both cropland and remnant grassland

Table 1. Richness (s), evenness (J), diversity (H), Hutcheson's (1970) T-test for diversity intra-community (D, '*' indicates $p < 0.05$), and Horn's (1966) similarity index intra-community (SIM) for the four habitats examined.

		Cropland					Ditch					Grassland					Pasture				
		s	J	H	D	SIM	s	J	H	D	SIM	s	J	H	D	SIM	s	J	H	D	SIM
	D	2	0.25	0.08		—	-	—	—		—	4	0.60	0.36		—	—	—	—		—
1981	J	—	—	—		—	-	—	—		—	-	—	—		—	-	—	—		—
	F	1	0	0		0.98	-	—	—		—	4	0.68	0.41		0.91	-	—	—		—
	M	-	—	—		—	-	—	—		—	-	—	—		—	-	—	—		—
	A	1	0	0		1.00	-	—	—		—	4	0.80	0.48		0.97	-	—	—		—
	M	1	0	0		1.00	-	—	—		—	5	0.66	0.46		0.97	-	—	—		—
	J	1	0	0		1.00	-	—	—		—	6	0.65	0.50		0.93	-	—	—		—
	J	4	0.66	0.40	*	0.85	-	—	—		—	6	0.69	0.54		0.97	-	—	—		—
	A	1	0	0	*	0.85	-	—	—		—	6	0.70	0.55		0.97	-	—	—		—
	S	1	0	0		1.00	-	—	—		—	5	0.91	0.63		0.95	-	—	—		—
	O	-	—	—		—	4	0.89	0.54		—	-	—	—		—	3	0.67	0.32		—
	N	3	0.38	0.18		0.96	5	0.75	0.52		—	5	0.79	0.55		0.84	1	0	0		—
	D	1	0	0	*	0.94	5	0.72	0.50		0.90	5	0.77	0.54		0.96	3	0.30	0.14		0.97
1982	J	-	—	—		—	5	0.75	0.52		0.99	-	—	—		—	1	0	0		0.96
	F	2	0.17	0.05		—	5	0.62	0.43		0.94	5	0.88	0.61		—	1	0	0		0.99
	M	1	0	0		0.99	5	0.55	0.39		0.99	4	0.97	0.58		0.97	2	0.20	0.06		1.00
	A	-	—	—		—	5	0.50	0.35		0.99	-	—	—		—	2	0.20	0.06		0.99
	M	2	0.14	0.04		—	6	0.51	0.40		0.97	5	0.87	0.61		—	2	0.39	0.12		1.00
	J	2	0.21	0.06		0.97	7	0.68	0.58		0.94	6	0.86	0.67	*	0.98	2	0.24	0.07		0.95
	J	3	0.33	0.16		0.94	9	0.68	0.65		0.96	7	0.88	0.74		0.92	2	0.76	0.23	*	0.97
	A	2	0.58	0.17		0.98	9	0.66	0.63		0.97	4	0.88	0.53	*	0.89	3	0.45	0.21		0.95
	S	2	0.58	0.17		1.00	6	0.72	0.56		0.95	5	0.85	0.59		0.95	3	0.40	0.19		0.95
	O	3	0.43	0.20		0.97	5	0.66	0.46		0.95	7	0.79	0.66		0.98	3	0.36	0.17		0.99

Table 1. Continued

		Cropland					Ditch					Grassland					Pasture				
	N	3	0.56	0.27		0.99	5	0.65	0.45		1.00	4	0.99	0.60		0.96	3	0.22	0.10		0.96
	D	2	0.90	0.27		0.95	6	0.57	0.44		0.99	4	0.84	0.51	*	0.94	4	0.27	0.16		0.97
1983	J	–	—	—		—	–	—	—		—	–	—	—		—	–	—	—		—
	F	2	0.49	0.15		0.95	3	0.56	0.27	*	0.55	3	0.86	0.41		0.97	2	0.19	0.06		1.00
	M	2	0.47	0.14		1.00	4	0.58	0.35		0.93	4	0.79	0.47		0.96	2	0.18	0.05		0.98
	A	3	0.44	0.21		0.96	3	0.63	0.30		0.93	3	0.63	0.30	*	0.95	2	0.49	0.15		0.93
	M	5	0.36	0.25		0.96	5	0.73	0.51	*	0.85	5	0.59	0.41		0.93	4	0.54	0.33		0.90
	J	1	0	0	*	0.93	5	0.83	0.58		0.87	6	0.75	0.59		0.92	3	0.46	0.22		—
	J	2	0.31	0.09		0.97	6	0.82	0.64		0.95	3	0.91	0.44		0.96	2	0.83	0.25		0.78
	A	2	0.45	0.14		0.93	6	0.87	0.68		0.96	4	0.83	0.50		0.96	2	0.65	0.20		0.78
	S	1	0	0	*	0.95	7	0.80	0.67		0.94	4	0.76	0.46		0.85	3	0.53	0.25		0.97
	O	1	0	0		1.00	5	0.70	0.49		0.84	5	0.68	0.48		0.84	3	0.63	0.30		0.89
MEAN		1.97	0.27	0.11		0.96	5.45	0.69	0.50		0.93	4.76	0.80	0.52		0.94	2.42	0.37	0.15		0.95

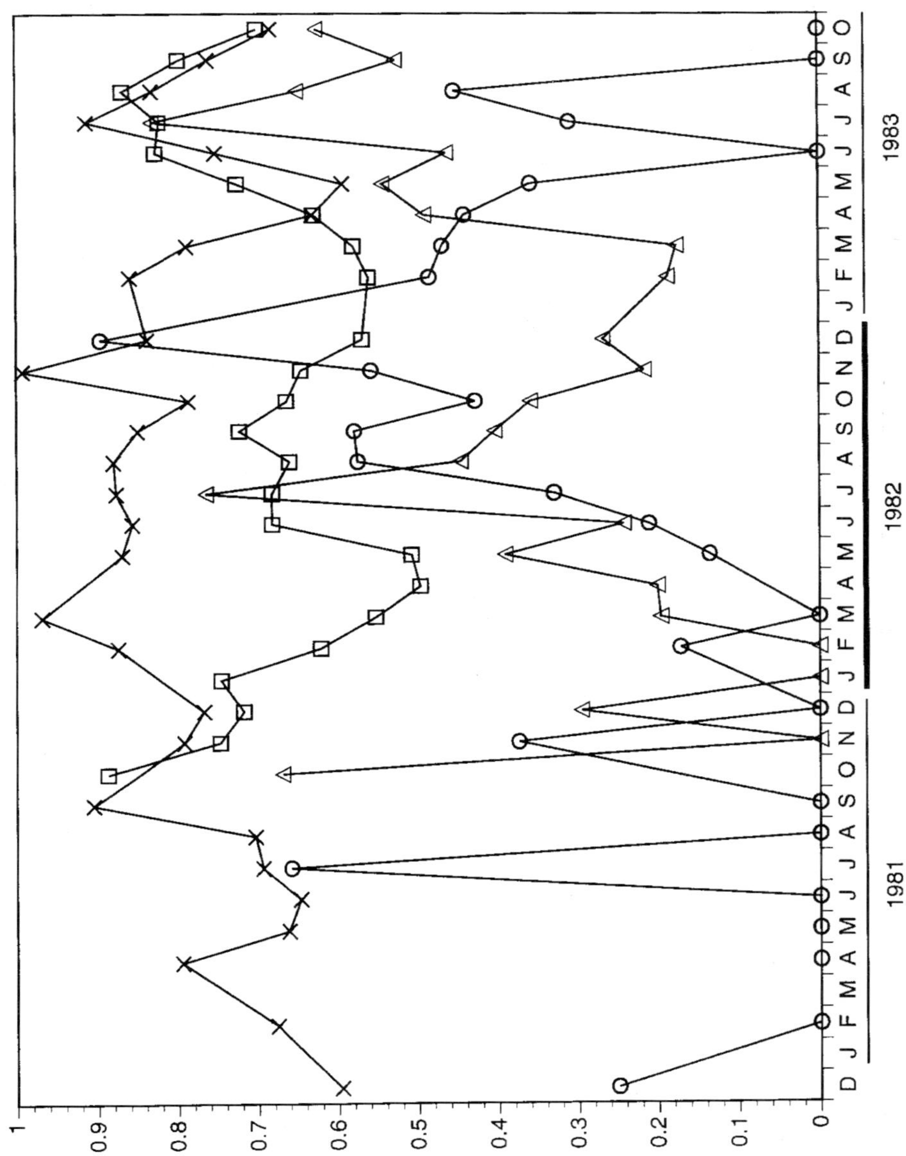

Fig. 4. Evenness of each habitat (circle = cropland, square = ditch, X = remnant grassland, triangle = pasture).

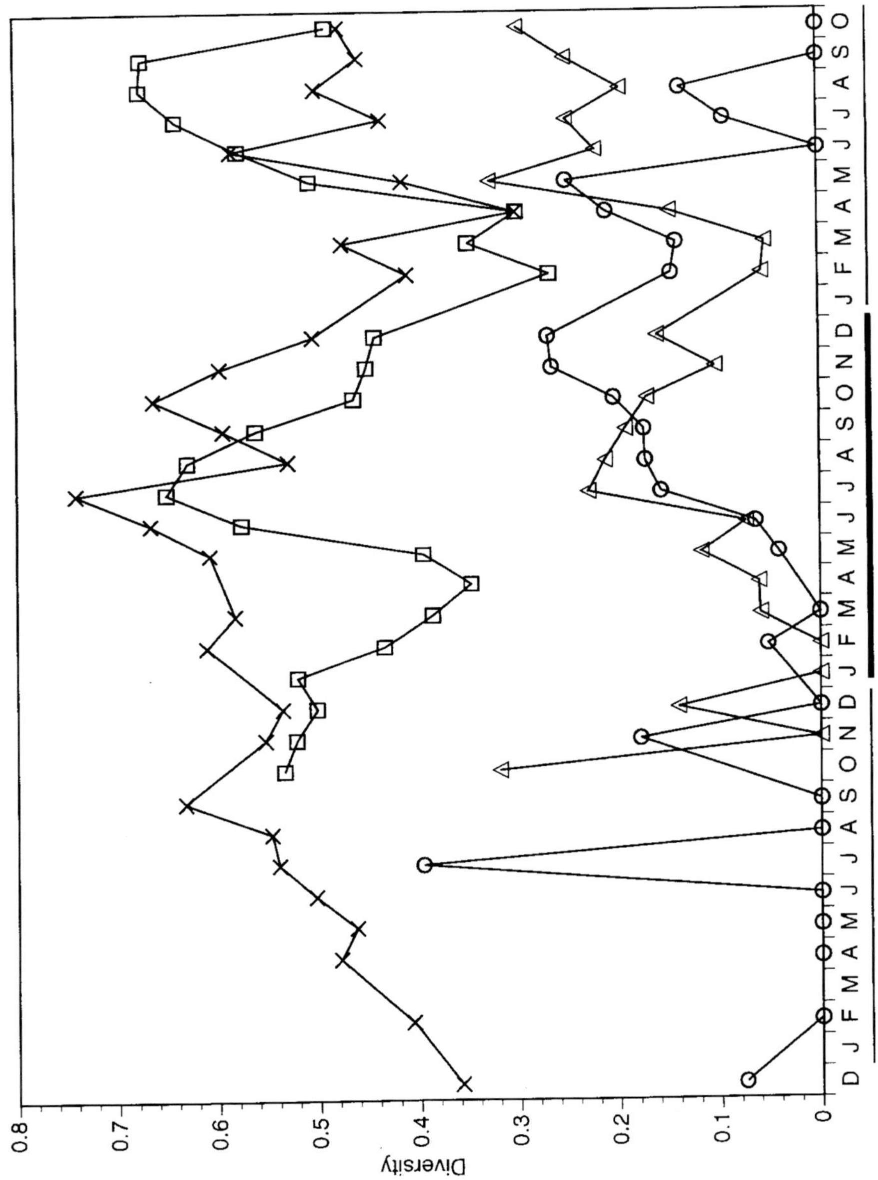

Fig. 5. Diversity of each habitat (circle = cropland, square = ditch, X = remnant grassland, triangle = pasture).

Table 2. Inter-community comparisons, Horn's (1966) similarity index inter-community (SIM) and Hutcheson's (1970) T-test for diversity intra-community (D, '*' indicates $p < 0.05$), of the four habitats.

		Cropland vs. Ditch		Cropland vs. Grassland		Cropland vs. Pasture		Ditch vs. Grassland		Ditch vs. Pasture		Grassland vs. Pasture	
		SIM	D	SIM	D	SIM	D	SIM	D	SIM	D	SIM	D
	D	—		0.88	*	—		—		—		—	
1981	J	—		—		—		—		—		—	
	F	—		0.60	*	—		—		—		—	
	M	—		—		—		—		—		—	
	A	—		0.58	*	—		—		—		—	
	M	—		0.50	*	—		—		—		—	
	J	—		0.52	*	—		—		—		—	
	J	—		0.53		—		—		—		—	
	A	—		0.59	*	—		—		—		—	
	S	—		0.55	*	—		—		—		—	
	O	—		—		—		—		0.52		—	
	N	0.48	*	0.59	*	0.95	*	0.77		0.49	*	0.57	*
	D	0.77	*	0.48	*	0.96		0.73		0.78	*	0.55	*
1982	J	—		—		—		—		0.74	*	—	
	F	0.86	*	0.65	*	0.99		0.82		0.86	*	0.63	*
	M	0.87	*	0.53	*	0.98		0.70		0.86	*	0.52	*
	A	—		—		—		—		0.89		—	
	M	0.85	*	0.40	*	0.95		0.58		0.84	*	0.44	*
	J	0.78	*	0.55	*	0.96		0.76		0.74	*	0.49	*
	J	0.78	*	0.53	*	0.84		0.83		0.70	*	0.43	*
	A	0.63	*	0.48	*	0.86		0.82		0.67	*	0.46	*
	S	0.60	*	0.48	*	0.87		0.88		0.70	*	0.52	*
	O	0.50	*	0.53	*	0.89		0.81	*	0.58	*	0.58	*
	N	0.50	*	0.49	*	0.86		0.74	*	0.54	*	0.48	*
	D	0.56	*	0.39	*	0.85		0.76		0.58	*	0.46	*
1983	J	—		—		—		—		—		—	
	F	0.84		0.34	*	0.93		0.30	*	0.87	*	0.31	*
	M	0.82	*	0.26	*	0.93		0.47		0.87	*	0.23	*
	A	0.91		0.13		0.88		0.10		0.80	*	0.11	*
	M	0.79	*	0.41		0.87		0.31		0.69		0.41	
	J	0.62	*	0.36	*	0.91		0.48		0.65	*	0.46	*
	J	0.69	*	0.53	*	0.83		0.63		0.57	*	—	
	A	0.70	*	0.39	*	0.87		0.66		0.77	*	0.46	*
	S	0.68	*	0.73	*	0.90		0.80		0.70	*	0.77	*
	O	0.71		0.81	*	0.87		0.71		0.91		0.71	
MEAN		0.71		0.51		0.90		0.64		0.72		0.48	

and 2 of 22 that were different in both ditch and pasture (Table 2). None of the periods of significant difference coincided in any of the habitats.

Horn's (1966) similarity indices also were used to make intra-community comparisons between consecutive months. High similarity values within a habitat indicate stability in species composition. The highest mean similarity in a habitat was 0.96 for the cropland, followed closely by the pasture (0.95), remnant grassland (0.94), and ditch (0.93) (Table 1). A closer examination of the similarity values shows that the drought had little effect on the composition of species within the cropland as evinced by a mean similarity of 0.96 before October 1982 and 0.96 after. The composition of species within the remnant grassland also was not appreciably affected by the drought (0.94 prior and 0.93 after). Both ditch and pasture, however, showed large differences in species composition before and during the drought. The mean similarity in the ditch after October 1982 was 0.89, whereas before that time it was 0.96. This was mirrored in the pasture, which showed a decline from 0.97 to 0.92 for the same period. The lowest intra-community similarity (0.55) was registered in the ditch in February 1983.

Peromyscus maniculatus (deer mice) was the only permanent species found in all four communities and had the greatest habitat breadth (Table 3). Deer mice were most common in the ditch (152.8 MNA/1,000 trap nights), followed by the cropland (140.8), pasture (126.2), and remnant grassland (73.1), where it was the second most abundant species. Populations of *P. maniculatus* slumped during the drought in all habitats. Populations of deer mice in the cropland were found to be correlated with those in the ditch ($r = 0.498$, $p = 0.022$) and remnant grassland ($r = 0.522$, $p = 0.004$). Populations of *P. maniculatus* in the ditch were correlated with those in the pasture ($r = 0.488$, $p = 0.015$).

Sigmodon hispidus (hispid cotton rat), the second most numerous species as indicated by relative density, had the second greatest habitat breadth (Table 3). In the remnant grassland, the hispid cotton rat was the third most abundant species and was classed as permanent (Fig. 6). In the ditch it was a semi-permanent species, establishing resident populations from October 1981 through December 1982, when it became the second most plentiful species (Fig. 7).

Populations of *S. hispidus* and *Microtus ochrogaster* (prairie vole) in the ditch were negatively correlated ($r = -0.642$, $p < 0.001$). Populations of *Mus musculus* (house mouse) in the ditch were positively correlated with those of *S. hispidus* ($r = 0.407$, $p = 0.048$), whereas populations of *S. hispidus* and *B. hylophaga* (Elliot's short-tailed shrew) were positively correlated in the remnant grassland ($r = 0.592$, $p = 0.001$) and ditch ($r = 0.703$ $p = 0.001$).

The western harvest mouse (*Reithrodontomys megalotis*) was the most plentiful species in the remnant grassland, where it was classed as permanent (Table 3); populations were high in 1981 and 1983 but slumped in 1982 (Fig. 6). *R. megalotis* occurred in all three of the other habitats but only as a transient species (Table 3). Populations of *R. megalotis* were not correlated with any other permanent or semi-permanent species.

Microtus ochrogaster was a semi-permanent species that established resident populations in the remnant grassland from November 1981 through July 1982 and also had resident individuals in the ditch from February 1983 until September 1983 (Table 3 and Fig. 7). During the drought, the population of prairie voles in the remnant grassland fell sharply and remained low. No prairie voles were captured in the cropland or pasture. Only populations of *Sigmodon hispidus* in the ditch showed a significant but negative correlation with those of *M. ochrogaster*.

The hispid pocket mouse (*Chaetodipus hispidus*) had an unusually high habitat breadth considering that it was never more than a transient in any of the communities (Table 3). Hispid pocket mice were relatively rare in all habitats before the drought, with only a few scattered individuals captured. As the drought deepened in the spring of 1983, however, more individuals were caught in all four habitats. By the end of the study, considerable numbers of *C. hispidus* were being captured in all habitats even though resident populations were never established.

House mice were classed as a semi-permanent species in the ditch, where a resident popu-

Table 3. **The species captured, their habitat breadth, status in each habitat (P = permanent, S = semi-permanent, and T = transient), and relative density (#/1,000 trapnights) in each habitat.**

	Mean Habitat Breadth	Cropland	Ditch	Grassland	Pasture
Peromyscus maniculatus	3.03	P (140.8)	P (152.8)	P (73.1)	P (126.2)
Sigmodon hispidus	1.44	T (0.4)	S (93.1)	P (55.0)	—
Reithrodontomys megalotis	1.17	T (0.4)	T (9.5)	P (105.9)	T (0.7)
Microtus ochrogaster	0.87	—	S (16.7)	S (18.8)	—
Chaetodipus hispidus	0.82	T (0.4)	T (4.9)	T (4.6)	T (1.6)
Mus musculus	0.81	T (2.9)	S (40.0)	T (0.8)	T (1.4)
Spermophilus tridecemlineatus	0.80	T (0.6)	T (1.4)	T (1.9)	S (8.1)
Blarina hylophaga	0.77	—	T (4.6)	S (12.5)	—
Onychomys leucogester	0.57	S (5.6)	T (0.8)	—	T (0.9)
Reithrodontomys montanus	0.07	—	T (0.2)	—	T (0.2)
Peromyscus leucopus	0.03	—	—	T (0.6)	—

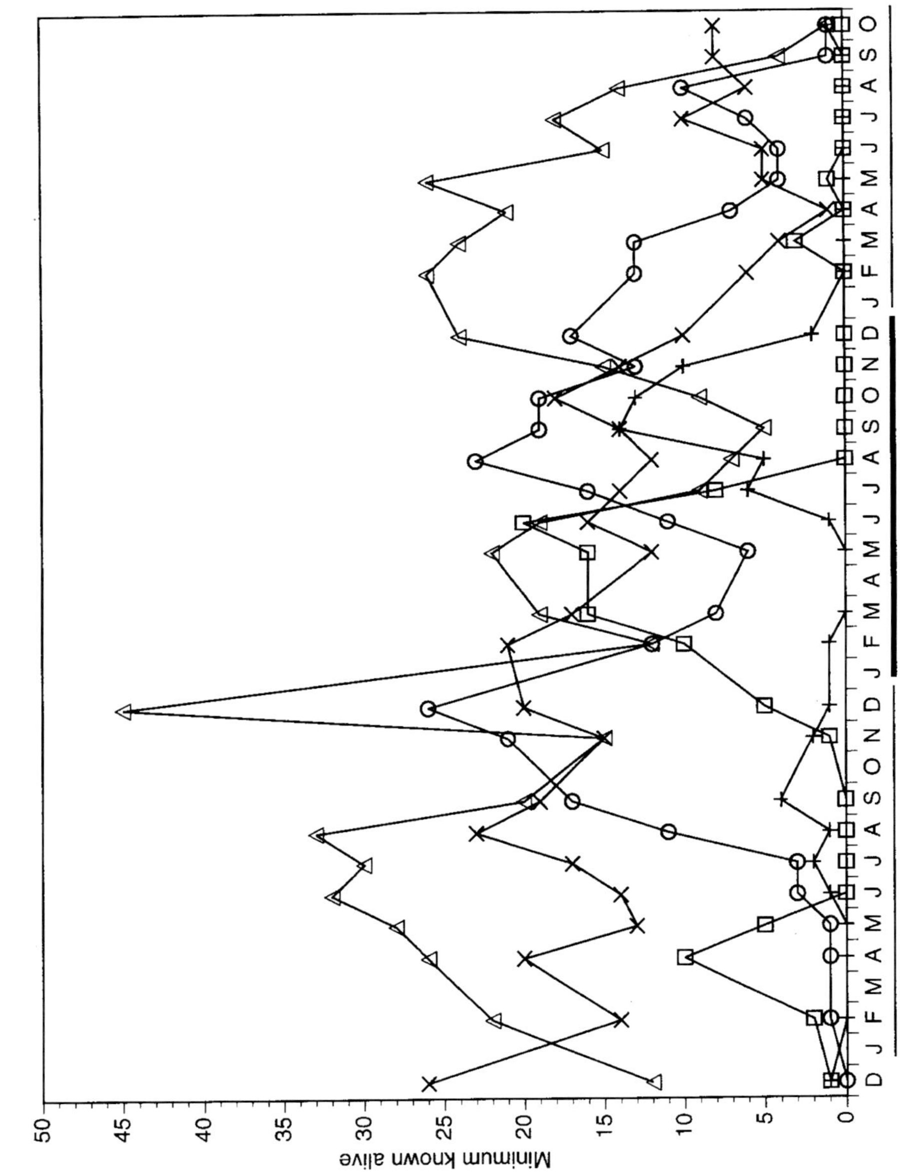

Fig. 6. Minimum number known to be alive of permanent and semi-permanent species in the remnant grassland (X = *Peromyscus maniculatus*, circle = *Sigmodon hispidus*, triangle = *Reithrodontomys megalotis*, square = *Microtus ochrogaster*, and cross = *Blarina hylophaga*).

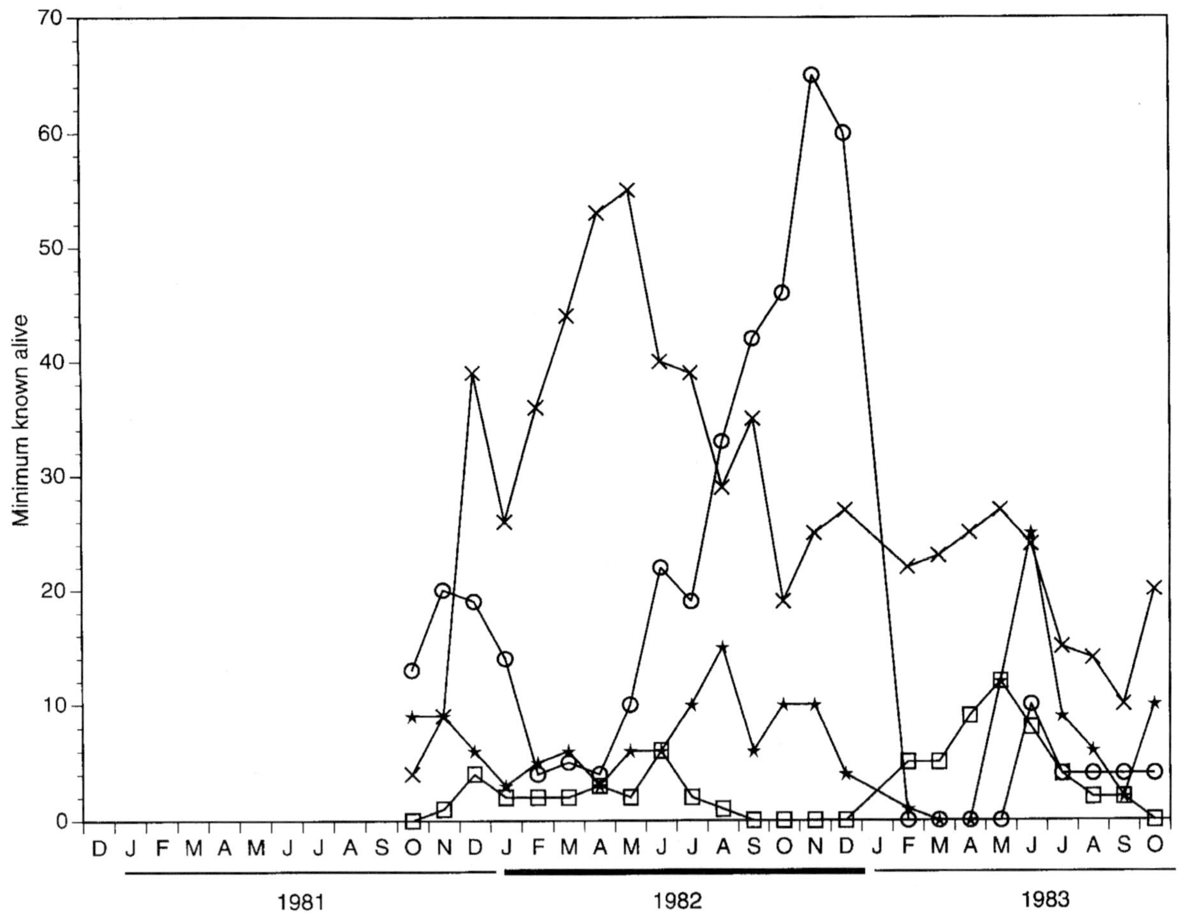

Fig. 7. Minimum number known to be alive of permanent and semi-permanent species in the ditch (X = *Peromyscus maniculatus*, circle = *Sigmodon hispidus*, square = *Microtus ochrogaster*, and semicircle = *Mus musculus*.

lation was established from February 1982 through September 1982 (Table 3). Transient individuals were caught in the other three habitats. *Sigmodon hispidus* was the only species whose population was correlated with that of *Mus musculus* in the ditch.

The thirteen-lined ground squirrel *(Spermophilus tridecemlineatus)* was a semi-permanent species in the pasture, where it formed a resident population, from May through August 1982. In the other three habitats, it was merely a transient. *S. tridecemlineatus* also was captured in high numbers from July through August 1983 in the pasture, but no individuals became residents. Insufficient numbers of *S. tridecemlineatus* were caught in enough trapping periods to justify statistical analysis.

Blarina hylophaga formed semi-permanent populations in the remnant grassland from July through December during 1981 and 1982 and was a transient species in the ditch. Nevertheless, it had relatively high habitat breadth. Populations of *B. hylophaga* were correlated to those of *S. hispidus* in remnant grassland, as mentioned above.

Onychomys leucogaster, the northern grasshopper mouse, occurred at very low numbers in the cropland, ditch, and pasture, but succeeded in forming a resident population in the cropland from July 1982 through March of 1983. During their tenure in the above habitats, populations of *O. leucogaster* were eclipsed in number by transient species. Insufficient numbers of northern grasshopper mice were captured to justify statistical work.

The plains harvest mouse (*Reithrodontomys montanus*) was captured as a transient in the ditch and pasture, whereas the white-footed mouse (*Peromyscus leucopus*) was captured only as a transient in the remnant grassland. Numbers of both of these species were insufficient to justify statistical analysis.

DISCUSSION

Richness, Evenness, and Diversity

In order to maintain a high diversity of small mammals, a complex community structure is required that provides animals with nesting areas, cover, and food resources. If these are present, a complex assemblage of small mammals can be supported as manifested by relatively high richness and evenness values, resulting in a high diversity index. Conversely, if the ecosystem lacks the vegetative structure to provide the necessities of life, low richness, evenness, and diversities can be expected. There is, however, another possibility. If less complex vegetative communities are adjacent to more complex ones, individuals residing in the latter occasionally might venture into the former during periods of high density or in search of food. This would result in a relatively high richness value but low evenness and diversity indices occurring in less complex habitats. In the mosaic of habitats sampled for small mammals in west–central Kansas, examples of all three of these possibilities were found.

The vegetatively complex communities of the remnant grassland and roadside ditch support large assemblages of small mammals, as indicated by high diversity values. The lush, relatively dense vegetation in the remnant grassland and ditch hold prairie snows and thereby provide soil moisture for a luxuriant spring and summer growth of vegetation that provides food, litter, and cover throughout the year. Additionally, the sloping nature of ditch banks and the rolling topography of the remnant grassland increase the ease of burrow construction for nests. It is not surprising that richness, evenness, and diversity values show these two habitats to be similar.

High mean richness values in the ditch suggests that "waste areas" are important habitats for some small mammals (Fleharty and Navo 1983; Yahner, 1983; Getz and Brighty, 1986; Kaufman and Kaufman, 1989) as well as dispersal routes for others (Getz et al., 1978; Kaufman and Kaufman, 1990), as suggested by high numbers of semi-permanent and transient species (Table 3). Dispersing species increase the richness and diversity of the ditch, but the seasonality and variable numbers associated with dispersal cause the small mammal assemblage to appear unstable when compared with other habitats.

Diversity in the ditch generally was very high (Fig. 5) and exhibited a regular pattern in the two years that it was sampled. Peak diversities in the ditch in 1982 and 1983, although similar in timing, shape, and amplitude, were driven by different factors. The peak diversity in 1982, prior to the drought, was driven by increased richness (Table 1), whereas the peak in 1983, during the drought, was driven by increased evenness (Fig. 4), primarily because of reduced populations of *P. maniculatus* and *S. hispidus*.

The remnant grassland was the only habitat in which *P. maniculatus* was not the only permanent species or the most abundant species (Table 3). As a consequence of the reduced numbers of deer mice and increased numbers of other permanent species, *R. megalotis* and *S. hispidus*, the remnant grassland had the greatest mean evenness value. The remnant grassland had more permanent and semi-permanent species than the other habitats, but it had fewer transients than the ditch. This combination resulted in a lower mean richness in the remnant grassland than that of the ditch (Table 3). Even though the mean richness of the ditch was greater than that of the remnant grassland, the latter had the greatest mean diversity because of the consistently large evenness values.

Grazed pastures and croplands have less complex vegetative structure, which results in reduced diversity. Croplands throughout much of the year provide little cover and lack the vegetative complexity to provide permanent habitat for large assemblages of small mammals. Grazed pastures, although providing nearly complete cover by cropped short grasses, lack the structural complexity and variety of plants found in either the roadside ditch or remnant grassland. Additionally, there is almost no litter; soils are compacted by the trampling of cattle, which is less conducive to burrowing and nest construction. As expected, diversity values were lower in these two habitats.

Low richness, evenness, and diversity values have been found to characterize cultivated ecosystems (Fleharty and Navo, 1983; Navo and Fleharty, 1983; Castrale, 1984 and 1985; Kaufman and Kaufman, 1989 and 1990); our results are no exception. The habitat alteration necessary for cultivation is too drastic for most species to endure (Navo and Fleharty, 1983; Kaufman and Kaufman, 1990). However, for those that are capable of inhabiting this immature ecosystem, large populations often result (Castrale, 1985).

Richness values in the cropland were much lower than those in the ditch, and there was little evidence of use of the cropland by individuals dispersing from the ditch. However, this may have been a function of the distance between the two traplines: that in the cropland was 200 m from the ditch habitat. Consequently, we did not see the increase in richness suggested by Kaufman and Kaufman (1989). Won (1990), however, set traplines in a wheatfield that bordered a ditch and, as expected (Kaufman and Kaufman, 1989), richness values between ditch and cropland were similar. Kaufman and Kaufman (1989) sampled fencerows between cultivated fields and grazed prairies and found that the fencerow habitat was an important one for nesting and burrowing for a relatively large number of small mammal species that could use both grazed prairie and cropland for food. Kaufman and Kaufman (1989) speculate that these incursions would result in high richness values even though diversity and evenness were low.

Richness was higher in the pasture than in the cropland, but was still relatively low when compared to the remnant grassland or ditch. Evenness in the pasture, although low, increased during the drought (Fig. 4) because the population of deer mice decreased, while those of hispid pocket mouse, thirteen-lined ground squirrels, and northern grasshopper mice increased. This increase in evenness resulted in an increased diversity in the pasture (Fig. 5). The drought impacted the population of *P. maniculatus* in the pasture more than in the other habitats. Declines in the population of deer mice were offset partially by increases in other species, so only a slight decrease in the total number of small mammals was apparent (Fig. 3).

Total Numbers

Peromyscus maniculatus is the dominate species in many types of cropland in the United States (Whitaker, 1967; Houtcooper, 1978; Navo and Fleharty, 1983; Castrale, 1984 and 1985; Wooley et al., 1985; Clark and Young, 1986; Getz and Brighty, 1986; Kaufman and Kaufman, 1990) and is largely unaffected by agricultural practices (Castrale, 1985; Wooley et al., 1985; Getz and Brighty, 1986; Wegner and

Merriam, 1990). The disturbances associated with farming operations in the cropland studied herein likewise had little impact on *P. maniculatus* populations. Large populations of deer mice were found in all stages of the three-year crop rotation and were often greater than those in the other habitats. It is arguable that the conversion of native prairie habitats to cultivations has actually increased population sizes of small mammals such as deer mice that are able to co-exist with cultivation (Fleharty and Navo, 1983; Kaufman and Kaufman, 1990).

M. musculus and *O. leucogaster* are also noted as occurring in cropland habitats (Hanson, 1944; Whitaker, 1967 and 1968; Fleharty and Navo, 1983; Navo and Fleharty, 1983; Castrale, 1985; Clark and Young, 1986; Getz and Brighty, 1986; Won, 1990). *Mus musculus* was caught only as a transient within the cropland, which supports the suggestion that this species persists in agricultural habitats by being nomadic (Whitaker, 1967). *M. musculus* enters croplands only when there is ample overhead cover or human artifacts present and disperses quickly after harvest (Whitaker, 1967 and 1968; Houtcooper, 1978; Fleharty and Navo, 1983; Getz and Brighty, 1986). *Onychomys leucogaster* favors disturbed habitats with little soil litter or overhead cover (Choate and Terry, 1974). Of the habitats herein sampled, only the cropland had those characteristics and northern grasshopper mice were captured most frequently here.

The ditch generally supported greater numbers of animals than the other three habitats (Fig. 3). The luxuriant, dense vegetation in the ditch creates a vegetatively complex community that provides better cover and food than the cropland and pasture for a large number of resident individuals in addition to dispersers.

Numbers of small mammals in the pasture remained modest but relatively stable over the study period compared to the other habitats where the drought caused significant declines.

Intra-Community Similarity

Mean intra-community similarity is a measure of the stability of species composition. High values indicate little change from one trapping period to the next, whereas low values indicate appreciable changes in species composition. The cropland, which was the least diverse habitat, had the greatest mean intra-community similarity. Having fewer species, the cropland had fewer possible sources of variability and thus a greater stability in species composition. The low intra-community value in the cropland in June 1981 signaled the appearance of a relatively large population of *O. leucogaster*, and the continued low similarity the next month reflected the immediate loss of the same population.

The ditch, which had the greatest mean richness, had the lowest mean intra-community similarity (0.55). This resulted from the species composition of the ditch relying on the dynamics of many species, a change in any one of which lowered the intra-community similarity. This low value was recorded in February 1983 when richness fell from six species in the preceding trapping period (December 1982) to three. Although the loss of three species considerably altered the community, the loss of one of those species, *S. hispidus*, was more important than the others.This was because in December 1982 the population of 64 cotton rats comprised more than half all of the small mammals found there (Fig. 7).

The low value recorded in the ditch in April and May 1983 was the result of a rapidly increasing population of *M. ochrogaster*, perhaps because of the population crash of *S. hispidus*. Had fewer species from the ditch been transients and more been permanent or semi-permanent, as was true in the grassland, the stability in the ditch would have been greater. Although the presence of transients in the ditch increases the richness and diversity, the seasonality and unstable numbers of transients decreases intra-community similarity.

The remnant grassland exhibited consistently high intra-community similarity values until the end of the study, when there was a precipitous decline in the number of *S. hispidus*. Whether this decline was real or an artifact from cessation of trapping is unknown.

The low intra-community similarity values in the pasture in July and August 1983 represents a rapid increase in the number of *S. tridecemlineatus* captured. This increase coincided with record high temperatures and suggests that these diurnal ground squirrels were altering their behavior to avoid the high temperatures during the day by becoming crepuscular and thus more susceptible to trapping in the evening.

Inter-Community Similarity

Although the statistical procedures performed on richness, evenness, and diversity indicated that the cropland most closely resembled the pasture and the remnant grassland most closely resembled the ditch, Horn's (1966) similarity index (inter-community similarity) presented a different perspective. The results of these inter-community comparisons show that the actual species composition that forms the small mammal community in the ditch most resembles that of the pasture, even though the structure of the ditch (as represented by the community characters) resembles that of the remnant grassland.

The similarity of the ditch to the remnant grassland was altered by the drought. Before the drought, the species composition of the ditch resembled the species compositions of the other habitats. After the drought, the species composition of the ditch was less like that of the remnant grassland because of a decrease in populations of *S. hispidus* and *M. ochrogaster*. This shifted the species composition in the ditch away from that in the remnant grassland and toward that in the pasture, where those species were rare or absent.

HISTORICAL EFFECTS ON SMALL MAMMALS

In discussions on biodiversity, it commonly is asked whether the activities of humans have increased or decreased the biodiversity of a particular region. Because detailed biotic analysis on the Great Plains did not take place until after settlement by ranchers and farmers, we can only speculate whether the activities of these initial agriculturalists altered the small mammal/vegetation relationship that existed. This speculation necessitates a brief historical review.

The military installation named Fort Hays was established on 4 July 1867 on a hill above Big Creek, less than three miles from all four of the habitats studied herein. Historical accounts of the native mammals of Ellis County exist in the medical records kept at Fort Hays during the period 1867–1889. These records indicate that "The common prairie dog [*Cynomys ludovicianus*] are also in large numbers throughout this section." Other mammals present include the "Jack-ass rabbit [*Lepus* sp.], a species of large size" and "the families Antelopidae and Bovidae, both appearing in large herds, the latter in the buffalo are frequently seen in herds of several thousand . . . generally lingering in small parties in the vicinity grazing and are generally found within a short distance throughout the year." Additionally, the "American Elk" [*Cervus elaphus*] and "striped gopher" [*Spermophilus* sp.] were recorded in an entry dated 20 June 1889.

The wapiti (*Cervus elaphus*) was common in the region prior to settlement by European man. Mr. J.R. Mead, a hunter, referred to droves "of 1,000 more or less" and said they were particularly numerous in the 1850s and 1860s in the area, where they preferred broken country with timber and streams (Hoffmeister, 1947).

The prairie was both grazed and disturbed by the trampling of hooves and the wallows of bison as well as by prairie dog excavations that extended over vast areas. Some springs existed, as did a few streams and smaller rivers. Some were permanent, some ephemeral, and they probably supported riparian communities consisting of moister, more luxuriant vegetation and, in places, shrubs and a few trees. J.A. Allen (1871) visited Fort Hays in 1871 and in his description of the area states, "The 'Plains' are here, as usual, somewhat rolling broad level plateaus, being separated by low ridges, or broken by sharp ravines and hollows. They are covered with short grass, usually but two or three inches high, except in the hollows and near the streams where it often grows to the height of one or two feet."

Human activities that altered these prairie ecosystems doubtlessly caused some species of mammal to decrease in abundance (or even become extirpated), some to increase in abundance, and some, heretofore not present, to move into the area. We believe that agricultural activities in this region did all three. In the large mammal category, bison, elk, and pronghorn (*Antilocapra americana*) were eliminated. Likewise, black-footed ferrets (*Mustela nigripes*) and prairie dogs have largely been displaced.

Of the 11 species of small mammals that we captured, two species, *M. musculus* and *S. hispidus*, were not present until European man arrived on the scene. House mice probably were introduced accidentally with the possessions of the early settlers, whereas the neotropical hispid cotton rat invaded this area only in the latter half of this century. In both cases, but especially with *S. hispidus*, the establishment of roadside ditches, remnant grasslands, and waste areas around farmsteads facilitated the dispersal and survival of these species.

Blarina hylophaga, *M. ochrogaster*, and *R. megalotis* undoubtedly were present in limited numbers before agriculture, occupying habitats of dense, luxuriant vegetation in riparian habitats or around springs. As above, the construction of ditches, remnant grasslands, and waste areas contributed to their dispersal and survival and, we believe, to an increased abundance.

Spermophilus tridecemlineatus was adversely affected by breaking up some of the prairie for cropland. Cropland, remnant grasslands, or heavily vegetated roadside ditches, although occasionally used, are not considered preferred habitat for this species. Thirteen-lined ground squirrels probably are not as abundant as they once were, although they do well in grazed rangeland and artificially created habitats such as golf courses and school yards.

Peromyscus leucopus primarily inhabits riparian areas in western Kansas. It frequently disperses along roadside ditches or through one area of woody vegetation to another.

The effect of agricultural perturbations on three species—*C. hispidus*, *R. montanus*, and *O. leucogaster*—is less clear. *R. montanus* is thought to prefer shortgrass communities (Kaufman and Fleharty, 1974) and has been noted in dry, sandy soils (Choate and Fleharty, 1975); however, there is a paucity of information available about the ecology of this species on the Great Plains. *C. hispidus* has been reported living in areas characterized by short grass (Kaufman and Fleharty, 1974) and sandy soils (Choate and Fleharty, 1975), and increasing in abundance during the drought in the 1930s (Wooster, 1939). Nevertheless, accurate and detailed information on this species is also lacking.

Choate and Terry (1974) speculated that *O. leucogaster* might have been present in the preagricultural prairie, especially in sandy areas where non-human activities such as floods, fires, heavy grazing, and trampling may have helped create the habitat that the northern grasshopper mouse prefers. Bison wallows and excavations by prairie dogs probably should be added to that list. The addition of cropland in the Great Plains has added prime habitat for *O. leucogaster*, especially if that cropland is established on sandy soil (Fleharty and Navo, 1983). We postulate that crop production activities have had a positive effect on grasshopper mice, providing them with loose soil for dusting (Egoscue, 1960) and a ready source of food in the form of insects that are attracted by crops. This positive effect has allowed an increase in abundance of northern grasshopper mice.

Peromyscus maniculatus, the generalist species of the group, seemingly is not affected adversely by human activities. This species appears to be equally well adapted to survival on grazed rangeland, roadside ditches, remnant grassland, and even croplands in this vicinity.

All habitats we sampled have been affected by human activities. Even the remnant grassland, which one naively might believe to be somewhat like the native prairie, is artificial due to the lack of grazing and attendant increased depth of mulch and height of grasses and forbs. Closing this remnant prairie to grazing, mowing, and burning has created a very different ecosystem than the one present when bison, pronghorn, and other native herbivores roamed the prairie. Of the four habitats sampled, however, the moderately grazed pasture probably comes closest to approximating the native prairie habitat that was found within the region prior to inhabitation by the European settlers.

ACKNOWLEDGMENTS

Thanks go to the following individuals who tended traplines: Jerry Stanfield, Tricia Teller, the late Bruce Wendt, Shirley Valek, and Peg Svoboda. We are especially appreciative of the pre-publication review and suggestions provided by Dr. J.R. Choate.

I (Fleharty) give a special thanks to my major professor, Dr. James S. Findley. I was fortunate to be his field assistant for two summers, tramping and trapping the montane areas of New Mexico and Arizona. I have many fond memories of the long discussions that took place during that time, and I probably learned more biology during those summers than in any formal classroom.

LITERATURE CITED

Albertson, F.W., and G.W. Tomanek. 1965. Vegetation changes during a thirty-year period in grassland communities near Hays, Kansas. Ecology, 46:714-720.

Allen, J.A. 1871. Notes of an ornithological reconnaissance of portions of Kansas, Colorado, Wyoming, and Utah. II. Lists of birds observed in the vicinity of Fort Hays, Kansas, from May 26 to July 3, 1871; with annotations. Bulletin of the Museum of Comparative Zoology, 3:131-142.

Castrale, J.S. 1984. Wildlife use of cultivated fields set aside under the payment in kind (PIK) program. Indiana Academy of Science, 93:173-180.

Castrale, J.S. 1985. Responses of wildlife to various tillage conditions. Transactions of the North America Wildlife and Natural Resource Conference, 50:142-155.

Choate, J.R., and E.D. Fleharty. 1975. Synopsis of native, Recent mammals of Ellis County, Kansas. Occasional Papers, The Museum, Texas Tech University, 37:1-80.

Choate, J.R., and D.M. Terry. 1974. Observations on habitat preference of *Onychomys leucogaster* (Rodentia: Muridae) on the central Great Plains. Transactions of the Kansas Academy of Science, 76:263-265.

Clark, W.R., and R.E. Young. 1986. Crop damage by small mammals in no-till cornfields. Journal of Soil and Water Conservation, 41:338-341.

Darton. N.H. 1916. Guidebook of the western United States. Part C. The Santa Fe Route Bulletin 613, Government Printing Office, Washington, DC, 200 pp.

Egoscue, H.J. 1960. Laboratory and field studies of the northern grasshopper mouse. Journal of Mammalogy, 41:99-110.

Elliott, R.S. 1883. Notes taken in sixty years. R. P. Studley and Co., St. Louis, MO, 336 pp.

Evans, R.A., and R. M. Love. 1957. The step-point method of sampling—A practical tool in range research. Journal of Range Management, 10:208-212.

Fleharty, E.D. 1972. Some aspects of small mammal ecology in a Kansas remnant prairie. Proceedings of the Midwest Prairie Conference, 2:97-103.

Fleharty, E.D., and M.A. Mares. 1973. Habitat preference and spatial relations of *Sigmodon hispidus* on a remnant prairie in west-central Kansas. The Southwestern Naturalist, 18:21-29.

Fleharty, E.D., and K.W. Navo. 1983. Irrigated cornfields as habitat for small mammals in the sandsage prairie region of western Kansas. Journal of Mammalogy, 64:367-379.

Fleharty, E.D., and D.L. Stadel. 1968. Distribution of *Peromyscus leucopus* (woods mouse) in western Kansas. Transactions of the Kansas Academy of Science, 71:231-233.

Frydendall, M.J. 1969. Rodent populations on four habitats in central Kansas. Transactions of the Kansas Academy of Science, 72:213-222.

Getz, L.L., and E. Brighty, 1986. Potential effects of small mammals in high-intensity agricultural systems in east-central Illinois, U.S.A. Agriculture, Ecosystems and Environment, 15:39-50.

Getz, L.L., F.R. Cole, and D.L. Gates. 1978. Interstate roadsides as dispersal routes for *Microtus pennsylvanicus*. Journal of Mammalogy, 59:208-212.

Hansen, C.M., and E.D. Fleharty. 1974. Structural ecological parameters of *Peromyscus maniculatus* in west-central Kansas. The Southwestern Naturalist, 19:293-303.

Hanson, H.C. 1944. Small mammal censuses near Prairie du Sac, Wisconsin. Wisconsin Academy of Sciences, Arts and Letters, 36:105-129.

Hilborn, R., J.A. Redfield, and C.J. Krebs. 1976. On the reliability of enumeration for mark and recapture census of voles. Canadian Journal of Zoology, 54:1019-1024.

Hoffmeister, D.F. 1947. Early observations on the elk in Kansas. Transactions of the Kansas Academy of Science, 50:75-76.

Horn, H.S. 1966. Measurement of "overlap" in comparative ecological studies. The American Naturalist, 100:419-424.

Houtcooper, W.C. 1978. Distribution and abundance of rodents in cultivated ecosystems. Indiana Academy of Science, 87:434-437.

Hulett, G.K., J.H. Brock, and J.E. Lester. 1972. Community structure and function in a remnant Kansas prairie. Proceedings of the Second Midwest Prairie Conference, 2:104-112.

Hutcheson, K. 1970. A test for comparing diversities based on the Shannon formula. Journal of Theoretical Biology, 29:151-154.

Kaufman, D.W., and E.D. Fleharty. 1974. Habitat selection by nine species of rodents in north-central Kansas. Southwestern Naturalist, 18:443-452.

Kaufman, D.W., and G.A. Kaufman. 1989. Nongame wildlife management in central Kansas: Implications of small mammal use of fence rows, fields, and prairie. Transactions of the Kansas Academy of Science, 92:198-205.

Kaufman, D.W., and G.A. Kaufman. 1990. Small mammals of wheat fields and fallow wheat fields in north-central Kansas. Transactions of the Kansas Academy of Science, 93:28-37.

Navo, K.W., and E.D. Fleharty. 1983. Small mammals of winter wheat and grain sorghum croplands in west-central Kansas. The Prairie Naturalist, 15:159-172.

Shannon, C.E., and W. Weaver. 1949. The mathematical theory of communication. University of Illinois Press, Urbana, IL, 117 pp.

Stout, I.J., and D.E. Sonenshine. 1973. A convenient bait for small mammal livetrapping studies. Acta Theriologica, 18:123.

Watson, L. 1872. Watson to directors, [1872]. Jennings Collection, MSS Division, Kansas State Historical Society, Topeka, KS.

Walker, F. 1872. A compendium of the ninth census. Government printing office, Washington, DC.

Wegner, J., and G. Merriam. 1990. Use of spatial elements in a farmland mosaic by woodland rodent. Biological Conservation, 54:263-276.

Whitaker, J.O., Jr. 1967. Habitat relationships of four species of mice in Vigo County, Indiana. Ecology, 48:867-872.

Whitaker, J.O., Jr. 1968. Relationship of *Mus*, *Peromyscus* and *Microtus* to the major textural classes of soils of Vigo County, Indiana. Proceedings of the Indiana Academy of Science, 77:206-212.

Whittaker, R.H., and S.A. Levin, eds. 1975. Niche theory and application. Benchmark papers in Ecology. Vol. 3, Halstead Press, Stroudsburg, PA, 448 pp.

Won, C.M. 1990. Ecological characteristics of small mammals under the influence of dryland crop production in west-central Kansas. Unpublished M.S. thesis. Fort Hays State University, 27 pp.

Wooley, J.B., L.B. Best, and W.R. Clark. 1985. Impacts of no-till cropping on upland wildlife. Transactions of the North American Wildlife and Natural Resource Conference, 50:157-168.

Wooster, L.D. 1939. The effects of drought on rodent populations. Turtox News, 17:1-2.

Yahner, R.H. 1983. Population dynamics of small mammals in farmstead shelterbelts. Journal of Mammalogy, 64:380-386.

Zar, J.H. 1984. Biostatistical Analysis. Prentice-Hall Inc., Englewood Cliffs, NJ, 718 pp.

E.D. Fleharty, Department of Biological Sciences & Allied Health, Fort Hays State University, Hays, Kansas 67601, and Rob Channell, Department of Zoology, University of Oklahoma, Norman, Oklahoma 73019

Mammal Species of Concern in New Mexico

Clyde Jones and C. Gregory Schmitt

Abstract

Sixteen taxa of mammals are considered as threatened or endangered in New Mexico, eight species have been extirpated in recent times, and at least 13 species have been introduced into the state. Comments are presented with regard to the status of these species in the region. Statements are included about some recent nomenclatorial changes in the names of these species of mammals. In addition, a brief history is presented about some legislative actions and other pertinent activities that affect these mammal species of concern in New Mexico.

INTRODUCTION

In the Old World, concerns for certain species of mammals extended back to Greek and Roman times, when the control of animals was the responsibility of the landowner on whose property they occurred. In the Middle Ages, it was accepted that mammals were the property of the monarchs who controlled forests, preserves, and estates. In the New World, concerns for certain species of mammals (and other animals) were expressed by some of the early human inhabitants of the area, such as the frequently cited quotation by Seattle, Chief of the Suguamish: "What is there to life if man cannot hear the lonely cry of the whippoorwill or the arguments of the frogs around a pond at night? What is man without the beasts? If all the beasts were gone, men would die from a great loneliness of spirit." With occupancy of North America by Europeans, numerous activities took place in connection with expressions of concerns about mammals and other kinds of wildlife in North America. These activities have resulted in the modern attitudes and corresponding legislative actions that control wildlife in general and mammals in particular. The purpose of this paper is to review briefly some of the historical events and legislative actions with regard to species of mammals of concern in general, and to present an annotated checklist of mammal species of concern (defined as species whose distribution is peripheral or has been drastically reduced) in New Mexico in particular. Updated nomenclature and comments on regional status are included as appropriate.

A comprehensive chronology of representative legislation affecting wildlife in North America was presented by Matthiessen (1987); a modified version is presented in Table 1. Several significant steps in legislative actions affecting mammals warrant some discussion.

Numerous important actions were taken early in the history of the development of territories and states (Table 1). A significant step, however, was the decree in 1896 by the U.S. Supreme Court (Geer vs. Connecticut) that game (and other wildlife, by extension) was the property of the state rather than the landowner (Matthiessen, 1987). This action strengthened the ability of the states to control wildlife within their boundaries.

A major milestone in wildlife legislation was the passage of the Lacey Act of 1900 (Table 1). The most important provision of this legislation was the prohibition of market hunting and interstate transportation of "any wild animals or birds" that were taken as violations of state laws. In other words, the principle thrust of the Lacey Act was to provide for the aid of the federal government in the enforcement of wildlife laws of the states. The Lacey Act was authored mostly by John F. Lacey, a Republican congressman from Iowa who was an avid sportsman (Reiger, 1986).

Life Among the Muses: Papers in Honor of James S. Findley. Edited by T.L. Yates, W.L. Gannon & D.E. Wilson.

Table 1. Partial list, in chronologic order, of some important events with regard to concerns about mammals of North America in general and New Mexico in particular (modified from Matthiessen, 1987).

Date	Event
1629	The West India Company grants hunting rights to persons planting colonies in the New Netherlands.
1630	Massachusetts Bay Colony authorizes the first bounty system of payment of one penny per wolf.
1694	Massachusetts declares the first closed hunting season on deer.
1718	Massachusetts establishes the first closed term of three years on deer.
1739	Massachusetts develops the first system of game wardens.
1776	Closed seasons on deer established in all colonies except Georgia.
1830	Maine declares a closed hunting season on moose.
1852	California establishes closed hunting seasons on pronghorn and elk.
1859	U.S.–Mexican boundary survey.
1861	Nevada declares closed hunting seasons on mountain goats and mountain sheep.
1864	Idaho establishes closed hunting seasons on bison, deer, elk, pronghorn, mountain goats, and mountain sheep.
1869	New York establishes a closed term of five years on moose.
1870	Maine declares a closed hunting season on caribou.
1871	New York authorizes a bounty system of thirty dollars per Adirondack wolf.
1872	Establishment of the first U.S. national park (Yellowstone).
1874	Legislation "to prevent the useless slaughter of buffaloes [sic] within the Territories of the United States" is passed by Congress, but is tabled by President Grant.
1878	State game departments and commissions established by California and New Hampshire.
1879	Michigan declares a ten-year closed term on elk.
1885	Establishment of the U.S. Division of Economic Ornithology and Mammalogy under the supervision of C. Hart Merriam. Establishment of the first national park in Alberta, Canada (Banff).
1890	Wyoming declares a ten-year closed term on bison.
1894	Park Protection Act provides protection for wildlife in national parks of the U.S.
1895	System of resident and non-resident hunting licenses established in several states.
1896	The U.S. Supreme Court decrees that game is the property of the state rather than the landowner.
1897	Montana adopts a law providing a two-year prison term (felony) for the killing of bison.
1900	Lacey Act prohibits importation of foreign wildlife without permit, as well as interstate traffic in wildlife killed in violation of state laws.
1901	Nevada adds game protection to the curriculum of public school education.
1902	Alaska initiates a law for the protection of brown bears.
1903	Establishment of the first U.S. national wildlife refuge (Pelican Island, Florida).
1905	California adopts a law providing a two-year prison term (felony) for the killing of elk.
1906	Northwest Game Act requires the licensing of fur traders in Canada.
1908	President Roosevelt established the U.S. National Conservation Commission under the chairmanship of Gifford Pinchot.
1909	Antelope removed from the list of game animals everywhere in its geographic range, except Arizona (which adopted this regulation in 1911) and Canada.
1910	U.S. legislation established to prohibit the taking of sea otters in Alaskan waters. New York adopts the Bayne Bill prohibiting the sale of wild game.
1911	Great Britain, Russia, and the U.S. sign the Seal Treaty controlling the take of fur seals and sea otters. Mexico establishes protection of elephant seals.
1912	New Mexico State Legislature creates the New Mexico Department of Game and Fish.
1913	Game protective associations, composed of hunters and fishermen, established in Albuquerque, New Mexico, by Aldo Leopold.

Table 1 continued

1913	Vernon Bailey publishes *Life Zones and Crop Zones of New Mexico.*
1914	U.S. Congress authorizes funds for the control of wolves, prairie dogs, and other injurious animals.
1916	Passage of the Migratory Bird Treaty Act.
1925	Alaska establishes a Game Commission, under the authority of the Alaska Game Law, to provide protection for mammals and birds.
1932	*Mammals of New Mexico* published, written by Vernon Bailey.
1934	Game and fish sanctuaries authorized for suitable areas in U.S. national forests.
1937	International Whaling Convention establishes seasons, whaling grounds, and full protection for gray whales and right whales. Mexico-U.S. treaty provides for protection of migratory birds and mammals.
	Pittman-Robertson Act establishing a 10% excise tax on sporting arms and ammunition to be allocated to states for restoration and maintenance of wildlife habitat, management of wildlife areas and resources, and research into problems of wildlife management.
1940	Bureau of Biological Survey (Department of Agriculture) merged with the Bureau of Fisheries (Department of Commerce) to form the U.S. Fish and Wildlife Service (Department of Interior).
1955	Comprehensive program in research on mammals established at the University of New Mexico by James S. Findley.
1957	U.S. Fish and Wildlife Service separated into the Bureau of Commercial Fisheries (Department of Commerce) and U.S. Fish and Wildlife Service (Department of Interior).
1966	Endangered Species Preservation Act, in which the Secretary of Interior was directed to "carry out a program in the United States of conserving, protecting, restoring, and propagating selected species of native fish and wildlife."
1969	Endangered Species Act provided for protection of species in danger of "worldwide extinction" and strengthened prohibitions on importation and interstate commerce of such animals and their parts.
1972	Marine Mammal Protection Act provides federal programs for protection and research on whales, pinnipeds, polar bears, sea otters, and sirenians.
1973	Convention on International Trade in Endangered Species of Wildlife Fauna and Flora (CITES) provided for restrictions in international commerce in species believed to be threatened by economic exploitation.
	Endangered Species Act establishes comprehensive federal program for protection and research on animals and plants threatened with extinction.
	Aday and Gennaro publish *Mammals of New Mexican Llano Estacado and Its Adjacent River Valleys.*
1974	New Mexico Department of Game and Fish given authority to study all indigenous, non-domestic vertebrate species (fishes, amphibians, reptiles, birds, mammals), plus crustaceans and mollusks for establishing recommendations to be presented to the State Game Commission concerning their status as endangered species.
1975	*Mammals of New Mexico* published, authored by James S. Findley, Arthur H. Harris, Don E. Wilson, and Clyde Jones.
1980	Fish and Wildlife Conservation Act provides federal programs for grants to states for conservation activities and research on non-game wildlife. Alaska National Interest Lands Conservation Act provides protection for mammals and other wildlife within the national parks, national wildlife refuges, and national wilderness preservation areas.
1987	*Natural History of New Mexican Mammals* published, written by James S. Findley.
1989	North American Wetlands Conservation Act provides federal funds to be matched with non-federal funds for acquisition and restoration of wetlands in the United States, Canada, and Mexico.

The Migratory Bird Treaty Act of 1916 was an important piece of early federal legislation that placed responsibilities concerning migratory birds on the federal governments of the United States and Canada (Table 1). The Endangered Species Preservation Act of 1966 was federal legislation designed to protect non-game species of animals and plants (Table 1). In accordance with this Act, the Secretary of the Interior was directed to "carry out a program in the United States of conserving, protecting, restoring, and propagating selected species of native fish and wildlife." The Department of the Interior was given authority to work with other federal departments with large land holdings, such as Agriculture and Defense, in carrying out provisions of the Act, including the acquisitions of lands. The Secretary of the Interior was directed to determine a species as threatened with extinction "after consultation with the affected states that its existence is endangered because its habitat is threatened with destruction, drastic modification, or severe curtailment, or because of over exploitation, disease, predation, or because of other facts, and that its survival requires assistance."

The Endangered Species Act of 1969 (Table 1) was more extensive than the previous Act in that it provided protection for species that might be in danger of "worldwide extinction," prohibited importation and interstate commerce in protected species (except for educational, medical, and related purposes), and broadened coverage to species of "wild mammal, fish, wild bird, amphibian, reptile, mollusk, or crustacean." In addition, this Act called for an international meeting of government officials to deal with the problems relating to enforcement of laws to protect threatened and endangered species of wildlife on a worldwide basis.

As a result of an international conference held in Washington, D.C., in early 1973, the Convention on International Trade in Endangered Species of Wildlife Fauna and Flora (CITES) was signed by representatives from many countries. A major intent of this convention was to restrict international commerce in species of animals and plants believed to be endangered by economic exploitation.

Later in the year, the Endangered Species Act of 1973 was approved by the U.S. Congress (Table 1). This comprehensive Act included provisions for combination of U.S. and foreign lists of threatened and endangered species, strengthened definitions of threatened and endangered categories, and implementation of the provisions of the CITES treaty. It further required all federal agencies to undertake programs for conservation of threatened and endangered species, and established protocols for cooperative agreements with states utilizing matching federal funds for protection of threatened and endangered species. Although there were amendments in 1978, 1982, and 1988, the Endangered Species Act of 1973 has remained more or less intact for the past 20 years.

The Endangered Species Act of 1973 expired officially in 1992. As of this writing, reauthorization of the Endangered Species Act of 1973 is pending, and is generating serious discussions among a wide array of special interest groups, government officials, and representatives from some industries, as well as conservationists and biologists. Although no significant wildlife legislation has been introduced for more than a decade (Matthiessen, 1987), there is evidence of environmental awareness and appropriate concerns by the current administration (Gore, 1992).

Although not involved directly with mammal species of concern in New Mexico, the Marine Mammal Protection Act of 1972 warrants some discussion because of some unique provisions of this particular legislation. For example, this Act removed all authority by states over all marine mammals and implemented a comprehensive federal program. Provisions were set up, however, for the appropriate states to regain control of research and management programs on marine mammals with federal assistance and under federal guidelines. Perhaps one of the most important and interesting parts of the Marine Mammal Protection Act of 1972 was the subtle change in management philosophy from one of maximum sustainable yield to a guideline to achieve optimal sustainable populations of marine mammals.

The North American Wetlands Conservation Act of 1989 provided federal funds to be matched with non-federal funds in cooperative ventures to protect wetland habitats in the United States, Canada, and Mexico.

Some important milestones concerning conservation of wildlife in New Mexico include: 1) approval of payment of bounties on certain predators in 1887 by the Territorial Legislature; later in that year the state legislature passed laws limiting periods of legal hunting and created other restrictions; 2) in 1903, the legislature directed the Governor to appoint "an officer to be known as the game and fish warden of the Territory of New Mexico"; 3) in 1909, the Legislature empowered the Territorial Warden to appoint "one chief deputy who shall also act as clerk and bookkeeper . . ."; 4) in 1912, the New Mexico legislature created the Department of Game and Fish; the Governor was directed to appoint "by and with the consent of the senate . . . some person skilled in matters relating to game and fish, to be Game and Fish Warden, who shall be the head of the game and fish department"; 5) the 1919 legislature relieved the chief deputy warden of clerical duties (or some of them) by empowering the State Game Warden to appoint one office clerk; 6) the 1921 New Mexico Legislature created the State Game Commission; 7) the 1925 Legislature transferred to the State Game commission the duty of employing a State Game Warden; 8) in 1931, the State Game Commission was "authorized and directed" to define game birds, animals, and fish; and to determine "when, to what extent, if at all, and by what means" game animals, birds, and fish were to be taken; 9) the 1939 Legislature authorized the State Game Commission to "receive any moneys to which the State of New Mexico may become entitled" under an Act of Congress (i.e., the 1937 Pittman-Robertson Act), which provides that the United States shall aid the states in wildlife restoration projects; 10) in 1945, the State Game Commission was enlarged by Legislative action, from three members to five, not more than three to be of the same political party at the time of appointment; and 11) beginning in 1955, the hiring of conservation officers, clerks, and other employees was the function of the Director.

In 1974, the New Mexico Department of Game and Fish (NMDGF) was given authority to study all indigenous, non-domestic species of vertebrates, including fishes, amphibians, reptiles, birds, and mammals, as well as crustaceans and mollusks, by passage of the Wildlife Conservation Act (17-2-37 to 17-2-46 NMSA 1978); Table 1. Beginning in 1975, through the authority and jurisdiction of the State Game Commission, several kinds of mammals were listed as endangered (Tables 2 and 3). Since that time, at approximately two-year intervals, NMDGF completes a review of species listed as endangered in New Mexico and makes recommendations concerning modifications to the listing of endangered species for consideration by the State Game Commission. Recommendations and background information on the biennial review, most recently completed in 1996, is provided to a wide variety of the public for their review and comment. The Wildlife Conservation Act was amended by the New Mexico State Legislature in 1995 (17-2-37 to 17-2-46 NMSA [1995 Repl.]). Amendments to this act required: 1) the State Game Commission to review and act on the biennial review of threatened and endangered species; 2) increased public involvement concerning adding and removing species from the threatened and endangered species list; 3) development of recovery plans for threatened/endangered species; and 4) that decisions to list species must be based solely on biological information (socio-economic information could only be used in development of recovery plans). Additionally, previously used terms of G2 and G1 were replaced with threatened and endangered, respectively.

Many of the modern concerns about mammals (and other wildlife) in New Mexico may have germinated from some of the activities initiated by Aldo Leopold in 1913 when he organized game protective associations in Albuquerque with one of the major purposes "to promote the protection and enjoyment of wild things." At least, it seems that these early experiences in New Mexico influenced the philosophy of Aldo Leopold in later years when he wrote, "The last word in ignorance is the man who says of an animal or plant: 'what good is it?'. . . If the biota, in the course of eons, has built something we like but do not understand, then who but a fool would discard seemingly useless parts? To keep every cog and wheel is the first precaution of intelligent tinkering" (Nash, 1982).

In the accounts that follow, scientific names, common names, and sequences of listings follow those of Jones et al. (1992) unless noted otherwise.

MAMMALS THREATENED OR ENDANGERED IN NEW MEXICO

Currently, 16 taxa of mammals are considered by the NMDGF and U.S. Fish and Wildlife Service (USFWS) as threatened or endangered; NMDGF lists nine species as threatened and six species as endangered; the USFWS lists four species of mammals as endangered (Table 2). Additionally, seven species of felids, six of which are not native to New Mexico, are listed as restricted species by the NMDGF. This list included species threatened or endangered in New Mexico; definitions of these rankings are presented in Table 3. A brief history of the rankings of each species that has been listed by the NMDGF is summarized in Table 3.

Sorex arizonae (Arizona Shrew)

This insectivore is considered as endangered (Tables 2 and 3) in New Mexico. It was first listed by the state on 10 February 1978 (Table 3); this shrew is not included on the federal list (Table 2), but is included in the USFWS list of species of concern (Table 4). The Arizona shrew is known in New Mexico from only one locality in the Animas Mountains (Conway and Schmitt, 1978; Cook, 1986). *Sorex arizonae* is known, however, from a total of about a dozen specimens from the Huachuca, Santa Rita, and Chiricahua mountains of southeastern Arizona (Hoffmeister, 1986). It is also known from portions of northern Chihuahua (Caire et al., 1978). The apparent rarity, extremely limited distribution, and probable small population size make this species highly vulnerable to any adverse habitat alterations.

Cryptotis parva (Least Shrew)

This species is included on the state list as threatened (Tables 2 and 3). The least shrew was first listed by New Mexico on 28 March 1985 (Table 3); it is not listed by the USFWS (Table 2). In New Mexico, extant populations of least shrews are known from several localities adjacent to the Llano Estacado and south plains in east-central New Mexico, including near Tucumcari, Quay County (Hoditschek et al., 1985), near Arch (Grulla National Wildlife Refuge), Roosevelt County (Owen and Hamilton, 1986), and near Roswell, Chaves County (Bitter Lake National Wildlife Refuge). Subsequent investigations to locate this species near Santa Rosa, Carlsbad, and Conchas Lake did not result in documentation of the species from these areas (Shuster, 1988). Although common throughout much of its geographic range, the least shrew has been considered rare on the Llano Estacado and adjacent south plains. There is some suggestion that *C. parva* may be more common in the south plains of Texas than thought previously (Jones et al., 1993). Whitaker (1974) provided a summary of the natural history of this species. Similar to many other organisms in the Southwest that require mesic habitats, this species is vulnerable to habitat loss resulting from water diversion, agriculture, and grazing within its limited habitat.

Leptonycteris curasoae (Southern Long-nosed Bat)

This bat is included as a threatened species by the NMDGF (Tables 2 and 3). It was first placed on the list of endangered mammals of New Mexico on 24 January 1975, delisted on 10 February 1978, and added to the list on 30 November 1990 (Table 3). This species is ranked as endangered on the federal list (Table 3). Arita and Humphrey (1988) designated all of the southern long-nosed bats from the mainland of North America as *L. c. verbabuenae*. These bats are known as summer residents of southeastern Arizona (Hoffmeister, 1986) and have been found in the Animas Mountains (Findley et al., 1975; Cook, 1986; Hoyt et al., 1994) and Guadalupe Canyon (Findley et al., 1975; Baltosser, 1980). Hoyt et al. (1994) conservatively estimated that they netted 150 to 200 individuals of *Leptonycteris* sp. in the proportion of 2:1 *L. curasoae* to *L. nivalis* in the Animas Mountains. Population size, trends, and threats to this species are not fully understood; however, the recent discovery of this bat in the Animas Mountains suggests that review of its listing as an endangered species may be warranted. Threats to this species include distur-

Table 2. Species of New Mexican mammals listed as threatened or endangered by the New Mexico Department of Game and Fish (1996) and the U. S. Fish and Wildlife Service (1995).

	RANKING	
SPECIES	NMDGF	USFWS
Sorex arizonae	endangered	not listed
Cryptotis parva	threatened	not listed
Leptonycteris curasoae	threatened	endangered
Leptonycteris nivalis	endangered	endangered
Lasiurus xanthius	threatened	not listed
Euderma maculatum	threatened	not listed
Lepus callotis	threatened	not listed
Tamias minimus atristriatus	endangered	not listed
Tamias quadrivittatus australis	threatened	not listed
Thomomys umbrinus	threatened	not listed
Microtus montanus arizonensis	endangered	not listed
Zapus hudsonius	threatened	not listed
Canis lupus	endangered	endangered
Mustela nigripes	not listed	endangered
Martes americana	threatened	not listed
Ovis canadensis mexicana	endangered	not listed

bance at roosting sites and destruction of habitat (e.g., food plants such as agaves) (Wilson, 1985), but see Cockrum and Patryszyn (1991).

Leptonycteris nivalis (Mexican Long-nosed Bat)

This phyllostomid is considered as endangered by NMDGF and USFWS (Tables 2 and 3). It was first placed on the list of endangered mammals of New Mexico on 24 January 1975, delisted on 10 February 1978, and added to the list on 30 November 1990 (Table 3). *Leptonycteris nivalis* is considered as endangered throughout its geographic range by the U.S. Fish and Wildlife Service (Table 3) and was first listed as endangered on 30 September 1988 (U.S. Fish and Wildlife Service, 1995). The U.S. Fish and Wildlife Service (1994) has prepared a recovery plan for *L. nivalis* and *L. curasoae.* This cave-dwelling species is known in the United States as a summer resident only from the Chisos Mountains, Brewster County, and the Chinati Mountains, Presidio County, Texas (Hensley and Wilkins, 1988; Schmidly, 1991), as well as in southern Hidalgo County, New Mexico (Arita and Humphrey, 1988; J. Scott Altenbach, pers. comm.). Hoyt et al. (1994) conservatively estimated that they netted 150 to 200 individuals of *Leptonycteris* sp. in the proportion of 2:1 *L. curasoae* to *L. nivalis* in the Animas Mountains. It appears that this species has experienced population declines. Disturbance in roost sites, including maternal colonies and destruction of habitat components (e.g., food plants such as agaves) are some of the significant reasons for concern for the status of this species, but see Cockrum and Petryszyn (1991).

Lasiurus xanthinus (Western Yellow Bat)

This bat is listed as threatened (Tables 2 and 3) by the NMDGF. It was first listed (as *Lasiurus ega xanthinus*) on 24 January 1975 and changed to *Nycteris ega* on 28 January 1985 (Table 3). The southern yellow bat is not included on the federal list (Table 2). *Lasiurus xanthinus* is the accepted name for this bat (Baker et al., 1988). This is mostly a neotropical species that ranges into southern California and Arizona, as well as in southwestern New Mexico (Hoffmeister, 1986), specifically in the Animas

Table 3. History of ranking of New Mexican mammals listed as endangered or threatened by the New Mexico Department of Game and Fish (1975-95). Taxonomy and nomenclature follows usage in the respective regulations and does not necessarily reflect current usage. Definitions are as follows: G1 = endangered, group 1, any species or subspecies whose prospects of survival or recruitment in New Mexico are in jeopardy; G2 = endangered, group 2, any species or subspecies whose prospects of survival or recruitment in New Mexico are likely to be in jeopardy within the foreseeable future; uplisted = moved from G2 to G1; delisted = removed from endangered species list; down-listed = moved from G2 to G1; NC = no change in ranking.

Species	Ranking	NMGF Regulation	Date First Listed	Subsequent Ranking	Date of Change	Reason for Listing or Change
Sorex arizonae	G2	563 amended (order 4-78)	10 Feb 1978			1
		657		uplisted to G1	9 Jan 1988	2
Cryptotis parva	G2	624	28 Mar 1985			3
Choeronycterus mexicana	G2	563	24 Jan 1975			4
		563 amended (order 4-78)		delisted	10 Feb 1978	5
Leptonycterus sanborni nivalis	G2	563	24 Jan 1975			4
		563 amended (order 4-78)		delisted	10 Feb 1978	5
Leptonycterus nivalis	G1	682		added to list	30 Nov 1990	6
Leptonycterus sanborni	G2	682	30 Nov 1990			6
Euderma maculatum	G2	657	9 Jan 1988			7
Lasiurus ega xanthinus	G2	563	24 Jan 1975			4
Nycteris ega	NC	624 amended			28 Mar 1985	8
Eumops perotis	G2	599	25 May 1979			9
		624		delisted	28 Mar 1985	10
Lepus callotis gaillardi	G1	563	24 Jan 1975			11
	G2	657		downlisted	9 Jan 1988	11
Eutamias minimus atristriatus	G1	624	22 Jul 1983			12
Eutamias quadrivittatus australis	G2	624	22 Jul 1983			13
	NC			NC	9 Jan 1988	14
Cynomys ludovicianus arizonensis	G2	563	24 Jan 1975			15
		563 amended (order 4-78)		delisted	10 Feb 1978	16

Table 3 continued

Cynomys ludovicianus ssp.	NC	563 amended (order 4-78)			10 Feb 1978	17
		657		delisted	9 Jan 1988	18
Thomomys umbrinus emotus	G2	563	24 Jan 1975			19
Perognathus nelsoni canescens	G2	563 amended (order 4-76)	21 May 1976			20
		624		delisted	28 Mar 1985	21
Peromyscus pectoralis laceianus	G2	563 amended (order 4-76)	21 May 1976			22
		563 amended (order 4-78)		delisted	10 Feb 1978	23
Microtus montanus arizonensis	G2	599	25 May 1979			24
	G1	682		uplisted	30 Nov 1988	25
Microtus ochrogaster ssp.	G2	563	24 Jan 1975			26
		563 amended (order 4-78)		delisted	10 Feb 1978	27
Zapus hudsonicus	G2	624	22 Jul 1983			28
Canis lupus baileyi	G1	563 amended (order 4-76)	21 May 1976			29
Canis lupus	NC	599			25 May 1979	30
Nasua narica molaris	G2	563	24 Jan 1975			31
		624		delisted	28 Mar 1985	32
Martes americana origenes	G2	563	24 Jan 1975			33
Mustela vison energumenos	G2	563	24 Jan 1975			34
		624		delisted	28 Mar 1985	35
Mustela nigripes	G1	563	24 Jan 1975			36
		657		delisted	9 Jan 1988	37
Lutra canadensis	G1	563	24 Jan 1975			38
		624		delisted	28 Mar 1985	39
Felis onca arizonensis	G1	563	24 Jan 1975			40
		599		delisted	25 May 1979	41

Table 3 continued

	G3	682 amended (order 2-91)			25 Jul 1991	42
Ovis canadensis mexicana	G1	599 amended (order 5-80)	24 May 1980			43
		624		NC	9 Jan 1985	44
		682		NC	30 Nov 1990	45

1 Known only from one specimen in a very limited locality in the Animas Mountains, Hidalgo Co.
2 After additional investigations, known only from one restricted site in the Animas Mountains, Hidalgo Co.
3 Known only from a few sites in central-eastern border region (e.g., Tucumcari, Quay Co.) where it appears confined to mesic habitats that are vulnerable to adverse alteration; limited range and presumed small population size thought to be significant threats.
4 Only a few maternal colonies are known in its limited New Mexican range, habitat deterioration thought to be a significant threat.
5 Numbers, range, and habitats were thought to be extensive enough and threats to them limited enough that their survival was not thought to be in actual or potential jeopardy.
6 Added to list as G1 to parallel federal listing (30 Sept. 1988).
7 Surveys revealed very few of these bats in areas where they were more regular (e.g., Jemez, San Mateo, and Mogollon Mountains).
8 Replaced *Lasiurus* with *Nycteris*.
9 Rare species known only from Guadalupe Canyon, Hidalgo Co., where habitat alteration or destruction could affect its survival in New Mexico.
10 Known only from one occasion in Guadalupe Canyon, Hidalgo Co., in 1969; its solitary occurrence, failure to locate species during subsequent surveys were reasons for delisting species.
11 Numbers, range, and habitats were known to be limited enough and threats to them significant enough that their survival was thought to be in actual or potential jeopardy; downlisted on 9 January 1988 because it was thought that threats to habitats and species were less than previous assessments.
12 Restricted to a few canyon areas in the Sacramento Mountains, Otero Co., where numbers appeared to be small and recent surveys failed to detect this species.
13 Southernmost population of a widespread species restricted to the Organ Mountains, Doña Ana Co.; limited range and habitat.
14 Listing amended to include populations in the Oscura Mountains, Socorro and Lincoln Cos., as these populations are most similar to populations in the Organ Mountains.
15 Populations suffered drastic declines as a result of extensive poisoning campaigns.
16 Subspecies *arizonensis* thought to be extirpated from New Mexico.
17 The "Tularosa" black-tailed prairie dog (*C. ludovicianus* ssp.), an undescribed subspecies, now confined to the Tularosa Basin in Lincoln, Otero, and possibly Doña Ana Cos., is declining and probably qualifies as threatened under the Endangered Species Act of 1973; listing included only prairie dogs in the Tularosa Basin.
18 Population had not been formally distinguished as distinct from non-endangered populations of this species.
19 Occurs as only one isolated population in the Animas Mountains, Hidalgo Co., and any habitat alteration could seriously threaten this species in New Mexico.
20 Listing amended to include species because of its extreme rarity and local occurrence (e.g., known from only one locality in Carlsbad Caverns National Park, Eddy Co.).
21 Solitary occurrence of species and lack of verification of the reported specimen.
22 Limited distribution along eastern flank of Guadalupe Mountains, Eddy Co.
23 Its numbers, range, and habitats are extensive enough and threats limited enough that its survival in New Mexico was not thought to be in actual or potential jeopardy.
24 Recently discovered small population in Catron Co. in only one very small sedge meadow that is extremely vulnerable.
25 Threats to the extremely limited habitat of this species warrant uplisting to G1.
26 Known only from a single disjunct population near Cimarron, Colfax Co.; habitat subject to alteration.
27 Additional populations were discovered near Cimarron, Colfax Co., and near Wagon Mound, Mora Co.
28 Species restricted to lowland and canyon wetlands which have been severely diminished by reclamation, grazing, and other human activities; continued loss of habitat jeopardize this species.

Table 3 continued

29 Thought to be extirpated from New Mexico due to extensive eradication programs.
30 Mexican wolf (*Canis lupus baileyi*) changed to gray wolf (*Canis lupus*) to conform to federal usage.
31 Thought to occur in such few numbers that disease or indiscriminate shooting and trapping could threaten its continued survival in New Mexico.
32 It appeared that this species had expanded its range in recent years and survival of species no longer appeared to be in question (it remained a protected species as a furbearer in NMSA 17-5-2 1978 compilation).
33 Species reaches its southern limits in the San Juan and Sangre de Cristo mountains and known from only a few records where indiscriminant shooting and trapping, in conjunction with habitat alterations, could threaten its continued survival in New Mexico.
34 Species at its southern limits in northern New Mexico, where it was rare and only seldom reported; habitat alteration and incidental killing were thought to threaten the survival of this species in New Mexico.
35 Species was delisted because it had not been reliably reported in more than a decade and there was no evidence that the species was extant (it remained a protected furbearer under NMSA 17-5-2 1978 compilation).
36 Species appeared to have declined, apparently due to extensive poisoning campaigns directed toward prairie dogs, and its survival in New Mexico was believed to have been severely jeopardized.
37 No known reliable reports of this species since 1970 and no confirmed record since 1934 (still protected as a furbearer under NMSA 17-5-2 1978 compilation).
38 Habitat degradation and adverse changes in food supplies were thought to have been responsible for extirpation of this species from most of its limited range.
39 Species had not been reliably reported since 1953 and there was no evidence that suggested this species was extant (remained a protected furbearer under NMSA 17-2-5 1978 compilation).
40 Added to list to conform to federal lists; species had not been substantiated in New Mexico in the wild since the early part of the century.
41 Species was delisted on the basis that no recent records of wild animals and its reoccurrence in New Mexico was thought to be unlikely.
42 Listing revised to include seven species of cats defined as those species or subspecies as set forth in section 4 of the Endangered Species act of 1973 under title 50 of the Code of Federal Regulations, Part 17, and as incorporated in whole or in part as listed in Chapters 2 through 8. These species were ranked as restricted species (= G3) and included the following: *Neofelis nebulosa* (clouded leopard), *Panthera pardus* (leopard), *P. unica* (snow leopard), *P. onca* (jaguar; the only species native to New Mexico in this ranking), *P. tigris* (tiger), *Felis concolor coryi* (Florida panther), and *F. pardalis* (ocelot). NMGF involvement with these taxa appears limited to matters of taking, possession, transportation, exportation, processing, sale or offering for sale, or shipment.
43 Species listed (non-captive animals only) due to significant declines in populations.
44 Listing revised to include only non-captive animals of the desert race only.
45 Listing revised to include all wild individuals and populations except the population in the Peloncillo Mountains, Hidalgo Co., and all stock in captivity.

Mountains (Cook, 1986) and Guadalupe Canyon (Mumford and Zimmerman, 1963; Baltosser, 1980). Loss or alteration of southwestern riparian areas and associated habitats are major threats to this peripheral species in New Mexico. This species may be expanding its geographic range in some areas of the United States in accordance with the widespread use of ornamental palm trees in landscaping (Schmidly, 1991).

Euderma maculatum (Spotted Bat)

This unique bat is considered as threatened in New Mexico (Tables 2 and 3) and was added to the list on 9 January 1988 (Table 3). The spotted bat is not listed by the USFWS (Table 2), but is included in the USFWS list of species of concern (Table 4). This vespertilionid has a large geographic range; it is thought to be uncommon in some areas, but it may be locally abundant in some places (Watkins, 1977). Although it seems that more spotted bats have been observed in Big Bend National Park than at any other single place (Easterla, 1973), *E. maculatum* is listed as a threatened species by the Texas Department of Parks and Wildlife (Jones, 1993). Limiting factors for this species are largely unknown. Surveys for spotted bats are continuing; use of increasingly sophisticated equipment (e.g., AnaBat bat detectors) and listening for their audible echolocation calls by the trained ear may ultimately reveal the presence of this species in new localities.

Lepus callotis (White-sided Jackrabbit)

In New Mexico, this hare is considered as threatened (Tables 2 and 3). It was originally placed on the state list of mammals by the NMDGF as a group 1 species on 24 January 1975, and on 9 January 1988, it was down-listed to group 2 (Table 3). This jackrabbit is not included on the federal list (Table 3), but is included in the USFWS list of species of concern (Table 4). *Lepus callotis* acquired its endangered status in New Mexico primarily because it is known only from grasslands in an area of about 120 km^2 in southern Hidalgo County (Best and Henry, 1993). This is an elusive species that was reported only a few times after its discovery in 1892 along the Mexican–U.S. Border (Mearns 1896) until two were collected in the Playas Valley in 1931 (Anderson and Gaunt, 1962) and a subsequent specimen was taken by Bogan and Jones (1975) in 1974. Later studies by Conway (1975), Conley and Brown (1977), and Bednarz and Cook (1984) confirmed previous statements that it is a desert grassland specialist. The number of sightings of this species on surveys during various periods between 1975-96 reveal a consistently downward trend (e.g., Bednarz and Cook, 1984; Mehlhop, 1995). Loss or degradation of grassland habitat within its restricted New Mexico range is thought to be the primary threat to this species. Status of this species in its far broader Mexican range requires further study.

Tamias minimus atristriatus (Least Chipmunk)

Originally listed as *Eutamias minimus atristriatus,* the least chipmunk from the highlands of Otero County is considered as endangered (Tables 2 and 3). This subspecies was added to the list of endangered New Mexican mammals on 22 July 1983 (Table 3). The least chipmunk is not included on the federal list of endangered mammals (Table 2). There has been no confirmation of this chipmunk in James or Peñasco canyons since 1969 (Conley, 1970) in spite of extensive sampling efforts during 1981-82 (Yates, 1982). Unverified reports, however, of least chipmunks in areas of former occupancy (James P. Ward, Lincoln National Forest, pers. comm.) warrant renewed surveys. A small population of least chipmunks apparently remains extant in a limited area on Sierra Blanca (Sullivan, 1985; Sullivan and Petersen, 1988). Alteration of suitable habitats due to increased human disturbances pose continued threats to this chipmunk. Use of *Tamias* as the generic name for these sciurids was discussed by Jones et al. (1992).

Tamias quadrivittatus australis (Colorado Chipmunk)

The populations of these chipmunks from the Organ and Oscura mountains are listed as threatened (Tables 2 and 3). This chipmunk was first listed by the NMDGF as *Eutamias quadrivittatus australis* on 22 July 1983 (Table 3). This subspecies of the Colorado chipmunk is not included on the federal list (Table 2), but

Table 4. New Mexican mammals listed as proposed endangered, candidate species, and species of concern by U.S. Fish and Wildlife Service, New Mexico Ecological Services Field Office, Albuquerque, New Mexico, 8 July 1996. Common names are those used by USFWS in this listing.

Proposed Endangered
- *Panthera onca* (jaguar)

Candidate Species
- *Vulpes velox* (swift fox)

Species Of Concern
- *Sorex arizonae* (Arizona shrew)
- *Choeronycteris mexicana* (Mexican long-tounged bat)
- *Myotis ciliolabrum* (western small-footed myotis)
- *Myotis evotis* (long-eared myotis)
- *Myotis lucifugus occultus* (occult little brown bat)
- *Myotis thysanodes* (fringed myotis)
- *Myotis velifer* (cave myotis)
- *Myotis volans* (long-legged myotis)
- *Myotis yumanensis* (Yuma myotis)
- *Euderma maculatum* (spotted bat)
- *Plecotus townsendii pallescens* (pale Townsend's big-eared bat)
- *Idionycteris phyllotis* (Allen's big-eared bat)
- *Nyctinomops macrotis* (big free-tailed bat)
- *Eumops perotis californicus* (greater western mastiff bat)
- *Ochotona princeps nigrescens* (Goat Peak pika)
- *Lepus callotis gaillardi* (white-sided jackrabbit)
- *Tamias canipes* (gray-footed chipmunk)
- *Tamias quadrivittatus australis* (Colorado chipmunk [Organ Mountians, NM])
- *Cynomys ludovicianus arizonensis* (Arizona black-tailed prairie dog)
- *Thomomys umbrinus guadalupensis* (Guadalupe southern pocket gopher)
- *Thomomys umbrinus mearnsi* (Mearns' southern pocket gopher)
- *Thomomys umbrinus paguatae* (Cebolleta southern pocket gopher)
- *Geomys bursarius arenarius* (desert pocket gopher)
- *Sigmodon ochrognathus* (yellow-nosed cotton rat)
- *Neotoma micropus leucophaea* (White Sands woodrat)
- *Ondatra zibethicus ripensis* (Pecos River muskrat)
- *Zapus hudsonius luteus* (New Mexican meadow jumping mouse)
- *Lutra canadensis sonorae* (Southwestern otter)

is included in the USFWS list of species of concern (Table 4). Populations of these chipmunks still exist in the Organ Mountains (Patterson, 1980) and chipmunks morphologically similar to this subspecies are known from the Oscura Mountains (Robert M. Sullivan, pers. comm.). These small, isolated populations are extremely vulnerable to habitat loss or alteration. Additionally, unregulated take and diseases (e.g., sylvatic plague) may pose threats to these relic populations.

Thomomys umbrinus (Southern Pocket Gopher)

This gopher is listed as threatened (Tables 2 and 3) by the NMDGF. This species was first placed on the endangered species list on 24 January 1975 (Table 3). The federal list of threatened and endangered wildlife does not include this species (Table 2), but it is included in the USFWS list of species of concern (Table 4). In New Mexico, this particular mammal (*T. u. emotus*) is restricted to the Animas Mountains (Findley et al., 1975; Hinseley and Thaeler, 1977; Cook, 1986), Hidalgo County. The endemism and restricted distribution are major reasons for its threatened status in the state. *Thomomys umbrinus* is known from at least five isolated mountain ranges in southeastern Arizona (Hoffmeister, 1986) and has an extensive geographic distribution in Mexico (Hall, 1981).

Microtus montanus arizonensis (Montane Vole)

The subspecies *M. m. arizonensis* is considered as endangered in New Mexico (Tables 2 and 3). This mammal was included on the list on 25 May 1979 in the Group 2 ranking and on 30 November 1990, it was ranked G1 (Table 3). The montane vole is not listed by the USFWS (Table 2). Although the species is widespread and common, this particular subspecies in New Mexico is known from only two localities, Centerfire Bog (Hubbard et al., 1983) and adjacent to Jenkins Creek (Frey et al., 1995), Catron County. This form ranges in the White Mountains of Apache and Greenlee counties in adjacent parts of Arizona (Hoffmeister, 1986). The restricted range of this species combined with its affinity for limited mesic habitats in this portion of southwestern New Mexico, make it extremely vulnerable to habitat deterioration.

Zapus hudsonius (Meadow Jumping Mouse)

This species is listed as threatened in New Mexico (Tables 2 and 3) by the NMDGF. On 22 July 1983, this mammal was added to the list of endangered species of the state (Table 3). The federal list does not include the meadow jumping mouse (Table 2), but the species is included in the USFWS list of species of concern (Table 4). The species is widespread and common; however, the subspecies, *S. h. luteus*, is restricted to Arizona and New Mexico. In Arizona, the meadow jumping mouse occurs in the White Mountains in Apache and Greenlee counties (Hoffmeister,1986). In New Mexico, this mammal is known from the San Juan, Jemez, Sangre de Cristo, and Sacramento mountains (Findley et al., 1975). Populations along the Rio Grande Valley were thought to have been extirpated due to changes in the ecology by modifications to the river for irrigation practices (Findley et al., 1975). Following its rediscovery in 1976 (Hafner et al., 1981) in the Rio Grande Valley, Morrison (1992) verified the presence of meadow jumping mice in all of the above areas and found them in several additional localities from which they had not been previously recorded. Whitaker (1972) provided a summary of the natural history of this species. This species may not be limited to naturally occurring habitats and, to some degree, may utilize localized habitats adjacent to agricultural waterways. Alteration of montane meadows and adjacent mesic grassy areas pose potential threats to this species.

Canis lupus baileyi (Mexican Wolf or Lobo)

In New Mexico, this species is listed as endangered by the NMDGF (Tables 2 and 3). The gray wolf was added to the state list on 21 May 1976 (Table 2). The USFWS listed gray wolves as endangered in the conterminous lower 48 states on 11 March 1967 (Minnesota populations listed as threatened; populations in Wyoming and parts of Idaho listed as nonessential experimental) (Table 2) (U.S. Fish and Wildlife Service, 1995). Bogan and Mehlhop (1983) and Nowak (1996) revised the taxonomy of wolves and discussed revisions relevant to the current listing of *C. l. baileyi* in the Southwest. García–Moreno et al. (1996) presented genetic data that show captive populations of Mexican wolves are distinct from dogs and northern gray wolves, and that three captive populations would be suitable to augment the genetic diversity of the certified lineage of captive Mexican wolves proposed for reintroduction into the wild. There are no known Mexican wolves remaining in the wild, and as of March 1996, the total certified captive population in the three lineages stood at 139 animals; 114 are held at 23 facilities, mostly zoos and wildlife sanctuaries, in the U.S. and 25 are held at five facilities in Mexico. The USFWS has proposed to reintroduce Mexican wolves, classified as nonessential experimental, into the Blue Range wolf recovery area (all of the Apache and Gila National Forests in east–central Arizona and west–central New Mexico) or the White Sands wolf recovery area (all of the White Sands Missile Range [WSMR], White Sands National Monument, Holloman Air Force Base, lands adjoining the western boundary of the missile range, including the Jornada Experimental Range and the San Andres National Wildlife Refuge) in south–central New Mexico. The history of Mexican wolves in New Mexico is discussed in the recovery plan on Mexican wolves (Mexican Wolf Recovery Team, 1982; currently being revised) and in the draft environmental impact statement released by the USFWS in 1995. Mech (1974) provided a summary of this species in North America.

Martes americana (American Marten)

This species is considered as threatened by the NMDGF (Tables 2 and 3). This mustelid was first placed on the list of endangered mammals on 24 January 1975 (Table 3) and is not included on the list presented by the USFWS. *Martes americana* has a limited distribution in New Mexico and is known only from the San Juan and Sangre de Cristo Mountain ranges (Findley et al., 1975). The presence of American martens from these mountain ranges was recently reconfirmed by salvage and preservation of two road-killed specimens—one from the San Juan Mountains, Rio Arriba County, in November 1995, and one found near Taos Ski Valley, Taos County, in July 1994 (NMDGF, unpubl. data). A reported sighting of this species in the Jemez Mountains in August 1990 by Marylin J. Altenbach and Charles W. Painter (pers. comm.), suggest this species may be extant in this range; however, subsequent surveys (i.e., use of tracking plates, hair snags, and visual searches for other types of evidence) for this species in the same area did not reveal any physical evidence of martens (NMDGF, unpubl. data). Clark et al. (1987) provided a summary of the natural history of this species.

Ovis canadensis (Mountain or Bighorn Sheep)

The state list includes *O. c. mexicana* (frequently referred to as desert bighorn sheep) as an endangered species (Tables 2 and 3). The species was first listed on 24 May 1980 (Table 3). This bovid is not included on the federal list (Table 3). Although previously widespread in New Mexico—i.e., perhaps as many as 16 areas were inhabitated by desert bighorn sheep as far north as the Zuni and Magdalena mountains and as far east as the Jicarilla and Guadalupe mountains. By 1955, native populations of desert bighorn sheep apparently survived only in the San Andres and Big Hatchet Mountains; they survived in the Guadalupe Mountains of southeastern New Mexico until 1946 (New Mexico Department of Game and Fish, 1995).

Currently, this subspecies occurs in only seven areas (three of which are a result of recent reintroductions and one is a captive population at Red Rock). In 1994, about 360 desert bighorn sheep were estimated to occur in New Mexico; approximately 100 at the Red Rock captive facility in Hidalgo County; approximately 250 free-ranging in these mountain ranges: San Andres (N = 35), Hatchet (N = 80), Peloncillo (N = 80), Alamo Hueco (N = 25; however, a June 1995 survey revealed only 3 males), Ladron (N = 35), and Fra Cristobal (N = 30). Reintroductions of desert bighorn sheep were made in the Hatchet Mountains in 1979 and 1982, Alamo Hueco Mountains in 1986, Ladron Mountains in 1992-93, Peloncillo Mountains in 1981-82, 1991, and 1993 (Fisher, 1994; New Mexico Department of Game and Fish, 1995), and in the Fra Cristobal Mountains in 1995. Desert bighorn sheep in the Chihuahuan Desert ecotype are the least secure population of bighorn sheep in the United States (Bailey, in press). The San Andres population has not recovered from a severe decline caused by an outbreak of scabies mites (*Psoroptes* sp.) in the late 1970s. In the smallest populations, inbreeding and high predator–prey ratios may limit population increases. All of New Mexico's populations are almost certainly losing genetic diversity (National Research Council 1995:107). Any contact with domestic sheep poses a serious threat of disease to desert bighorn sheep.

The current list of threatened and endangered species of mammals in New Mexico includes a total of 16 taxa. Of these, six are regarded as endangered and nine are listed as threatened (New Mexico Department of Game and Fish, 1996). Three of these species are ranked as endangered by the USFWS. Another species is classified as endangered by the USFWS that is not listed by the NMDGF (Table 2). By comparison, six taxa of land mammals currently are considered as endangered in Texas; four of them are included on the federal list (Jones, 1993). In Arizona, 14 species of mammals were formerly listed as endangered and three species were formerly listed as threatened. More recently, the Arizona Department of Game and Fish has apparently adopted a listing "Wildlife of Special Concern In Arizona" (Arizona Department of Game and Fish, 1996) in which 24 species of mammals are included; three of these species are also included on the federal list (U. S. Fish and Wildlife Service, 1995).

MAMMALS EXTIRPATED IN NEW MEXICO

Several subspecies and species of mammals have been driven to extinction within the state either directly or indirectly by the activities of humans. A listing of these mammals, along with a brief account of each, is presented below.

Cynomys ludovicianus arizonensis (Arizona Black-tailed Prairie Dog)

Black-tailed prairie dogs were formerly abundant and widespread east of the Rio Grande and in grasslands of the southwestern part of New Mexico (Findley et al., 1975). Bailey (1932) reported vast numbers of this species in Luna and Hidalgo counties; however, this species was not found during many trips to these areas between 1955 and 1972 (Findley et al., 1975). Arizona black-tailed prairie dogs (*C. l. arizonensis*) were eliminated from southwestern New Mexico as a result of extensive control measures (e.g., Bailey, 1932; Hubbard and Schmitt, 1984; Findley, 1987). Black-tailed prairie dogs are still extant in a few localities in the Otero Basin and in localized areas in the eastern grasslands of New Mexico (Hansen, 1977; Hubbard and Schmitt, 1984). This species also has been extirpated in Arizona (Hoffmeister, 1986). Isolated colonies still occur in west Texas (Jones and Jones, 1992). The status of prairie dogs in New Mexico is poorly understood and deserves renewed attention by researchers and land managers.

Sigmodon fulviventer goldmani ([Hot Springs] Tawny-bellied Cotton Rat)

Findley and Jones (1963) provided information that suggested this subspecies may be extinct and apparently replaced ecologically by *S. hispidus*. In spite of subsequent searches for *S. fulviventer* in the vicinity of this subspecies, no evidence of its continued existence has been documented. Reasons for the possible extirpation of this population are unknown; however, significant alteration of its habitat by over-grazing, desiccation, and other man-caused disturbances may have played an important role.

Canis lupus (Gray Wolf)

Although still included on the list of endangered species in New Mexico (see the previous discussion), there is no evidence that any wild wolves still occur in the the state. Some possible explanations of continuous reports of sightings of wolves in the area were discussed by Findley (1987). Gray wolves were reported as declining in the state since the turn of the century (e.g., Bailey, 1932); the last record of a wolf taken (specimen confirmed vs. sight records without photographic or other kinds of evidence) in the state was in 1965 in Hidalgo County (U.S. Fish and Wildlife Service, 1982). Wolok (1994) summarized details on reports of wolves in New Mexico and other parts of the Southwest. Brown (1983a) provides a thorough discussion on wolves in the Southwest. This species is considered to be extinct also in Arizona (Hoffmeister, 1986) and in Texas (Jones, 1993). As mentioned previously, captive breeding programs for this canine have been successful and reintroduction of Mexican wolves into the wild is now under consideration by the USFWS.

Ursus arctos (Grizzly or Brown Bear)

Although apparently widespread throughout the state until the turn of the century (Bailey, 1932), this species has been eliminated from the area. It seems that the last grizzly bear known to be killed in the state was in the decade of 1930 (Findley et al., 1975; Findley, 1987). *Ursus arctos* also is extinct in Arizona (Hoffmeister, 1986) and in Texas (Jones and Jones, 1992). A 1979 record of a grizzly bear from a remote part of the San Juan Mountains of southern Colorado (Fitzgerald et al., 1994 [1995]) was adjacent to relatively similar habitat in northern New Mexico.

Mustela nigripes (Black-footed Ferret)

Black-footed ferrets were first listed as an endangered species by the NMDGF on 24 January 1975 and removed from the list on 9 January 1988 (Table 3). This mustelid was associated closely with the vast colonies of prairie dogs that once occupied the central and western portions of the country. Apparently the black-footed ferret is now extinct in the wild over its entire historic range. The last verified record of this species in New Mexico was in 1934 near Gallup, McKinley County (Hubbard and Schmitt, 1984). The last known wild ferrets occurred in Wyoming during the 1980s; unfortunately, the animals died out several years ago, apparently due to an outbreak of distemper (Thorne and Williams, 1988). Captive breeding of these mammals by the USFWS, Wyoming Department of Game and Fish, and several zoos has been successful. Approximately 1,200 black-footed ferrets have been born in captivity and nearly 300 of these have been reintroduced into the wild. Reintroductions of black-footed ferrets into towns of *Cynomys leucurus* have occurred in Wyoming (1991, 1993, and 1994). Reintroductions of this species into towns of *C. ludovicianus* have also occurred in Montana and South Dakota in 1994. The Arizona Department of Game and Fish is currently involved with preliminary stages of reintroducing black-footed ferrets into colonies of *C. gunnisoni* near Seligman, Coconino County. A recovery plan for the black-footed ferret has been prepared by the U.S. Fish and Wildlife Service (1988).

Mustela vison (Mink)

This species was first listed as an endangered species (Group 2) by the NMDGF on 24 January 1975 and delisted on 28 March 1985. Findley et al. (1975) reported mink along permanent watercourses in and adjacent to the Sangre de Cristo Mountains, upper San Juan and Animas rivers in San Juan County, as far south as the Rio Grande near Los Lunas, Valencia County, and suggested they may also occur in the upper Rio Grande and possibly in the Rio Chama drainage as well. Bailey (1932) summarized records of this species in the northern part of the state, including the Pecos River drainage. Mink were also reported from the Cimarron area, Colfax County (Hill, 1942). This species was reported by Berghofer (1967) to be common throughout the northern half of New Mexico and Gila drainage, and plentiful in the upper Rio Grande and Chama River drainages; however, Findley et al. (1975) were unable to verify records from the Gila drainage. Two road-killed specimens of this species from near Ute Park, Colfax County, in 1987 were preserved; nevertheless, it is uncertain if these specimens represented wild, free-ranging individuals or possibly escaped, pen-reared captives (NMDGF, unpubl. data). Also, E.H. Stucker collected two at Mora River, 3½ mi E. Watrous, Mora County, New Mexico, March 1967; in MSB collection (MSB 24807 and 24808), both males.

Lutra canadensis (River Otter)

River otters were first listed by the NMDGF as an endangered (Group 1) species 24 January 1975 and delisted on 28 March 1985 (Table 3). Sight records and hearsay reports of this species were reported by Bailey (1932) from the Gila, Canadian, and upper Rio Grande drainages. Only one specimen of the river otter has been preserved from New Mexico, a specimen taken on 6 March 1953 on the Gila River one mile south of Cliff, Grant County (McClellan, 1954). NMDGF has received reported sightings of this species from the upper Rio Grande and Canadian drainages; however, subsequent investigations in these areas (e.g., Stahlecker, 1986; Polechla et al., 1994) have not yielded any physical evidence confirming the presence of this species in New Mexico. River otters were extirpated from Colorado and beginning in 1976, transplants (animals from at least six different states or Canadian provinces) into the Colorado, Gunnison, Piedra, and Dolores rivers have been made (Fitzgerald et al., 1994 [1995]). Hoffmeister (1986) summarized records of this species in Arizona. Between 1981 and 1984, the Arizona Department of Game and Fish introduced 46 river otters into the Verde River of central Arizona (Spicer, 1987). Spicer (1987) also provided a status report of this species along the Colorado River in Arizona. Reasons for the scarcity of this species from New Mexico are unknown. Habitat alterations (e.g., adverse modification of bank configurations along riverine habitats and destruction of riparian vegetation), indiscriminate shooting, overtrapping of beaver, overgrazing, lack of minimum stream flow, erosion and subsequent siltation, decline of fish population, possible contamination of prey species due to pesticides and heavy metals, and other types of human disturbances may have contributed to their decline and ultimate extirpation.

Panthera onca (Jaguar)

The jaguar was first listed by the NMDGF as endangered (Group 1) on 24 January 1975 and delisted on 28 March 1985 (Table 3). Bailey (1932) summarized early reports of jaguars in New Mexico, most of which were reported from the southern and southwestern part of the state. At one time, this spotted cat apparently ranged in New Mexico as far north as Springer (Hill, 1942); however, the location of a jaguar that was reportedly killed (date indeterminate) near Springer is unknown. Halloran (1946) also mentions reports of this species from the San Andres and Caballo mountains, neither of which are verifiable by specimens. Bailey (1932) also mentions a report of this species in Santa Fe. There have been no verifiable reports of these mammals from the state since 1904 (Findley et al., 1975; Findley, 1987) until a jaguar was observed and photographed in the Peloncillo Mountains, Hidalgo County on 7 March 1996 (Warner Glenn, pers. comm., 28 May 1996; picture on file at NMDGF). Brown (1983b) presented information on the locations in New Mexico and Arizona where this species had been taken between 1900-80. Hoffmeister (1986) also discussed this species in Arizona, including the reported killing of one of these cats in Arizona in 1971. More recently, Girmendonk (1994) summarized sighting reports of this species in Arizona. This species is extinct in Texas (Jones, 1993).

Cervus elaphus merriami (Wapiti or Merriam's Elk)

Originally, it was thought that there were two species of elk in New Mexico (Bailey, 1932); however, what was formerly considered Merriam's elk (*C. merriami*), is now considered as a geographic race of *C. elaphus*, the Rocky Mountain elk (Anderson and Barlow, 1978). The former persisted until around 1900 in southern New Mexico (the last known sighting was in the Sacramento Mountains), and the latter was apparently extirpated by 1909 (Bailey, 1932; Gates, 1967; Findley et al., 1975). Populations of *C. e. merriami* are extinct (Hoffmeister, 1986; Findley et al., 1975). In New Mexico, restoration of *C. elaphus nelsoni* (the Rocky Mountain Elk) from animals originating in Wyoming, Colorado, South Dakota, and Oklahoma began in 1911 (Gates, 1967). Today, the elk has been reestablished into much of its historic New Mexico range, has expanded its range significantly, and has become a popular species with big game hunters. This species has also been reintroduced successfully in Arizona (Hoffmeister, 1986) and in Texas (Jones and Jones, 1992; Jones, 1993).

Bison bison (American Bison)

Although herds of these mammals once were abundant in New Mexico east of the mountains, they were eventually hunted to extinction and essentially gone from the state by 1884; the last recorded killing of a bison in the state was in 1884 (Findley, 1987; Findley et al., 1975). American bison now occur primarily on private ranches and on federal reservations. NMDGF established a herd of bison at Ft. Wingate, Cibola County, in 1966, 1967, and 1969 using stock from Montana, Oklahoma, and North Dakota (New Mexico Department of Game and Fish, 1993). These animals were periodically sold and even infrequently hunted. During February 1993, the NMDGF and U.S. Bureau of Land Management introduced bison to El Malpais National Conservation Area, Cibola County from Ft. Wingate (New Mexico Department of Game and Fish, 1993). These animals were removed from the area in 1995. Both of these introductions of bison were in areas outside of their primary native New Mexican range. For some explanations for discontinuing the use of the generic name *Bison* in favor of *Bos*, see Jones et al (1992). We use *Bison* here after Wilson and Reeder (1993).

Ovis c. canadensis (Rocky Mountain Bighorn Sheep)

Rocky Mountain bighorn sheep (*O. c. canadensis*) were formerly known from four to six different populations in northern New Mexico (Bailey, 1932). By the early 1900s these populations were extirpated from New Mexico because of indiscriminant hunting, competitive interactions with domestic livestock, and transmission of disease from domestic sheep (Buechner, 1960). Restoration of this subspecies began as early as 1932 (Sands, 1967) with reintroduction of six Rocky Mountain bighorn sheep (source of stock was not detailed) in the Sangre de Cristo Mountains, northeast of Santa Fe. Subsequently, reintroductions of this subspecies, from animals that originated in Banff National Park and Waterton Lakes, Canada, and Dubois, Wyoming, have been made in the following areas: Sandia and Manzano Mountains, Turkey Creek, San Francisco River, Pecos Wilderness, Wheeler Peak Wilderness, Latir Wilderness, Ft. Wingate, and Cimarron Canyon (Dunn, 1996). Currently, five populations, consisting of approximately 550 animals, occur in the following areas: Turkey Creek (N = 60), San Francisco River (N = 80), Pecos Wilderness (N = 350), Wheeler Peak Wilderness (N = 50), and Manzano Mountains (N = 25) (Dunn, 1996). Dunn (1996) has evaluated suitable ranges for Rocky Mountain bighorn sheep in New Mexico and estimated that low-elevation habitats could potentially support 300 to 500 animals and that high-elevation habitats may support up to 625-775 animals. Historically, desert bighorn sheep occurred in the Turkey Creek and San Francisco River areas, not Rocky Mountain bighorn sheep. As discussed previously, desert bighorn sheep (*O. c. mexicana*) once inhabited several arid mountain ranges in central and southern New Mexico. Evidence of their occupancy in 14 ranges is available (New Mexico Department of Game and Fish, 1995).

In summary, at least eight species and three subspecies of native mammals have been extirpated from either part or all of New Mexico within historical times. Two kinds (*Cervus elaphus* and *Ovis c. canadensis*) have been reintroduced to form viable populations at least in some parts of the original ranges. Two other species (*Canis lupus* and *Mustela nigripes*) could be reintroduced in the future. One extirpated species (*Bison bison*) exists only in managed herds primarily on private lands and small parcels of public lands.

MAMMALS INTRODUCED IN NEW MEXICO

At least 15 species of mammals have been introduced either accidentally or intentionally into the state since the first occupancy by European humans. These now occur in the wild in numbers sufficient to justify discussing them as components of the mammalian fauna of New Mexico. These species are of special concern because they sometimes usurp indigenous mammals and/or their habitat. Introduced species can also affect interactions among natural populations by interfering with the processes of competition, reproduction (e.g., hybrids), and may be a nuisance to ranchers, farmers, and city dwellers.

Didelphis virginiana (Virginia Opossum)

The presence of this species in New Mexico was mentioned by Bailey (1932). Bermudez et al. (1995) provided a thorough review of records and sightings of this species in New Mexico. Records of this species east of the Pecos River are thought to probably be the result of natural expansion (Bermudez et al., 1995). Human related introductions probably have had some effect on distribution of this species, including records from the Rio Grande Valley and adjacent drainages, as well as those from the lower Pecos Valley and the record from Hidalgo County (Bermudez et al., 1995).

Sciurus niger (Fox Squirrel)

This species has become established in New Mexico, most likely as a result of introduction. Aday and Gennaro (1973) and Findley et. al (1975) reported 1958 and 1966 records from Roswell, Chaves County. Fox squirrels have also been observed in Carlsbad, Eddy County, and Lovington and Hobbs, Lea County (NMDGF, unpubl. data).

Rattus norvegicus (Norway Rat)

As early as 1851, this species was known to be present in several settlements in the state (Bailey, 1932). In 1888, Norway rats were known to occur in Albuquerque (Findley, 1987). However, this introduced species of mammal is thought to be relatively uncommon in New Mexico at the present time (Findley, 1987). Recently (1996), specimens have been received by the Museum of Southwestern Biology from the Bernalillo County Health Department.

Rattus rattus (Black Rat)

This species has been present in New Mexico since at least 1914 (Bailey, 1932). This mammal probably occurs in association with human settlements, mostly in the lower Rio Grande Valley (Findley, 1987).

Mus musculus (House Mouse)

It seems that this introduced species has been present in New Mexico at least since 1851 (Bailey, 1932). These rodents occur in association with structures constructed by humans, but they also occur in numerous habitats along with native mice and rats. House mice are common and widespread in New Mexico (Findley, 1987).

Myocastor coypus (Nutria)

Nutria apparently escaped from captivity near Roswell in 1937 and were reported at Bitter Lake National Wildlife Refuge (Sooter, 1943). Currently, nutria are not known to occur at Bitter Lake National Wildlife Refuge (Sonia R. Najera, pers. comm.). This species was introduced at San Simon Cienega, Hidalgo County, by the NMDGF in the early 1950s in an attempt to keep water areas open for nesting of waterfowl (Berghofer, 1967). Findley et al. (1975) reported finding a skull of nutria at San Simon Cienega in 1958. Hoffmeister (1986) reported that nutria were thriving during the 1960s when he visited San Simon Cienega. This species has been absent from San Simon Cienega since at least 1974. Marsh habitats that were created there for Mexican ducks (*Anas platyrhynchos diazi*) are no longer present and the area is reverting to a more natural cienega, at least in ungrazed portions of the area, which is unsuitable for this non-native species. During the 1950s, nutrias were sold to fur farmers throughout the state and escapees of this species from fur farms became a nuisance along the Ruidoso River due to their habit of burrowing through ditch banks and under fruit trees (Berghofer, 1967). This species still persists along the Rio Hondo from an area approximately 18 miles west of Roswell, Chaves County, to near Hondo, Lincoln County (Rand French, pers. comm.), where they have been trapped periodically between 1989 to 1995. There are also recent unconfirmed reports of this species along drainage ditches in the Albuquerque area (NMDGF, unpubl. data).

Canis familiaris (Feral Dog)

These animals are common in many parts of the state, especially on the outskirts of the larger human settlements, such as Albuquerque and Las Cruces. Feral dogs are known to kill domestic livestock; their impact on native mammals remains poorly understood.

Felis cattus (Feral Cat)

Like feral dogs, these free-ranging mammals are found commonly near the larger human settlements in the state. Investigations on impacts of feral cats on New Mexican mammals and other vertebrates are lacking. Jurek (1994) provided a thorough bibliography on feral, stray, and free-roaming domestic cats in relation to wildlife conservation.

Equus asinus (Feral Ass)

Free-ranging feral asses were common formerly in some parts of New Mexico and even were hunted along the Rio Grande Valley (Findley et al., 1975). They also apparently occur in portions of Lincoln County (Findley, 1987). Feral asses formerly occurred in the Jemez Mountains into White Rock Canyon, Sandoval County (Findley et al., 1975). Feral asses were known from the Bandelier National Monument and adjacent areas from the latter half of the 1930s and by 1976, reached numbers there exceeding 140 (Sydoriak, 1995). Apparently, efforts directed at controlling their numbers in this area began in 1946 and continued periodically until 1987. This non-native mammal was last observed at Bandelier National Monument during the summer of 1993, when a single individual was seen. An aerial survey for this species at Bandelier National Monument during 1994 did not detect their presence (Sydoriak, 1995).

Equus caballus (Feral Horse)

The last native horses in New Mexico became extinct about 10,000 years before the present (Findley et al., 1975). Currently, herds of feral horses occur in Rio Arriba and San Juan counties on public lands administered by the U.S. Forest Service and Bureau of Land Management. Feral horses also occur in portions of the Tularosa Basin, specifically, on WSMR. Available information strongly suggests that WSMR horses are derived from ranch stock that was either abandoned or escaped capture during the 1940s (Advanced Sciences, Inc., 1995). During the late 1960s, numbers of feral horses at WSMR were estimated at 200 to 300 (Advanced Sciences, Inc., 1995). In recent years, the population of feral horses at WSMR increased to 1,700 to 1,800 in four subpopulations (P. Morrow, WSMR, pers. comm., *in* Holdermann et al., 1996). Concern about overpopulation of feral horses at WSMR resulted in four roundups (1970-89) that involved removal of 751 horses by the U.S. Army, and more recently, the U.S. Army has removed significantly more feral horses from WSMR (D.A. Holdermann, WSMR, pers. comm.). Dieoffs of feral horses there were documented in 1985, 1989, and 1994; since 1991, the U.S. Army has monitored feral horse populations at WSMR (Holdermann et al., 1996). Currently, there are approxiamtely 100 to 200 feral horses remaining on WSMR (D.A. Holdermann, pers. comm.). Additionally, approximately 50 feral horses persist in the Sierra Larga Range, east of Socorro, Socorro County (Carla Alford, pers. comm.).

Sus scrofa (Feral Pig)

Findley (1987) reported that feral hogs in the state are found mostly in Hidalgo and Lincoln counties. Feral hogs are most frequently seen in lowland areas up to the foothills of the Peloncillo and Animas mountains, Hidalgo County.

Oryx gazella (Gemsbok)

A total of 93 gemsbok was introduced by the NMDGF during 1969-73 and 1976-77 to the WSMR (Reid and Patrick, 1983). Wood et al. (1970) conducted investigations on this species at the NMDGF's Red Rock facility prior to its introduction in New Mexico. The most recent survey data available (1995) on this species suggest a current population of 1,000 to 1,300, up from the estimated 600 to 800 that were thought to occur in 1990 (Snyder, 1995a). A total of 1,782 oryx have been taken by hunters during annual hunts conducted for this species since 1974, including approximately 190 taken during limited depredation hunts that were administered between 1990-95 (Snyder, 1995a). In spite of these depredation hunts, which are held to help control the expansion of this species into new areas, oryx continue to increase in numbers and distribution. For example, this species has been observed in several areas beyond WSMR, including the San Andres National Wildlife Refuge, White Sands National Monument, Bureau of Land Management and other government lands west of the San Andres Mountains, and at least as far north as a few miles north of U.S. Highway 380 in Socorro County.

Capra hircus (Wild Goat)

Domestic goats have become established in several regions in New Mexico, including the Guadalupe Mountains and along portions of the western escarpment of the Sacramento Mountains. The wild goat referred to as Persian ibex and/or Bezoar goat (sometimes referred to as *C. aegagrus*, generally considered directly ancestral to, if not conspecific with *C. hircus)* was introduced by the NMDGF into the Florida Mountains, Luna County, on 9 December 1970 when 15 were released (Sutcliffe, 1973). Investigations of this species after its introduction have been conducted by several researchers (e.g, Sutcliffe, 1973; Bavin, 1975; Woodroof, 1979; Morrow, 1986); however, information on the impacts of this species on the native flora and fauna is lacking. During 11 surveys on the Florida Mountains for this species conducted between 1985 and 1995, the mean population was estimated to be 562, which ranged from a low of 315 (1995) to a high of 1,020 (1985) (Snyder, 1995b). Assessments of habitat in the Florida Mountains by the Bureau of Land Management resulted in a population goal of 400 of these animals in the Florida Mountains. Hunts for this species have been conducted for several years to provide hunting opportunity and to attempt to control numbers of this goat on the Florida Mountains and to control the expansion of this species to other mountain ranges. Numbers of this goat killed by hunters between 1989-90 to 1994-95 ranged from 187 in 1989-90 to a high of 542 in 1993-94 hunts; hunts for this species in areas beyond the Florida Mountains have resulted in at least 23 kills (1990-91 hunts). This species has been reported as far north the Cooks' Range, as far south as the Tres Hermanas Mountains, both in Luna County, and as far east as Mt. Riley in the Potrillo Mountains, Doña Ana County (NMDGF, unpubl. data). Findley (1987) also reported this species from the Alamo Hueco, Doña Ana, and Mimbres mountains.

Capra ibex (Ibex)

This species, also known as Siberian ibex, was introduced into the Canadian River Gorge in 1975 and 1977 and increased to an estimated 100 animals. Limited hunting of this non-native goat occurred annually between 1981 to 1991, during which approximately 11 Siberian ibex were killed by hunters (Catanach, 1994). The last NMDGF surveys for this species were conducted in 1989 and no Siberian ibex were found. There has been no verifiable confirmation of this species in New Mexico since that time. This non-native species is apparently extirpated from New Mexico.

Ammotragus lervia (Barbary Sheep)

This species was introduced intentionally into the Hondo Valley about 1940 and into the Canadian River gorge about 1950 (Findley, 1987). Apparently, information on the sources of Barbary sheep introduced into New Mexico is incomplete (Ogren, 1965). Barbary sheep introduced into the Canadian River Gorge in 1950 by the NMDGF originated from 12 animals obtained from a population that had been introduced in the Hondo Valley, Lincoln County, and 45 animals from an introduced population on the Hearst Ranch, California. In 1956, 21 Barbary sheep from Thousand Oaks, California, were introduced by the San Juan County Wildlife Federation in Largo Canyon. This species has been hunted in New Mexico since 1955—1,008 killed between 1955-78 (Morrison, 1980) and 1,142 between 1990-94 (Gleadle, 1995). As a result of introductions of this species in New Mexico and other areas in the Southwest, Barbary sheep continue to expand in areas significantly beyond original release sites (e.g., Simpson, 1980). The impact of Barbary sheep on the native flora and fauna is understudied, but is worthy of serious concern and should be a topic of careful study and evaluation.

SUMMARY

In addition to the 15 species discussed above, other non-native mammals have been, and in some cases still are, held in captivity in privately owned class A game parks. Class A game parks are privately owned properties in which non-native ungulates are raised for private hunting interests. These areas are approved by the New Mexico State Game Commission and licensed by the NMDGF. Presently, there are two class A game parks, one in Santa Fe County and one in Cibola County; formerly, there was one also in Lincoln County. Some of the species reportedly held in captivity in these class A game parks include: *Cervus axis* (axis deer), *C. dama* (fallow deer), *C. nippon* (sika deer), *Hemitragus jemlahicus* (tahr), *Taurotragus oryx* (eland), and *Musimon ovis* (mouflon). *Hemitragus jemlahicus* has escaped at least once from its enclosure at Mt. Taylor, Cibola County, where in 1978 or 1979, at least one was killed by hunters.

By comparison on a regional basis, it seems that New Mexico and Texas have approximately the same numbers of species of mammals that have been introduced (unintentionally or intentionally) and are living in viable numbers in the wild. A reasonable prediction is that this number of introduced mammals will continue to survive, and to some degree, expand in New Mexico, as well as elsewhere. For example, a total of 25 such taxa are known now to occur in North America north of Mexico (Jones et al., 1992). The impact of these non-native mammals on the flora and fauna of New Mexico is poorly understood and warrants serious attention from the conservation community at large. There are some serious threats of the spread of other exotic species in the United States in general and in New Mexico in particular. For example, *Mustela putorius* (European ferret) has been documented to occur in the wild now in Arizona (Hoffmeister,1986), some of these mammals have been killed by hunters in Mississippi (Cheri A. Jones, pers. comm.), and there are reports of this species that may represent established populations in New Mexico (NMDGF, unpubl. data).

ACKNOWLEDGEMENTS

We are grateful to the numerous biologists who have contributed to the ever-growing data base that was used as the source of information presented in this paper. Of particular importance are many contributions of James S. Findley, who provided guidance and encouragement not only to us but to numerous students at the University of New Mexico and other biologists and naturalists interested in conservation of New Mexican mammals. We also acknowledge contributions made by the New Mexico Department of Game and Fish and the U.S. Fish and Wildlife Service toward conservation of the natural resources of New Mexico. We also thank M.A. Bogan, J.K. Frey, and W.L. Gannon for helpful comments and suggestions on the manuscript.

LITERATURE CITED

Aday, B.J., Jr., and A.L. Gennaro. 1973. Mammals (excluding bats) of New Mexican Llano Estacado and its adjacent river valleys. Studies in Natural Sciences, Eastern New Mexico University, 1(5):1-33.

Advanced Sciences, Inc. 1995. Environmental assessment for feral horse management, White Sands Missile Range, New Mexico. White Sands Missile Range Environmental Services Division, Report Number EA-001-95.

Anderson, S., and A.S. Gaunt. 1962. A classification of the white-tailed jackrabbits of Mexico. American Museum Novitates, 2088:1-16.

Anderson, S., and R. Barlow. 1978. Taxonomic status of *Cervus elaphus merriami* (cervidae). Southwestern Naturalist, 23(1):63-70.

Arita, H.T., and S.R. Humphrey. 1988. Revisión taxonomica del los murcielagos magueyeros del género *Leptonycteris* (Chiroptera: Phyllostomidae). Acta Zoologica Mexicana, 29:1-60.

Arizona Department of Game and Fish. 1996. Wildlife of special concern in Arizona, public review draft 16 March 1996. Unpublished list, Arizona Department of Game and Fish, Phoenix, 23 pp.

Bailey, J.A. In press. Status of major ecotypes of bighorn sheep in the United States.

Bailey, V. 1913. Life zones and crop zones of New Mexico. North American Fauna, 35:1-100.

Bailey, V. 1932. Mammals of New Mexico. North American Fauna, 53:1-412.

Baker, R.J., J.C. Patton, H.H. Genoways, and J.W. Bickham. 1988. Genic studies of *Lasiurus* (Chiroptera: Vespertilionidae). Occasional Papers, Museum, Texas Tech University, 117:1-15.

Baltosser, W.H. 1980. A biological inventory of endangered species occurring in Guadalupe Canyon, Hidalgo County, New Mexico. New Mexico Department of Game and Fish, Contract 519-68-06, Final report:1-73.

Bavin, R.L. 1975. Ecology and behavior of Persian ibex in the Florida Mountains, New Mexico. M.S. Thesis. Colorado State University, Fort Collins, Colorado, 141 pp.

Bednarz, J.C., and J.A. Cook. 1984. Distribution and number of the white-sided jackrabbit (*Lepus callotis gaillardi*) in New Mexico. Southwestern Naturalist. 29(3):358-360.

Berghofer, C.B. 1967. Protected furbearers. Pp. 187-194, *in* New Mexico wildlife management. New Mexico Department of Game and Fish, Santa Fe, 250 pp.

Bermudez, F.C., J.N. Stuart, J.K. Frey, and R.Valdez. 1995. Distribution and status of the Virginia opossum (*Didelphis virginiana*) in New Mexico. Southwestern Naturalist, 40(3):336-340.

Best, T.L., and T.H. Henry. 1993. *Lepus callotis*. Mammalian Species, 442:1-6.

Bogan, M.A., and C. Jones. 1975. Observations of *Lepus callotis* in New Mexico. Proceedings of the Biological Society of Washington, 88(5):45-50.

Bogan, M.A., and P. Mehlhop, 1983. Systematic relationships of gray wolves (*Canis lupus*) in southwestern North America. Occasional Papers, Museum of Southwestern Biology, 1:1-21.

Brown, D.E. 1983a. The wolf in the Southwest: The making of an endangered species. University of Arizona Press, Tucson, 195 pp.

Brown, D.E. 1983b. On the status of the jaguar in the Southwest. Southwestern Naturalist, 28(4):459-460.

Buechner, H.K. 1960. The bighorn sheep in the United States, its past, present, and future. Wildlife Monograph, No. 4, 174 pp.

Carie, W., J.E. Vaughan, and V.W. Diersing. 1978. First record of *Sorex arizonae* (Insectivora: Soricidae) from Mexico. Southwestern Naturalist, 23(3):532-533.

Catanach, M. 1994. Siberian ibex population trends, distribution, and harvest information. New Mexico Department of Game and Fish. Unpublished final report, W-93-R, segments 32-36, 4 pp.

Clark, T.W., E. Anderson, C. Douglas, and M. Strickland. 1987. *Martes americana*. Mammalian Species, 289:1-8.

Cockrum, E.L., and Y. Petryszyn. 1991. The long-nosed bat, Leptonycteris: An endangered species in the Southwest? Occasional Papers, The Museum, Texas Tech University, No. 142, 32 pp.

Conley, W.H. 1970. Geographic variation in the least chipmunk, *Eutamias minimus* in New Mexico and Arizona. Journal of Mammalogy, 51:695-702.

Conley, W.H., and W. C. Brown. 1977. Habitat partitioning between the jackrabbits *Lepus callotis* and *L. californicus*. New Mexico Department of Game and Fish. Contract 516-65-23, Final report, 9 pp.

Conway, M.C. 1975. A rare hare. New Mexico Wildlife, 21(2):45-50.

Conway, M.C., and C.G. Schmitt. 1978. Record of the Arizona shrew *Sorex arizonae* from New Mexico. Journal of Mammalogy, 59:631.

Cook, J.A. 1986. The mammals of the Animas Mountains and adjacent areas, Hidalgo County, New Mexico. Occasional Papers, Museum of Southwestern Biology, 4:1-45.

Dunn, W.C. 1996. Long range plan for the management of Rocky Mountain bighorn sheep in New Mexico, 1996-2002. New Mexico Department of Game and Fish, unpublished draft, 32 pp.

Easterla, D.A. 1973. Ecology of 18 species of Chiroptera at Big Bend National Park, Texas. Northwest Missouri State University Studies, 34:1-165.

Findley, J.S. 1987. The natural history of New Mexican mammals. University of New Mexico Press, Albuquerque, xii + 164 pp.

Findley, J.S., and C. Jones. 1963. Geographic variation in the least cotton rat in New Mexico. Journal of Mammalogy, 44:307-315.

Findley, J.S., A.H. Harris, D.E. Wilson, and C. Jones. 1975. Mammals of New Mexico. University of New Mexico Press, Albuquerque, xxii + 360 pp.

Fisher, A.S. 1994. Status of desert bighorn sheep in New Mexico, 1993. Desert Bighorn Council Transactions, 38:33-34.

Fitzgerald, J.P., C.A. Meaney, and D.M. Armstrong. 1994 [1995]. Mammals of Colorado. Denver Museum of Natural History and University Press of Colorado, Niwot, Colorado, 467 pp.

Frey, J.K., J.H. Fraga, and F.C. Bermudez. 1995. A new locality of the montane vole (*Microtus montanus arizonensis*) in New Mexico. Southwestern Naturalist, 40(4):421-422.

García–Moreno, J., M.D. Matocq, M.S. Roy, E. Geffen, and R.K. Wayne. 1996. Relationships and genetic purity of the endangered Mexican wolf based on analysis of microsatellite loci. Conservation Biology, 10(2):376-389.

Gates, G.H. 1967. Elk. Pp. 31-41, *in* New Mexico wildlife management. New Mexico Department of Game and Fish, Santa Fe, 250 pp.

Girmendonk, A.L.1994. Ocelot, jaguar and jaguarundi sighting reports: Arizona and Sonora,

Mexico. Technical Report 35, Nongame and endangered wildlife program. Arizona Game and Fish Department, 15 pp.

Gleadle, B. 1995. Barbary sheep population trend and harvest information. Unpublished final report, W-930R, Segments 32-36, New Mexico Department of Game and Fish, 19 pp.

Gore, A. 1992. Earth in the balance: Ecology and the human spirit. Houghton Mifflin Company, New York, 407 pp.

Hafner, D.J., K.E. Petersen, and T.L. Yates. 1981. Evolutionary relationships of jumping mice (genus *Zapus*) of the southwestern United States. Journal of Mammalogy, 66:501-512.

Hall, E.R. 1981. The mammals of North America. John Wiley and Sons, New York, 2nd ed., 1:XV + 1-600 + 90 and 2:VI + 601–1181 + 90.

Halloran, A. 1946. The carnivores of the San Andres Mountains, New Mexico. Journal of Mammalogy, 27:154-161.

Hanson, D.H. 1977. Taxonomic status of the prairie dog subspecies *Cynomys ludovicianus ludovicianus* (Ord) and *Cynomys ludovicianus arizonensis*. M.S. thesis, Eastern New Mexico University, Portales, New Mexico, 29 pp.

Hensley, A.P., and K.T. Wilkins. 1988. *Leptonycteris nivalis*. Mammalian Species, 307:1-4.

Hill, J. 1942. Notes on mammals of northeastern New Mexico. Journal of Mammalogy, 23:75-82.

Hinesley, L.L. and C.S. Thaeler, Jr. 1977. Karyotype and distribution of the southern pocket gopher, *Thomomys umbrinus emotus* Goldman. Journal of Mammalogy, 58:235-237.

Hoditschek, B., J.F. Cully, Jr., T.L. Best, and C. Painter. 1985. Least shrew *Cryptotis parva* in New Mexico. Southwestern Naturalist, 30(4):600-601.

Hoffmeister, D.F. 1986. Mammals of Arizona. University of Arizona Press, Tucson, xx + 602 pp.

Holdermann, D.A., V.W. Howard, R. Valdez, R. Ditner, and B.C. Thompson. 1996. Feral horse relationships to vegetation communities and wildlife habitat in the Tularosa Basin, New Mexico. Unpublished report to White Sands Missile Range, USFWS, and NBS Cooperative Agreement 14-16-0009-1572, 50 pp.

Hoyt, R.A., J.S. Altenbach, and D.J. Hafner. 1994. Observations on long-nosed bats (*Leptonycteris*) in New Mexico. Southwestern Naturalist, 39(2):175-179.

Hubbard, J.P., and C.G. Schmitt. 1984. The black-footed ferret in New Mexico. Unpublished report to New Mexico Bureau of Land Management, Santa Fe, New Mexico, 118 pp.

Hubbard, J.P., C.S. Thaeler, Jr., and C.G. Schmitt. 1983. Notes on voles (*Microtus*, Family Cricetidae) in New Mexico. Nemouria, Occasional Papers, Delaware Museum of Natural History, 28:1-8.

Jones, J.K., Jr. 1993. The concept of threatened and endangered species as applied to Texas mammals. Texas Journal of Science, 45:115–128.

Jones, J.K., Jr., and C. Jones. 1992. Revised checklist of recent land mammals of Texas, with annotations. Texas Journal of Science, 44:53-74.

Jones, J.K., Jr., R.S. Hoffmann, D.W. Rice, C. Jones, R.J. Baker, and M.D. Engstrom. 1992. Revised checklist of North American mammals north of Mexico, 1991. Occasional Papers, Museum, Texas Tech University, 146:1-23.

Jones, J.K., Jr., R.W. Manning, F.D. Yancey, and C. Jones. 1993. Records of five species of small mammals from western Texas. Texas Journal of Science, 45:104-105.

Jurek, R.M. 1994. A bibliography of feral, stray, and free-ranging domestic cats in relation to wildlife conservation. California Department of Fish and Game, Nongame Bird and Mammal Program Report 94-95, 24 pp.

Matthiesson, P. 1987. Wildlife in America. Elisabeth Sifton Books, Viking Penguin Inc., New York, 332 pp.

McClellan, J. 1954. An otter catch on the Gila River in southwestern New Mexico. Journal of Mammalogy, 35:443-444.

Mearns, E.A. 1896. Preliminary description of a new subgenus and six new species and subspecies of hares, from the Mexican border of the United States. Proceedings of the U.S. National Museum, 17:551-565.

Mech, L.D. 1974. *Canis lupus*. Mammalian Species, 37:1-6.

Mehlhop, P. 1995. Status of the white-sided

jackrabbit in New Mexico with recommendations for a monitoring protocol. New Mexico Department of Game and Fish. Contract 95-516-32, Final Report, 27 pp.

Mexican Wolf Recovery Team. 1982. Mexican wolf recovery plan. U.S. Fish and Wildlife Service, Albuquerque, New Mexico, 115 pp.

Morrison, B.L.1980. History and status of Barbary sheep in New Mexico. Pp. 15-16, *in* Proceedings of the symposium on ecology and management of Barbary sheep. Texas Tech University, Lubbock, Texas, 112 pp.

Morrison, J.L. 1992. Persistence of the meadow jumping mouse, *Zapus hudsonius luteus*, in New Mexico. Southwestern Naturalist, 37(3):308-311.

Morrow, P. 1986. Evaluation of Persian wild goat carrying capacity determinations, Florida Mountains, New Mexico. M.S. Thesis. New Mexico State University, Las Cruces, 67 pp.

Mumford, R.E. and D.A. Zimmerman. 1963. The southern yellow bat in New Mexico. Journal of Mammalogy, 44:417-418.

Nash, R. 1982. Wilderness and the American mind. Yale University Press, New Haven, Connecticut, xvii + 425 pp.

National Research Council, Committee on Scientific Issues in the Endangered Species Act. 1995. Science and the Endangered Species Act. National Academy of Science, Washington, D.C., 162 pp.

New Mexico Department of Game and Fish. 1993. Fort Wingate bison move to El Malpais. New Mexico Wildlife, 38 (3):6-9.

New Mexico Department of Game and Fish. 1995. New Mexico's long-range plan for desert bighorn sheep management, 1995-2002. Final Report Federal Aid in Wildlife Restoration Project W-127-R10, Job 1, New Mexico Department of Game and Fish, Santa Fe, 40 pp.

New Mexico Department of Game and Fish. 1996. List of threatened and endangered species. 19 NMAC 33.1; 31 January 1996.

Nowak, R.M. 1996. Another look at wolf taxonomy. Pp. 375-398, *in* Ecology and conservation of wolves in a changing world (L.N. Carbyn, S.H. Fritts, and D.R. Seip, eds.). Canada Circumpolar Press Institute, University of Alberta, Edmonton, Alberta, Canada, 620 pp.

Ogren, H.A. 1965. Barbary sheep. New Mexico Department of Game and Fish, Santa Fe, Bulletin No. 13, 117 pp.

Owen, R.D. and M.J. Hamilton. 1986. Second record of *Cryptotis parva* (Soricidae: Insectorivora) in New Mexico, with review of its status on the Llano Estacado. Southwestern Naturalist, 31(3):403-405.

Patterson, B.D. 1980. A new subspecies of *Eutamias quadrivittatus* (Rodentia: Sciuridae) from the Organ Mountains, New Mexico. Journal of Mammalogy, 61:445-464.

Polechla, P.J., S. DesGeorges, R. Gardiner, B. Hayes, G. Long, and D. Storch. 1994. Field reconnaissance for river otters of the upper Rio Grande, Cow Paddy to Lee's Trail, Taos County, New Mexico. Unpublished report to Bureau of Land Management, Taos, New Mexico, 19 pp.

Reid, W.H. and G.R. Patrick. 1983. Gemsbok (*Oryx gazella*) in White Sands National Monument. Southwestern Naturalist, 28(1):97-99.

Reiger, J.F. 1986. American sportsmen and the origins of conservation. University Oklahoma Press, Norman, 316 pp.

Sands, J.L. 1967. Bighorn sheep. Pp. 69-82, *in* New Mexico wildlife management. New Mexico Department of Game and Fish, Santa Fe, 250 pp.

Schmidly, D.J. 1991. The bats of Texas. Texas A&M University Press, College Station, xvii + 188 pp.

Shuster, C.J. 1988. Distribution, relative population densities, and habitat of *Cryptotis parva* in New Mexico. New Mexico Dept. Game and Fish, Contract 516.6-75-19, Final Report, 1-20.

Simpson, C.D., ed. 1980. Proceeding of the symposium on ecology and management of Barbary sheep. Texas Tech University, Lubbock, Texas, 112 pp.

Snyder, P. 1995a. Oryx population trends, distribution, and harvest information. New Mexico Department of Game and Fish, unpublished final report, W-93-R, segments 32-36, 7pp.

Snyder, P. 1995b. Persian ibex population trends, distribution, and harvest information. New Mexico Department of Game and Fish, unpublished final report, W-93-R, segments 32-36, 10 pp.

Sooter, C. 1943. Nutria good diver. Journal of Mammalogy, 24:503-504.

Spicer, R.B. 1987. Status of the Arizona river otter, *Lutra canadensis sonora* (Rhoads), along the Colorado River in Arizona, unpublished report for U.S. Fish and Wildlife Service, Office of Endangered Species, Albuquerque, New Mexico, Contract 14-16-0002-82-216, 39 pp.

Stachlecker, D.W. 1986. A survey for the river otter (*Lutra canadensis*) in Taos and Colfax counties, New Mexico. New Mexico Department of Game and Fish, Contract 519-75-01, Final Report, 10 pp.

Sullivan, R.M. 1985. Phyletic, biogeographic, and ecologic relationships among montane populations of least chipmunks *Eutamias minimus* in the Southwest. Systematic Zoology, 34:419-448.

Sullivan, R.M., and K.E. Petersen. 1988. Systematics of southwestern populations of least chipmunks *Tamias minimus* reexamined: A synthetic approach. Museum of Southwestern Biology, Occasional Papers 5., 27 pp.

Sutcliffe, D.H. 1973. Post-release investigations of Iranian ibex in New Mexico and evaluation of a proposed release site. M.S. Thesis. New Mexico State University, Las Cruces, 38 pp.

Sydoriak, C. 1995. Resource management plan, Bandelier National Monument. Unpublished manuscript, National Park Service files, Los Alamos, New Mexico 87544.

Thorne, E.T., and E.S. Williams. 1988. Disease and endangered species: The black-footed ferret as a recent example. Conservation Biology, 2(1):66-74.

U.S. Fish and Wildlife Service. 1988. Black-footed ferret recovery plan. U.S. Fish and Wildlife Service, Denver, Colorado, 154 pp.

U.S. Fish and Wildlife Service. 1994. Mexican Long-nosed bat (*Leptonycteris nivalis*) recovery plan. U.S. Fish and Wildlife Service, Albuquerque, New Mexico, 91 pp.

U.S. Fish and Wildlife Service. 1995. Endangered and threatened wildlife and plants. Federal Register 50 CFR 17.11 and 17.12, Washington, D.C., 44 pp.

Watkins, L.C. 1977. *Euderma maculatum*. Mammalian Species, 77:1-4.

Whitaker, J.O., Jr. 1972. *Zapus hudsonius*. Mammalian Species, 11:1-7.

Whitaker, J.O., Jr, 1974. *Cryptotis parva*. Mammalian Species, 43:1-8.

Wilson, D. E. 1985. Status report, *Leptonycteris sanborni* Hoffmeister, Sanborn's long-nosed bat. U.S. Fish and Wildlife Service, Office of Endangered Species, Albuquerque, New Mexico, unpublished report, i-ii + 1-36.

Wilson, D.E., and D.M. Reeder (eds). 1993. Mammal species of the world: A taxonomic and geographic reference. Second ed. Smithsonian Institution Press, Washington, DC, 1206 pp.

Wolok, M. 1994. Reported wolf observations in New Mexico, Texas and Mexico. Unpublished report, U.S. Fish and Wildlife Service, Albuquerque, New Mexico, 10 pp.

Wood, J.E., R.J. White, and J.L. Durham. 1970. Investigations preliminary to the release of exotic ungulates in New Mexico. New Mexico Department of Game and Fish Bulletin No. 13, 58 pp.

Woodroof, W.O. 1979. Ecology and behavior of Persian wild goat (*Capra aegagrus*) in New Mexico. M.S. Thesis. New Mexico State University, Las Cruces, 95 pp.

Yates, T.L. 1982. Status of the Peñasco chipmunk. New Mexico Department of Game and Fish, Contract 519-71-01, Preliminary Report, 3 pp.

Clyde Jones, Department of Biological Sciences and The Museum, Texas Tech University, Lubbock, Texas 79409-3191, and C. Gregory Schmitt, New Mexico Department of Game and Fish, Endangered Species Program, Santa Fe, New Mexico 87504

Non-Human Mortality, Injuries, and Possible Cannibalism in Utah Black Bears

HAL L. BLACK

Abstract

Over the past eight years, my students and I have made occasional observations of non-human caused mortality, injuries, and apparent cannibalism in two black bear study sites in Utah. Here I report two possible cases of cannibalism, four cases of death at old age, and two cases of serious injuries. I suggest that human-caused mortality in bears may be in the final outcome, not unlike that which black bears experienced in the past from other sympatric carnivore species. Further, I suggest that bears are a robust, physically durable species that, during long lives, sustain rather extensive injuries.

INTRODUCTION

Beginning in 1987, the Utah Division of Wildlife Resources, the U.S. Forest Service, the Bureau of Land Management, private organizations, and individuals have, in conjunction with Brigham Young University, conducted research into the biology, behavior, and population ecology of black bears. The two regions of the state where I have conducted field work or am currently working are the LaSal Mountains of southeastern Utah and the Book Cliffs region of the East Tavaputs Plateau in eastern Utah. Various personnel associated with these studies have observed cases of non-human-caused mortality, severe injuries, and possible cannibalism, events rarely documented. I report these cases here to increase our understanding of these events as they relate to the biology, evolution, and ecology of black bears.

METHODOLOGY

Bears were captured with Aldrich leg snares or barrel traps using peanut butter, bacon ends, and carrion as bait. Selected individuals (primarily breeding-age females) were fitted with radio transmitters and monitored from fixed-wing aircraft and from 4x4 trucks and ATVs using directional antennas. A premolar was extracted for aging by cementum annuli (Matson's Laboratory, Missoula, MT, USA).

CASE STUDIES

Below are observations on eight individual bears. Two were probably cannibalized, four died presumably of old age, and two had severe injuries to limbs.

Gramps

An old male with well-worn teeth was trapped 16 August 1991 in the Book Cliffs. He was extremely thin, and his eyes were sunken deep into his head. The field crew was of the opinion that he was on the verge of dying and, being anxious to document natural mortality in an obviously old male, they fitted him with a radio-transmitter. To our surprise, he lived through the remainder of fall, denned, and successfully hibernated. On 8 March 1992, he was outside of his den in a bed at the base of a large Douglas fir when crews approached. Upon detection, he became rather cooperative by scurrying back into his den (under a large rock) where he was then anesthetized. His pelage was judged to be in good condition, although he was emaciated. Within two weeks his radio signal indicated mortality, and on our return visit he was found dead, sprawled on the ground near his den. Old, large male bears are rare in hunted

Life Among the Muses: Papers in Honor of James S. Findley. Edited by T.L. Yates, W.L. Gannon & D.E. Wilson.

populations; this one had lived long enough to no longer be considered a trophy (i.e., if weight or just general physical condition count). Yet, on the basis of a front pad width of 140 mm, he would have stirred the interest of hunters who frequented the area. In another sense, he was an evolutionary trophy. If he attained sexual maturity at four, he had probably gone through 14 breeding seasons. Assuming he was successful at demonstrating his biological fitness, he could have induced ovulation in, say, three females per summer, and assuming a litter size of two (about average for Utah bears), he could have produced eighty-four cubs before he apparently starved to death. He paid several prices for longevity. In addition to the extensive wear on his teeth (Fig. 1), he was missing two lower incisors, an upper first molar, premolars 3 and 4, and a lower molar. With the exception of one canine, which was well worn, all other canines had worn to the point that they had seemingly lost all function—both trophic and behavioral. He had broken both fibuli and one ulna, but all had healed with proper alignment (Fig. 2). Several metacarpels and metatarsals showed evidence of arthritis. He was missing a small portion of one ear and about 50 mm of his tail. He attained old age sharing his home range with cowboys, elk, and deer hunters, numerous active gas wells and their attendants, campers, and bear hunters and their dogs.

Auntie

On 21 April 1990, a 14-year-old female was found dead on a small snowbank in a ravine within 10 m of a bear bed at the base of a large Douglas fir. No den was found in the immediate vicinity. Careful examination of the carcass failed to determine any cause of death. There were no bullets or arrow wounds (the spring bear hunt was in progress) and no other evidence of injury. She appeared to have walked to the snowbank, lain down, and died. Her teeth showed wear comparable to the old male seen in Fig. 1, and she was missing three upper molars. Her lower canines were worn nearly to the

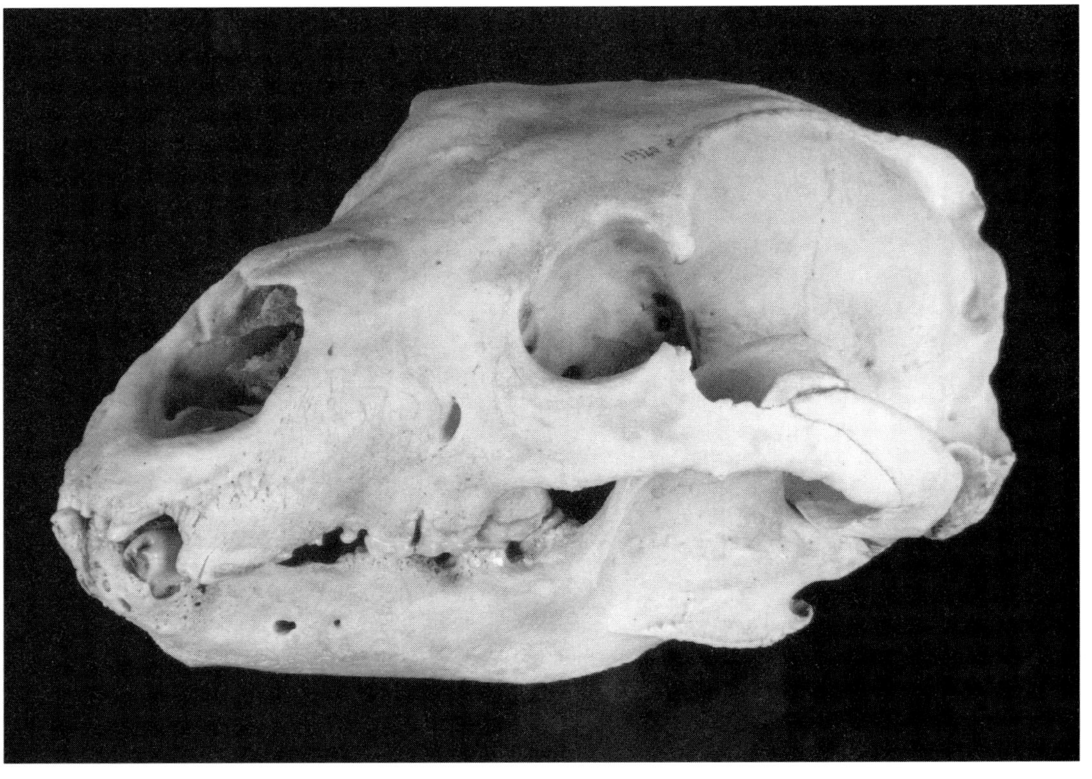

Fig. 1. Skull of a 17-year-old male bear (Gramps) that died of natural causes. Note the extensive wear of all teeth, especially the canines.

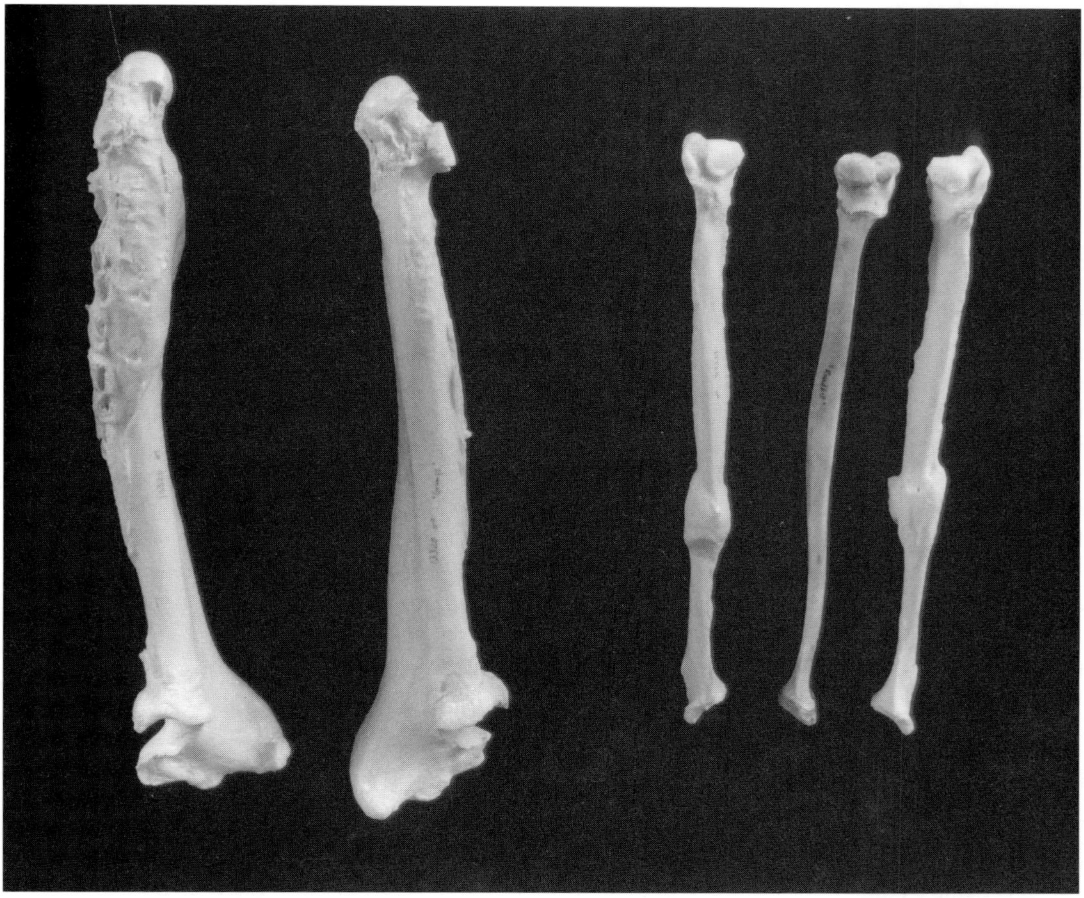

Fig. 2. (A) Normal ulna (right) and an apparently fractured and healed one (left). (B) Three fibuli showing two that were broken and subsequently mended. Fibula in the center is normal. (All but middle fibula in B are from Gramps).

level of the incisors, and the upper canines were perhaps one-third the length of those of a prime female bear.

Marty's Friends

We were "led" to the burial grounds of two adult bears on the LaSal Mountains by a radio-collared male named Marty. We had visited his den in early March to check his radio-collar and found him much too awake to do so. We returned in May to assess vegetation and take physical measurements of the den. On first inspection, the mound of soil piled up in front of the den entrance suggested a new den, but the size of the den interior suggested otherwise. Our suspicion that this was an old den was confirmed by the discovery of the remains of portions of two adult bear skeletons in the soil in front of the den. A partial skull and lower jaw and most of the major leg bones, along with scapuli from two different bears, were found. One interpretation is that these two bears on separate occasions had used this den prior to Marty and had died during hibernation of natural causes (there was no evidence that the cave ceiling had fallen on them before they were excavated from the den by Marty or some other bear). We suspect that their remains were excavated from the den at the same time, given their distribution in the soil. Bears will reuse dens (Beck, 1991), but it is unusual for two bears to have apparently died of natural causes in the same den. Another explanation is that the smaller bear (a mature female or a 5-6 year-old male) was killed by the larger one when discovered in the den and this was followed by the death of the larger bear.

Jackie Dee

On the LaSal Mountains 9 May 1987, a 54.5 kg, 2-year-old female was snared by the left front foot. In resisting the snare, she damaged the first digit. She was killed in July 1988, and a posthumous examination showed her foot to be stiff and probably dysfunctional. We salvaged the carcass and when we prepared the skeleton, we found that the right humerus had been broken subsequent to the initial capture (Fig. 3). It had healed atypically with a 75 mm overlap of the shaft elements. During the mending process, there appeared to have been extensive irritation of the ribs beneath the humerus since new bone had developed on and between several ribs (Fig. 3). Alternatively, the ribs may have been broken at the same time as the humerus. That these injuries had apparently taken their toll seemed evident since she had not gained weight in more than 14 months. Nevertheless, her pelage was in good condition and was salvaged along with the entire skeleton as a museum specimen.

Road Warrior

In mid-November 1993, a motorist struck and killed a 138.6 kg, 4-year-old male bear on a highway in mountains east of Salt Lake City. The Utah Division of Wildlife Resources retained the skin of this bear for display and donated the skull and skeleton to Brigham Young University. After skinning, a taxidermist removed 27.3 kg of subcutaneous fat from the specimen, about 20% of its total body weight. In preparing the skeleton, we found that the left femur had been broken in a previous accident and was in the early stage of mending. Callus had formed around the severely overlapped bone and some stabilization had occurred (Fig. 4). The atypical fusion left the femur a full 83 mm shorter than the right. The hair on the sole of the foot of the broken leg was nearly twice as long as that on the good foot, indicating that for some time the broken leg was carried or at least only lightly used. We judged this bear to be in good physical condition and surmised that the injury occurred 6–8 weeks prior to his death. This would have been in early fall during a time when natural foods, both hard and soft mast, and some carcasses and offal from deer and elk kills by hunters would have been readily available, perhaps reducing the need to travel extensive distances to forage.

Interstate 80, where this bear was killed, runs through elk and deer habitat, and road-kills, especially of deer, are regular events. In fact, there was a deer carcass about 50 m from where this bear was killed, suggesting that he may have been attracted to the road to feed. Expeditious removal of road kills from the roadside should be encouraged in general, but might be most advantageous in the fall when bears are in a hyperphagic mode in preparation for hibernation. At this time, adult males are more vulnerable since they remain active later into the fall than most other bears.

Chelsea

A 2½-year-old female, weighing 47.7 kg when caught on the LaSal Mountains on 5 July 1989, was radio-collared and released. She was lean, as is common for young bears at this time of year, but otherwise in good physical condition. She showed no injuries from the snare. She was radio-located several times by fixed-wing aircraft and from the ground in the two months following capture, but in mid-September her radio transmission indicated she had shed the collar or had died. When we reached her location on 30 September, we found her remains in a small grassy meadow in a mixed stand of Gambels oak and Ponderosa pine. About 70% of the skeleton, the skull and lower jaw, and perhaps 40% of her skin were found. Her remains appeared typical of those left by feeding black bears, with several bones uneaten and the skin inverted over large portions of the skeleton (LeCount, 1986). In the immediate vicinity were 25-30 bear scats containing bear hair, bone fragments, and acorn hulls. About 20 m from the remains was a bear bed at the base of a large pine. That bear(s) had fed extensively on acorns in the area was evidenced by numerous broken limbs and terminal stems on the oak. We suspect that, as is typical for bears, several may have been using this patch, and this young female was killed by a mature bear. Three radio-collared adult females had overlapping home ranges that included the site where this young female died and was eaten. Crowding and perhaps competition for food is thought to be the cause of high rates of cannibalism in an Is-

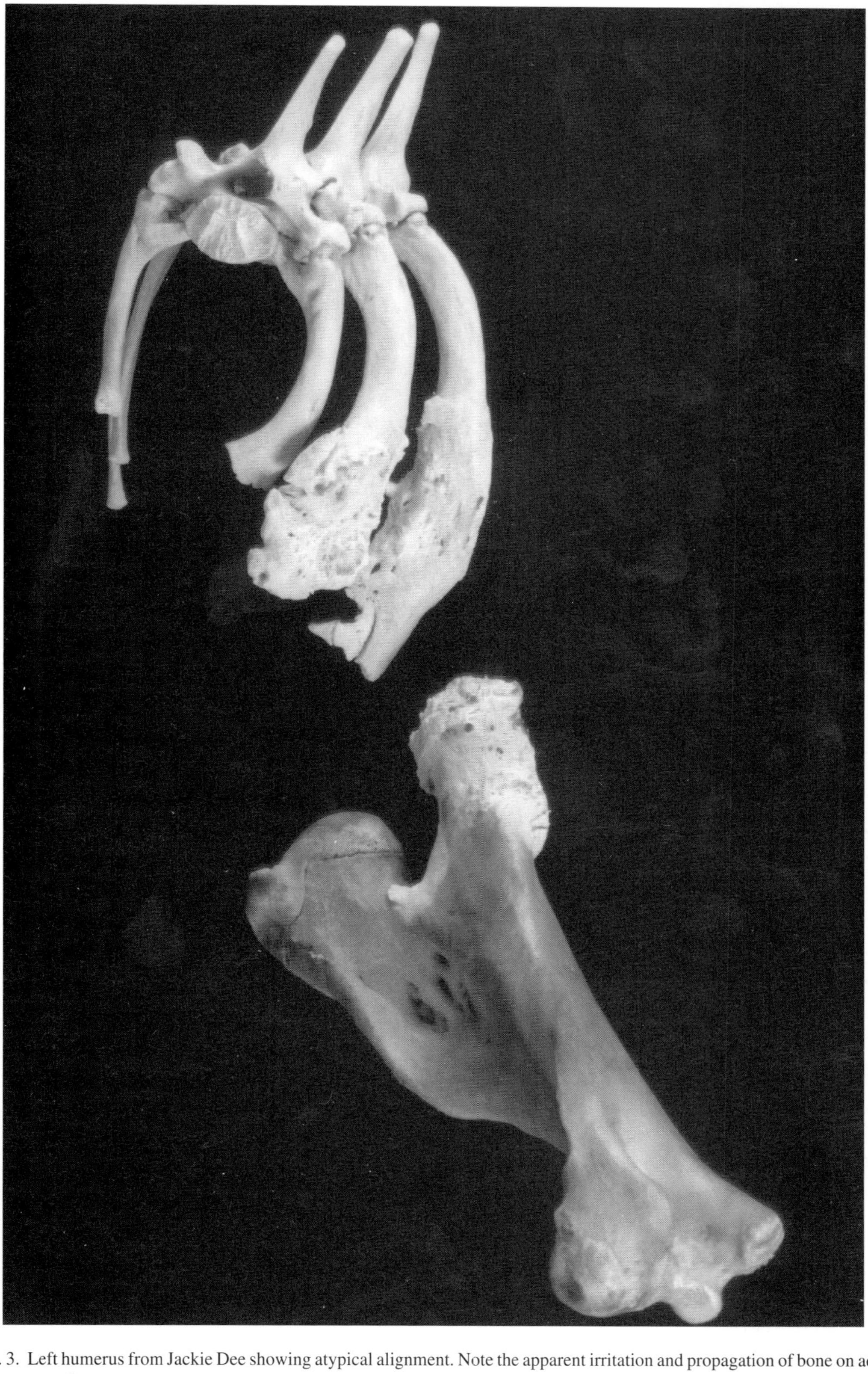

Fig. 3. Left humerus from Jackie Dee showing atypical alignment. Note the apparent irritation and propagation of bone on adjacent ribs.

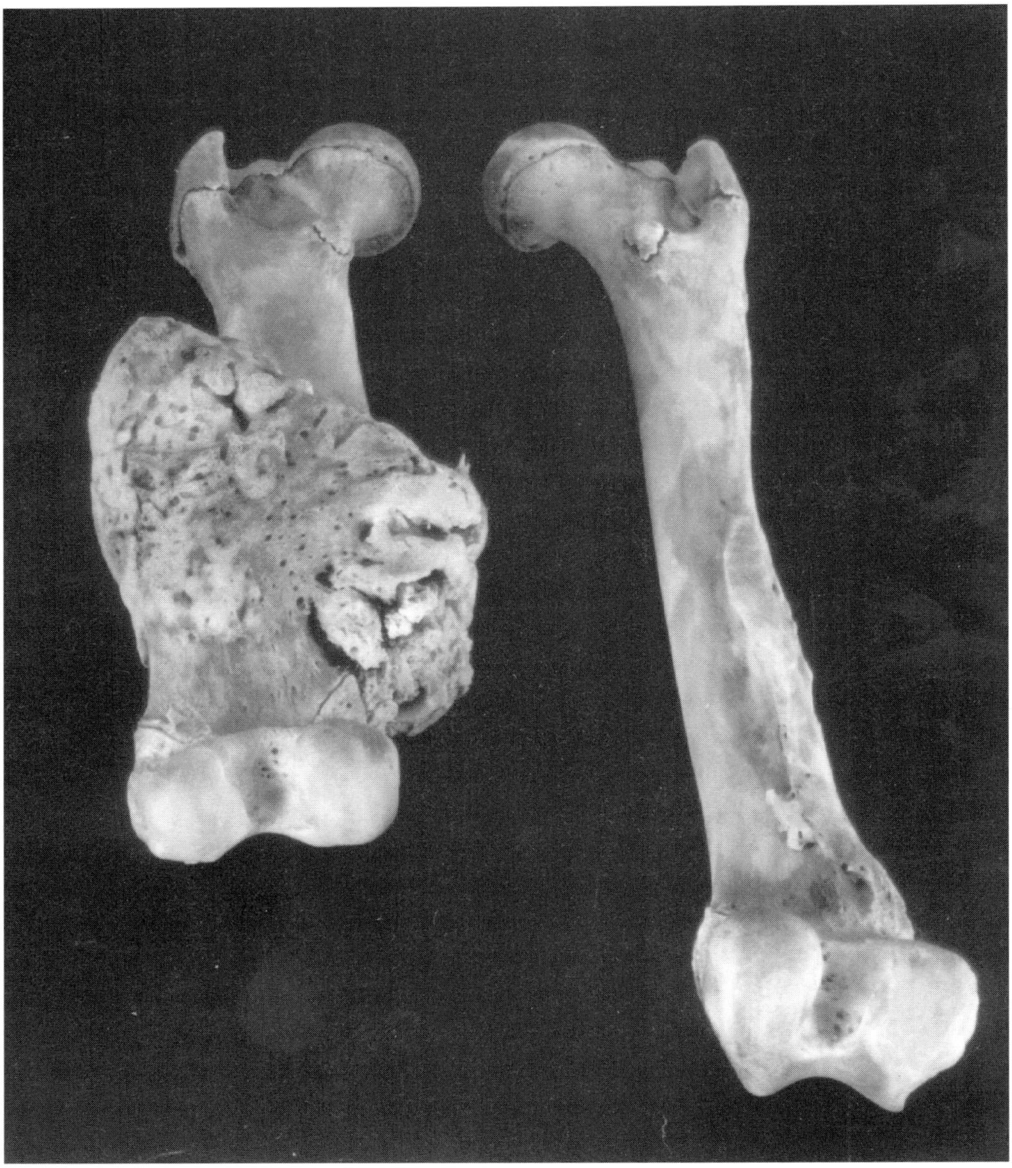

Fig. 4. Partially healed left femur shown with the right un-injured femur from a four-year-old bear (Road Warrior).

land population of bears in Wisconsin (Noyce, 1995). Mature males had been caught and tagged in the area as well. Of course, this bear may have been scavenged and not cannibalized.

Tina

This 7-year-old was located 17 March 1990 in a den at about 2958 m elevation on the LaSal Mountains. She weighed 65.9 kg, 2.3 kg less than she weighed when first caught in May of 1988. With the exception of some mange around both eyes, she was in good pelage. This bear is one of the few we captured with black hair (approximately 80% of the bears in our samples are shades of brown). Two female yearlings, each weighing 18.2 kg, were denned with her. Their pelage was sound and they appeared healthy. Upon returning to this den in summer, we found that one of the yearlings had died or was cannibalized. Only the lower jaw and skull, several carpel, and tarsal bones, and the proximal halves of the humeri, ulni, and radii remained. A portion of the right maxilla had been removed exposing the roots of the fourth premolar and the first molar. This appeared to be the result of a bite delivered while being killed and/or eaten by a den mate or some other carnivore, perhaps a cougar or coyote.

DISCUSSION

The low number of observed natural mortalities in Utah black bears is consistent with other studies throughout North America (Beck, 1991; Rogers, 1987; Schwartz and Franzmann, 1991, for example). The leading source of mortality is human caused, is persistent and in some ways, may mimic the predation black bears undoubtedly experienced before the elimination of the majority of their competitors and predators. Historically, brown bears and wolves occupied much of their range; during the Pleistocene, predators like saber-tooths (*Smilidon*), giant short-faced bears (*Arctodus*) and spectacled bears (*Tremarctos*) (Shaw and Cox, 1993) may have kept populations of black bears at densities below that seen at present. Geist (1993) suggests that both humans and black bears flourished when the larger *Arctodus* went extinct. Black bears may have reached their highest population densities before the advent of the more effective firearms of recent years. But their current populations, in spite of hunting, are considered by some to be in good shape and in no immediate danger of collapse (Garshelis, 1993).

That many bears die from causes unrelated to human activities seems evident from the large number of marked bears in our study areas that are never caught or seen a second time. We suspect that they simply "perform death alone," as Lewis Thomas (1973) puts it. Most of nature dies, as Thomas suggests, "behind things, under things," and we wouldn't have encountered the few deaths we reported here without radiotelemetry.

The injuries I report here may seem rather dramatic and can be viewed as alterations or departures from adaptive form (Frazzetta, 1975), leaving individuals at some disadvantage in competition and survival. However, the 4-year-old male (Road Warrior) seemed, by visual estimate of body fat, to have survived the broken femur rather well, and the 17-year-old male (Gramps) had rather extensive bone injuries and still lived much beyond average life expectancies for males in a hunted population. On the LaSal Mountains, we never caught, over a four-year period, any male five years of age or older. In the case of the small two-year-old female (Jackie Dee), however, a defective paw, broken humerus, and irritated ribs may have been too far beyond the threshold; she hadn't gained weight in more than 14 months. She may have survived her injuries, though, had she not been shot. Beck (1991) reports that an adult female bear lost, over a two-year period, one front paw and the arm up to the elbow on the other front leg, yet successfully raised cubs. This would seem to compromise a rock-turning, log-ripping, tree-climbing bear, and yet she apparently compensated for the injuries. This entire portrait of injuries suggests that wildlife managers and especially those dealing with threatened or endangered species should perhaps act with optimism when attempting to rehabilitate severely injured animals.

Chelsea is the only bear, in more than seven years of study, that we strongly suspect was cannibalized. This occurred in the fall in a patch of acorns where she may have been feeding or where she may have disturbed a feeding bear(s). We suspect that a large male killed her, but this is based on literature which suggests that when cannibalism occurs, it is usually done by males (Tietje et. al., 1986). Our experience and that of others (Beck, 1991) suggests that male bears make extensive movements in the fall to feed on patches of food prior to hibernation. This young female, then, may have been unrelated and thus decreases any apparent maladaptiveness associated with intra-specific predation. If the acorn crop and, for that matter, all soft and hard fall mast crops were poor, then the risk to this small bear would have been increased. Theory would suggest that bear-food-rich habitats should yield fewer cases of cannibalism than areas where one or two mast crops can dictate reproductive success (the Great Smoky Mountains, for example, where failure of hard mast crops have led to a nearly complete absence of cub production in a given year).

CONCLUSIONS

Black bears seem to be strong, durable animals with both males and females growing to old age and tolerating rather severe injuries. The only clear-cut case of multiple den use in more than 80 den visitations in our two study areas is the one reported here. Our observations differ markedly from those of Beck (1991), who found frequent reuse of dens. Perhaps some dens are of sufficiently high quality that bears seek them out. To my knowledge, however, this is yet to be a proven motivation for den reuse. While crowding may lead to extensive cannibalism in non-hunted populations, the hunted populations we are studying appear to experience cannibalism as a rare event. Again crowding or high population densities of many black bear populations are probably a result of human intervention, and densities may have been lower when black bears shared their range with a diverse group of large carnivores that may have regularly preyed on them and/or compromised recruitment by direct or indirect competition for food resources.

ACKNOWLEDGMENTS

I thank the following students and employees for their attention to detail, their hours, and even days of work without receiving adequate compensation, and for their listening politely to my stories, even though they had heard many of them on previous occasions: Janene Auger, Max "Grizz" Baker, Bert Frost, Annie Henry, Mark Hurley, Heidi Johnson, Richard Lance, Margaret Martin, Scott Mortensen, Gary Ogborn, Wade Paskett, Scott Richardson, Marc Seid, Dana Shurtleff, April "Kawasaki" Young, and Kevin Young. I thank private individuals, the Utah Division of Wildlife Resources, the Bureau of Land Management, the Forest Service, and Brigham Young University for financial support.

LITERATURE CITED

Beck, T.D.I. 1991. Black bears of west-central Colorado. Tech. Publ. No. 39, Colorado Division of Wildlife, 86 pp.

Frazzetta, T.H. 1975. Complex adaptations in evolving populations. Sinauer Assoc., Inc., Sunderland, Massachusetts, 267 pp.

Garshelis, D.L. 1993. Monitoring black bear populations: Pitfalls and recommendations. Pp. 123-144, *in* Proceedings of the Fourth western black bear workshop (J.A. Keay, ed.). National Park Service Report 93/12, 144 pp.

Geist, V. 1993. Bear and early humans. Pp. 24-25, *in* Bears: Majestic creatures of the wild (I. Stiring, ed.). Rodale Press, Emmaus, Pennsylvania, 240 pp.

Noyce, K. 1995. Island black bear cannibalism. International Bear News, 4(2):14.

LeCount, A.L. 1986. Black bear field guide: A manager's manual. Arizona Game and Fish Dept. Special Report 16, 127 pp.

Rogers, L.L. 1987. Effects of food supply and kinship on social behavior, movements, and population growth of black bears in northeastern Minnesota. Wildlife Monograph, 97:1-72.

Schwartz, C.C., and A.W. Franzmann. 1991. Interrelationship of black bears to moose and forest succession in the northern coniferous forest. Wildlife Monograph, 113:1-58.

Shaw, C.A., and S.M. Cox. 1993. The giant short-faced bear. Pp. 22-23, *in* Bears: Majestic creatures of the wild (I. Stirling, ed.) . Rodale Press, Emmaus, Pennsylvania, 240 pp.

Thomas, L. 1973. Death in the open. New England Journal of Medicine, 288:92-93.

Tietje, W.D., B.O. Pelchat, and R.L. Ruff. 1986. Cannibalism of denned black bears. Journal of Mammalogy, 67:762-765.

Hal L. Black, Department of Zoology, Brigham Young University, Provo, Utah 84602

Skeletal Architecture of the Forelimbs in Kangaroo Rats (Heteromyidae: Dipodomys): Adaptations for Digging and Food Handling

Kerry S. Kilburn

Abstract

The skeletal architecture of the forelimbs of kangaroo rats was investigated to examine potential trade-offs between efficiency of food handling and digging ability. Analyses of skeletal components were performed on nine specimens of *D. ordii* and 22 *D. merriami*, and compared with existing data on fossorial and non-fossorial mammals. High-speed cinematography was used to examine details of digging behavior. Together, the analyses show a division of labor among forelimb segments, with the antebrachium specialized for high-speed food handling, and the scapula and humerus adapted for digging. No trade-off in function appears to exist between these behaviors.

INTRODUCTION

The heteromyid rodents of North America have been the focus of a wide range of biological investigations throughout this century (Genoways and Brown, 1993). One area of the biology of heteromyids that largely has been neglected is functional morphology. Following the descriptive monographs of Howell (1932), Hatt (1932), and Wood (1935), this field lay virtually dormant until research on the auditory apparatus of heteromyids was published beginning in the early 1960s (e.g., Webster, 1962; Webster and Webster, 1971, 1975). More recently, investigators have focused on the functional morphology of hind limbs of heteromyids and their function in quadrupedal and bipedal locomotion (Nikolai and Bramble, 1983 and references therein; Thompson, 1985).

Little attention has been paid to the functional morphology of the forelimbs of heteromyids. Of special interest is the relationship between musculoskeletal architecture and function in the forelimbs of the bipedal kangaroo rats (*Dipodomys*). In this genus, as in other bipedal rodents, the forelimbs are reduced in size relative to the hind limbs (Howell, 1932; Wood, 1935; Eisenberg, 1963; Nikolai and Bramble, 1983). In kangaroo rats, the forelimbs play only a minor role in supporting the body during rapid bipedal locomotion (e.g., Howell, 1932; Eisenberg, 1963; Nikolai and Bramble, 1983). They are used, however, during slow quadrupedal locomotion (Nikolai and Bramble, 1983). In addition, members of this genus rely on the forelimbs for food handling and digging. Nikolai and Bramble (1983) and Thompson (1985) have suggested that bipedality in kangaroo rats has evolved in response to selection for both cursorial locomotion and the retention of a forelimb specialized for food handling and digging.

The ability to dig burrows and to handle food rapidly and efficiently are both critical to the ability of kangaroo rats to persist in open, hot, and arid environments. Moreover, interspecific differences in harvesting efficiency and digging ability have been proposed as possible mechanisms facilitating the coexistence of species of kangaroo rats. Finally, these behaviors potentially place different functional demands on the forelimb; the way in which those differences have been resolved over evolutionary time has never been addressed explicitly.

In this study, I consider each behavior in terms of its ecological significance, the poten-

Life Among the Muses: Papers in Honor of James S. Findley. Edited by T.L. Yates, W.L. Gannon & D.E. Wilson.

tial functional demands it places on the forelimb, and the skeletal adaptations that should result from selection for specialization for each. I then use data from skeletal analyses and high-speed cinematography of digging to address the question of whether specialization for either function has been achieved at the cost of reduction of the other.

Each limb segment (bones or combinations of bones) may be thought of as a lever moving about a fulcrum (joint) (Maynard Smith and Savage, 1956; Hildebrand, 1988). The force and velocity with which distal limb segments move (out-force, Fo, and out-velocity, Vo) is a function of several characteristics (Fig. 1): force and velocity applied to the limb segment by muscles moving that segment (in-force, Fi, or in-velocity, Vi); the perpendicular distance from the center of rotation of the joint to the line of action of the muscle (the effective in-lever arm, Li); and the linear distance from the center of rotation of the joint to the point on the limb at which the

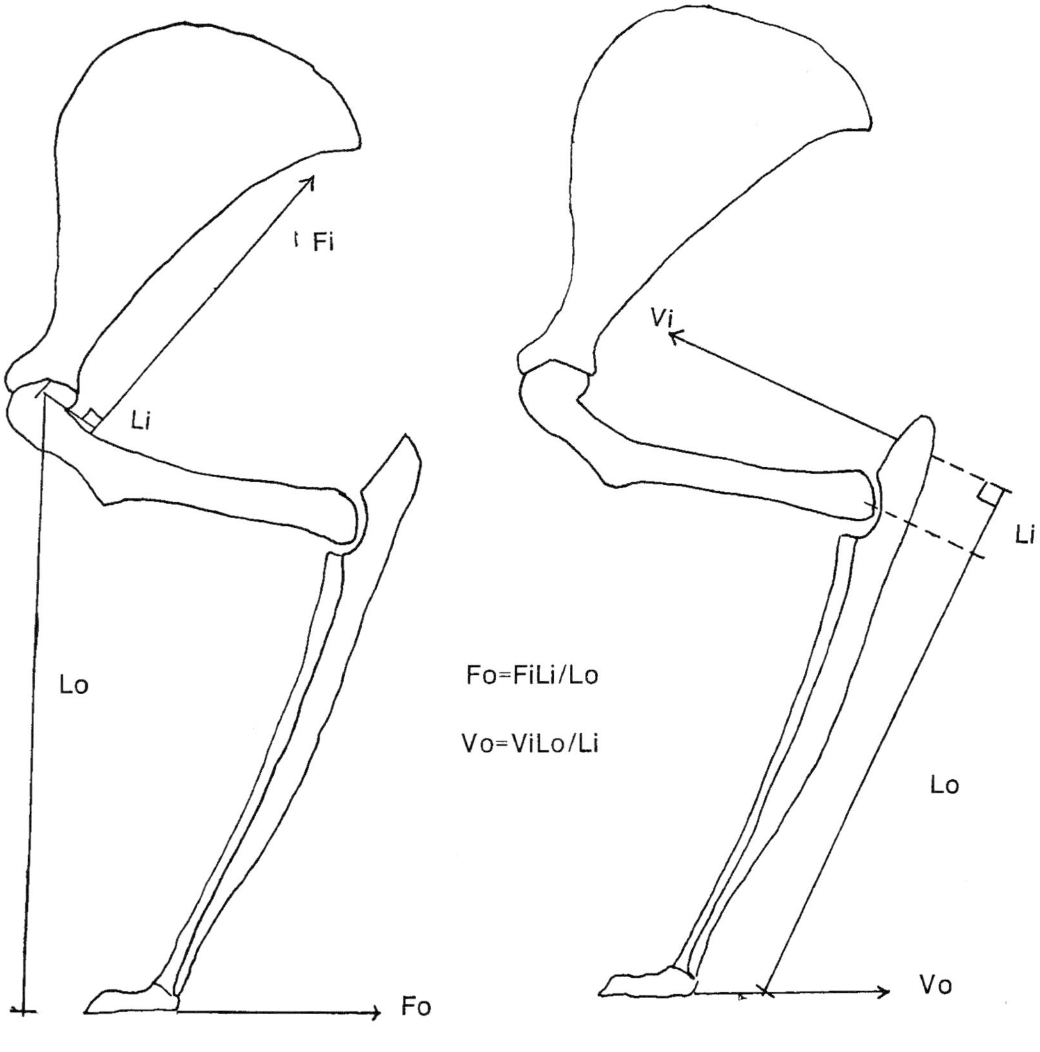

Fig. 1. Diagram of selected mechanical components of the forelimb and their relationships in the production of force and velocity. Left: lever arms and force vectors for the teres major. Fi = in-force, Fo = out-force, Li = effective in-lever arm, Lo = out-lever arm. Right: lever arms and velocity vectors for the triceps complex. Vi = in-velocity, Vo = out-velocity.

out-force or out-velocity is applied. These characteristics are related to each other according to the equations:

$$Fo = FiLi/Lo$$
$$Vo = ViLo/Li$$

Selection to increase out-force should result in an increase in the in-lever arm and/or a decrease in the out-lever arm. Conversely, selection to increase out-velocity should result in an increase in the out-lever arm and/or a decrease in the in-lever arm. Consequently, if selection acted only on lever arms, adaptation to increase out-force of a limb segment would result in a decrease in out-velocity of the same segment.

Digging and foraging behaviors appear to be highly stereotyped among species of kangaroo rats (*D. spectabilis*, Howell, 1932; *D. deserti, D. nitratoides, D. microps, D. ordii, D. ingens, D. panamintinus, D. heermanni, D. agilis, D. merriami*, Eisenberg, 1963; *D. deserti, D. merriami*, Nikolai and Bramble, 1983). Food handling consists of two activities: gleaning seeds from the soil and placing the seeds in external cheek-pouches. The behavior involved in gleaning seeds from the soil is not well-known. Nikolai and Bramble (1983) described high-speed pouching behavior in *D. deserti* collecting millet seeds from glass plates. A typical pouching cycle consists of: 1) extension of forelimb to substrate with simultaneous pronation and opening of hands; 2) hands closing on seeds and retracting toward the mouth (flexion of the forearm) with simultaneous supination of hand; 3) elevation of forearms such that fingers penetrate the openings of the cheek pouches, release of food, and extension of the forearm away from the mouth.

Selection to increase the speed and efficiency of this process should result in increasing the out-velocity at the hand during extension and flexion of the forearm. Predicted skeletal modifications should therefore include: 1) small hands and slender bones in the antebrachium to reduce weight at the distal limb segments (Nikolai and Bramble, 1983); and 2) a relatively long ulna and short olecranon to increase the ratio of the out-lever arm to the in-lever arm at the forearm.

Kangaroo rats are scratch-diggers, using a rapid to-and-fro movement of the hand against the substrate (Hildebrand, 1985). This behavior potentially involves three actions, each of which serves to move the claws and hand posteriad along or through the substrate (Maynard Smith and Savage, 1956; Lehmen, 1963; Hildebrand, 1985). These are: 1) flexion of the humerus on the scapula (humeral retraction) via contraction of the teres major and latissimus dorsi muscles; 2) extension of the forearm via contraction of the triceps complex; 3) flexion of the hands and digits via the action of the carpal and digit flexors.

Because digging requires that the limb be flexed against the soil with sufficient force to overcome the resistance of the substrate, the limbs of fossorial animals must be adapted to provide a powerful out-force (Maynard Smith and Savage, 1956; Hildebrand, 1985). The skeletal adaptations associated with high out-force production include the following: 1) a well-developed post-scapular process for the origin of a robust teres major; 2) a short, stout humerus with a prominent deltoid tubercle and a large medial epicondyle for the origin of well-developed pronators, carpal flexors, and digit flexors; and 3) a long olecranon and short ulna, which results in a large ratio of the in-lever arm to the out-lever arm and increases the mechanical advantage of the triceps complex (Howell, 1932; Wood, 1935; Maynard Smith and Savage, 1956; Lehmen, 1963; Hildebrand, 1985).

Comparing skeletal adaptations expected from specialization for these two behaviors shows at least one important conflict. For efficient high-speed seed pouching, the forearm (radius and ulna) should be long and slender with a short olecranon process. For digging, the forearm should be relatively short with a well-developed olecranon. Moreover, to resist bending while the forearm is being extended against the ground, radius and ulna should be robust rather than slender (Lanyon and Rubin, 1985).

Such conflicting selection pressures may result in a continuum of outcomes. At one extreme, we might see specialization for food handling at the expense of digging ability. This conclusion is explicitly stated for *D. merriami* by Hardy (1945) and Reynolds (1958) and is implicit in the work of other investigators

(Howell, 1932; Nikolai and Bramble, 1983). At least some species of kangaroo rats, however, are capable of constructing burrows in obdurate soils (*D. spectabilis*, Vorhies and Taylor, 1922, Bailey, 1931, Howell, 1932; *D. ordii*, Bailey, 1931, Eisenberg, 1963; *D. merriami*, Huey, 1942, Setzer, 1949). *D. ingens*, *D. heermanni* and *D. nitratoides* in the Carrizo Plains of California construct burrows in dry, packed soil that resists digging even with pick-axes; in the foothills of the San Mateo mountains of New Mexico, *D. spectabilis* constructs burrows in gravel (particle size >1 cm) (pers. obs.).

These observations support the hypothesis that selection for efficient food handling has led to no loss in digging ability in this genus. The question, then, is how such a loss is avoided. To answer this question, I have analyzed aspects of the forelimb skeleton in *D. ordii* and *D. merriami*, and compared my results to existing data in the literature from other species of *Dipodomys*, other scratch-diggers, and non-fossorial mammals. The context for the interpretation of these results is provided by high-speed films of stereotyped digging behavior in *D. ordii* and *D. merriami*.

Although the restriction of the analyses to two species is potentially problematic, I believe it is justified in this study. First, as described earlier, food handling and digging behaviors are highly stereotyped throughout the genus. Similarly, although skeletal characters may vary in detail among species, the overall architecture of the forelimb is conserved across the genus (Howell, 1932; Wood, 1935; Nikolai and Bramble, 1983). Finally, *D. ordii* and *D. merriami* are representatives of divergent lineages within the genus *Dipodomys* (Schnell et al., 1978). Therefore, I am confident that the behaviors and characters common to these two species are typical of the genus as a whole.

METHODS

Specimens Examined

I obtained nine skeletal specimens of *D. ordii* and 22 of *D. merriami* from the Museum of Southwestern Biology, University of New Mexico. Individuals used in the high-speed cinematography (one adult male of each species) were live trapped at a study site located 18.8 km (11.3 mi) E San Antonio, Socorro Co., New Mexico. Throughout the six-week training period, both animals were maintained in individual cages under a 12:12 light–dark cycle.

Skeletal Analyses

The ratio of width across humeral epicondyles to humeral length as an index of the degree of development of the pronators, carpal flexors, and digit flexors, and ratio of length of olecranon process to length of the ulna as a measure of the mechanical advantage (out-lever arm/in-lever arm) for the forearm extensors are commonly reported in comparisons of fossorial and non-fossorial mammals. Therefore, I measured: 1) greatest length of the humerus, from the most proximal point of head to distal-most extension of epicondyles; 2) width across the humeral epicondyles; 3) length of olecranon process from tip to center of the trochlear fossa (determined visually); and 4) total ulnar length, from tip of the olecranon to distal end of the bone. All measurements were made using dial calipers mounted on the stage of a dissecting microscope and were taken to the nearest 0.02 mm.

High-Speed Cinematography

One individual of each species was trained to dig in a glass aquarium containing ca. 4 cm of fine sand (particle size < 1 mm). Because the forelimbs are obscured when the animals dig in loose sand, I prepared a substrate composed of 50% sand and 50% finely-ground clay. This was wetted, placed in the digging chamber, and allowed to dry in a slope of approximately 45°. While training the animals to dig on this substrate, I observed no detectable qualitative differences between their digging behavior on the sand and on the sand/clay compound. I took 200 ft of film at 150 frames/s to get clear footage of three digging bouts for each animal.

Analyzing the footage frame-by-frame, I was able to record the position of the hand and elbow during the digging stroke. The humerus, however, was obscured by the animals' loose skin and long pelage. To determine the position of the humerus during the digging cycle, I anesthetized the two individuals used in the films, placed them on their sides, and recreated the positions of the hands and elbows at the points of maximum extension and flexion of the forelimb. I did this by tracing the surface of the substrate and the anterior-most and posterior-most positions of the hands from the high-speed film. I also traced body posture and the position of the elbow at each of these points. I then placed the tracing of substrate and hand positions on a piece of dissecting wax, laid the anaesthetized animal on its side, and anchored the hand in the position indicated from the tracing. Next, I adjusted the animal's posture and anchored the elbow into position according to the tracings. With the digging posture and limb position recreated, I palpated the forearm to determine the position of the humerus and of the scapulo-humeral joint. I repeated this procedure for both anterior and posterior positions, for a total of eight digging strokes for each individual.

RESULTS

Skeletal Analyses

The forelimb in both species is distinctly dichotomous with respect to adaptations for digging and food handling (Fig. 2). Both species possess a well-developed post-scapular process. The humerus is short and robust with a prominent deltoid crest and broad medial epicondyles, as would be expected of animals highly adapted for digging. The ulna, however, is long and slender with a relatively short olecranon. Results of the skeletal analyses and comparable information from other authors are shown in Table 1 (Wood, 1935; Lehmen, 1963; Hildebrand, 1985). The ratio of epicondlylar width to humeral length for both *D. merriami* and *D. ordii* falls within the range reported for other species of kangaroo rat (Wood, 1935; Nikolai and Bramble, 1983). Moreover, the ratio for *Dipodomys* is similar to that reported by Hildebrand (1985) for a variety of fossorial mammals (Table 1).

In contrast, the ulna possesses characteristics that would be expected to result from selection for high-speed handling of food, rather than for digging. The ratio of length of olecranon to length of the rest of the ulna is well below that reported for fossorial mammals by Hildebrand (1985). Ratios I obtained are more similar to those reported for the raccoon, which is not fossorial (Hildebrand, 1985; Table 1). To compare my results with those of Lehmen (1963), I computed the ratio of length of olecranon to total length of ulna; again, ratios for *D. merriami* and *D. ordii* are more similar to those reported for non-fossorial species than for fossorial species (Table 1).

The ratio of olecranon length to ulnar length provides an index of the ratio of the in-lever arm to the out-lever arm (Fig. 1); a low value, therefore, corresponds to a high value for the inverse ratio (out-lever arm to in-lever arm). This, in turn, should lead to high out-velocity at the distal limb segments, as would be expected if selection acted to increase the efficiency of food handling. The slender bones and small hands characteristic of kangaroo rats, and subsequent reduction of weight at the distal end of the forelimb, also increases out-velocity (Nikolai and Bramble, 1983).

Cinematography

Direct observation of six captive individuals of each species confirmed that digging involves two distinct phases. First, animals rapidly scratch the surface of the soil until a pile of loose sand has been accumulated. Second, they pull the limbs beneath the body and extend the forearms, pushing the sand to the rear beneath the body. This action is often completed by kicking the soil further to the rear using the hindlimbs (Eisenberg, 1963; Nikolai and Bramble, 1983). Because breaking through and loosening the surface of the soil places greater functional demands on the digging apparatus than does the removal of already-loosened soil, I have concentrated on this phase only.

Frame-by-frame analysis of the high-speed film clearly shows that the major action during the digging stroke is humeral retraction (Fig. 3). In those specimens, movement of the hand posteriad always was accompanied by movement of the elbow posteriad and dorsad, which means that the humerus was being flexed on the scapula. Moreover, in all positions analyzed for both species, the antebrachium was held at an almost constant angle of ca 90° to the humerus. Thus, extension of the antebrachium is not involved in powering the digging stroke;

Table 1. **Results of skeletal analyses and comparative data. EW = epicondylar width as a percentage of length of humerus; OLE(1) = length of olecranon to center of semi-lunar notch as a percentage of length of ulna from semi-lunar notch to distal ulna; OLE(2) = length of olecranon to center of semi-lunar notch as a percentage of total length of ulna. Data from present study are means +/- SD (n); superscripts indicate data from the literature.**

Taxon	Character		
	EW	OLE (1)	OLE(2)
D. merriami	30.99 ± 3.10 (22) 37[1]	15.79 ± 1.46 (14)	13.02 ±1.1 (14)
D. ordii	31.66 ± 1.37 (9) 26[1]	18.74 ± 2.94 (6)	15.70 ± 1.46 (6)
Dipodomys	30–33[2]		
Tachyorychtes	30[3]	42[3]	22[4]
Ctenomys	33[3]	31[3]	22[4]
Geomys			22[4]
Rattus, Liomys *Octodon,* *Brachyuromys*			15[4]
Cannomys	30[3]	34[3]	
Procyon		12[3]	
Thomomys	35[3]	30[3]	
Taxidea	34[4]	35[3]	

[1] Wood (1935)
[2] Nikolai and Bramble (1983)
[3] Hildebrand (1985)
[4] Lehmann (1963)

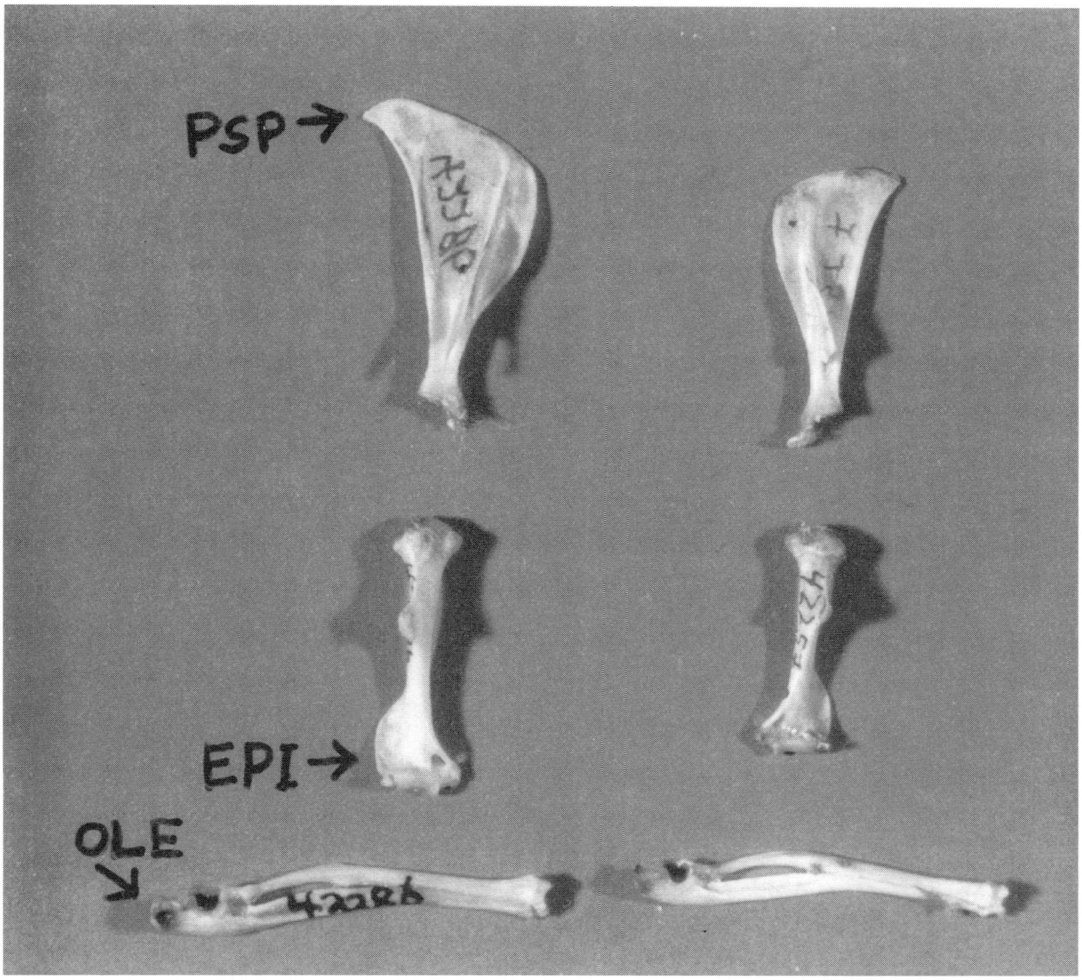

Fig. 2. Scapula, humerus, and ulna from *D. merriami* (right) and *D. ordii* (left). PSP = post-scapular process, EPI = humeral epicondyles, OLE = olecranon. The numbers on some of the bones are MSB catalog numbers.

rather, it appears to serve solely to stabilize the hand against the substrate.

DISCUSSION

With these results, a resolution of the question of the outcome of selection for conflicting functions in the forelimb of kangaroo rats is possible. Because humeral retraction is the major action involved in digging, the burden of force production is placed on the teres major and latissimus dorsi muscles. Thus, the scapula and humerus have been modified to enhance the force produced by these muscles. The forearm acts to transmit the tensional load (load along the long axis of the bone; Lanyon and Rubin, 1985) the muscles produce to the hand. As this limb segment does not add a large force component to the digging stroke, it is relatively free from selection for structural characteristics required to resist bending loads. The long, slender ulna and relatively short olecranon associated with food handling at high speed, therefore, place few mechanical constraints on digging ability.

Comparing anatomical characteristics of fossil heteromyids to those of living forms, Wood (1935) noted that *Dipodomys* possesses a short humerus and the shortest olecranon of five genera in the family. Because *Dipodomys* is considered to be the most highly-derived genus in its family (Wood, 1935; Hafner and Hafner,

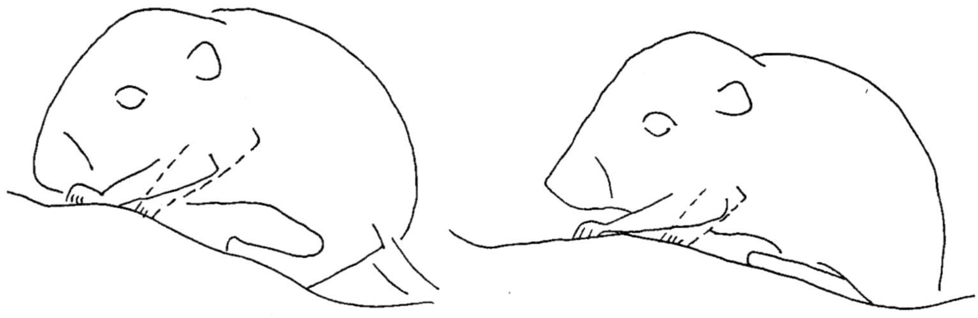

Fig. 3. High-speed film tracings of digging behavior in *D. ordii* (right) and *D. merriami* (left). Solid lines represent the position of the hand and forearm at the anteriormost position of the digging stroke; dashed lines, at the posteriormost position.

1983), a reasonable hypothesis is that the architecture of the forelimb comprises a combination of derived characters that have evolved in response to selection pressure for both digging ability and efficiency of food-handling, rather than a compromise or trade-off between two competing forces.

This hypothesis is consistent with the known importance of both food handling and digging to the ability of kangaroo rats to survive in their desert environments. Subterranean burrows, for example, provide critical refugia from daytime conditions of extreme heat and aridity (Schmidt–Nielsen and Schmidt–Nielsen, 1950a, 1950b; Kay, 1977; Kay and Whitford, 1978). Interspecific differences in digging ability have been proposed as a proximate mechanism mediating habitat segregation on the basis of the size of soil particles and facilitating coexistence among species of *Dipodomys* (Hardy, 1945; Findley et al., 1975).

The ability to collect and pouch seeds rapidly is presumed to reduce the time the animals must spend foraging outside the burrow, hence reducing exposure to predators, and to reduce locomotor expenditure during foraging by allowing an animal to make fewer trips to collect and store a given amount of food (Nikolai and Bramble, 1983). Interspecific differences in the efficiency with which different species of heteromyids harvest seeds from different types of seed patches has been proposed as a possible mechanism for microhabitat partitioning in communities of heteromyids (e.g., Price and Heinz, 1984; Price and Waser, 1985).

Although this study supports an adaptive hypothesis for forelimb morphology, architecture of the forelimb in this genus does not represent "perfect" adaptations to both digging and foraging behaviors. For example, the mechanical advantage of the teres major decreases as the distance from the scapulo-humeral joint to the ground increases (Fig. 1). This distance is, in part, a function of the length of the antebrachium and could be decreased with a decrease in length of forearm. However, length of forearm is likely constrained, not only by the functional demands of food handling, but also by its other roles (e.g., locomotion, environmental exploration, grooming; Howell, 1932; Eisenberg, 1963). This system is remarkable in that, although length of forearm may be constrained by the functional demands placed upon it, the resultant structure of the forearm places a minimum of constraints on the rest of the limb. In this genus, at least, strength and agility may not be mutually exclusive; rather, an elegant suite of anatomical and behavioral adaptations has permitted a division of labor among limb segments such that neither digging ability nor foraging efficiency is compromised.

ACKNOWLEDGEMENTS

J.S. Altenbach performed the high-speed cinematography throughout the course of the study. P. Glass provided useful suggestions in the construction of the filming chamber and in the training of the animals for filming. W.L. Gannon, Museum of Southwestern Biology, provided excellent access to specimens and other museum resources. T. Griffin and P. Nuchims, West Virginia State College, prepared photographs and drawings. This manuscript has been greatly improved by the editorial comments of S.D. Thompson, D.J. Hafner, T.L. Best, J.S. Altenbach, and two anonymous reviewers. The study was performed in partial fulfillment of the requirements for the Ph.D. degree at the University of New Mexico.

LITERATURE CITED

Bailey, V. 1931. Mammals of New Mexico. North American Fauna, 53:1-412.

Eisenberg, J.F. 1963. The behavior of heteromyid rodents. University of California Publications in Zoology, 69:1-100.

Findley, J.S., A.H. Harris, D.E. Wilson, and C. Jones. 1975. Mammals of New Mexico. University of New Mexico Press, Albuquerque, 359 pp.

Genoways, H.H., and J.H. Brown, eds. 1993. The biology of the Heteromyidae. American Society of Mammalogists, Special Publication #10, 719 pp.

Hafner, J.C., and M.S. Hafner. 1983. Evolutionary relationships of heteromyid rodents. Great Basin Naturalist Memoirs, 7:3-29.

Hardy, R. 1945. The influence of types of soil upon the local distribution of some mammals in southwestern Utah. Ecological Monographs, 15:71-108.

Hatt, R.T. 1932. The vertebral column of ricochetal rodents. Bulletin of the American Museum of Natural History, 63:599-738.

Hildebrand, M.H. 1985. Digging of quadrupeds. Pp. 89-109, *in* Functional vertebrate morphology (M.H. Hildebrand, D.M. Bramble, K.F. Liem, and D.B. Wake, eds.). Harvard University Press, Cambridge, Massachusetts, 430 pp.

Hildebrand, M.H. 1988. The analysis of vertebrate structure. Third ed. Wiley, New York, 701 pp.

Howell, A.B. 1932. The saltatorial rodent *Dipodomys*: The functional and comparative anatomy of its muscular and osseous systems. Proceedings of the American Academy of Arts and Sciences, 67:377-536.

Huey, L.M. 1942. A vertebrate faunal survey of the Organ Pipe National Monument, Arizona. Transactions of the San Diego Society of Natural History, 9:353-376.

Kay, F.R. 1977. Environmental physiology of the banner-tailed kangaroo rat. II. Influences of the burrow environment on metabolism and water loss. Comparative Biochemistry and Physiology, A. Comparative Physiology, 57:471-477.

Kay, F.R. and W.G. Whitford. 1978. The burrow environment of the banner-tailed kangaroo rat, *Dipodomys spectabilis*, in southcentral New Mexico. The American Midland Naturalist, 99:270-279.

Lanyon, L.E. and C.T. Rubin. 1985. Functional adaptations in skeletal structures. Pp. 1-25, *in* Functional vertebrate morphology (M.H. Hildebrand, D.M. Bramble, K.F. Liem, and D.B. Wake, eds.). Harvard University Press, Cambridge, Massachusetts, 430 pp.

Lehmen, W.H. 1963. The forelimb architecture of some fossorial rodents. Journal of Morphology, 113:59-76.

Maynard Smith, J. and R.J.G. Savage. 1956. Some locomotory adaptations in mammals. Zoological Journal of the Linnean Society (London), 42:603-622.

Nikolai, J.C. and D.M. Bramble. 1983. Morphological structure and function in desert heteromyid rodents. Great Basin Naturalist Memoirs, 7:44-64.

Price, M.V., and K. Heinz. 1984. Effects of body size, seed density and soil characteristics on rates of seed harvest by heteromyid rodents. Oecologia (Berlin), 61:420-425.

Price, M.V., and N.M. Wasser. 1985. Microhabitat use by heteromyid rodents: Effects of artificial seed patches. Ecology 66:211-219.

Reynolds, H.G. 1958. The ecology of the Merriam's kangaroo rat (*Dipodomys merriami* Mearns) on the grazing lands of Arizona. Ecological Monographs, 28:111-27.

Schmidt–Nielsen, B., and K. Schmidt–Nielsen. 1950a. Evaporative water loss in desert rodents in their natural habitat. Ecology, 31:75-85.

Schmidt–Nielsen, B., and K. Schmidt–Nielsen. 1950b. Pulmonary loss in desert rodents. American Journal of Physiology, 162:31-36.

Schnell, G.D., T.L. Best, and M.L. Kennedy. 1978. Interspecific morphologic variation in kangaroo rats (Dipodomys): Degree of concordance with genic variation. Systematic Zoology, 27:34-45.

Setzer, H.W. 1949. Subspeciation in the kangaroo rat, *Dipodomys ordii*. University of Kansas Publications, Museum of Natural History, 1:473-573.

Thompson, S.D. 1985. Bipedal hopping and seed-dispersion selection by heteromyid rodents: The role of locomotion energetics. Ecology, 66:220-229.

Vorhies, C.T., and W.P. Taylor. 1922. Life history of the kangaroo rat, *Dipodomys spectabilis spectabilis* Merriam. U.S. Department of Agriculture Bulletin, 1091:1-40.

Webster, D.B. 1962. A function of the enlarged middle-ear cavities of the kangaroo rat, *Dipodomys*. Physiological Zoology, 35:248-255.

Webster, D.B., and M. Webster. 1971. Adaptive value of hearing and vision in kangaroo rat predator avoidance. Brain, Behavior, and Evolution, 4:310-322.

Webster, D.B. and M. Webster. 1975. Auditory systems of Heteromyidae: Functional morphology and evolution of the middle ear. Journal of Morphology 146:343-376.

Wood, A.E. 1935. Evolution and relationships of the heteromyid rodents with new forms from the Tertiary of Western North America. Annals of the Carnegie Museum, 24:73-262.

Kerry S. Kilburn, Department of Biological Sciences, Old Dominion University, Norfolk, VA 23529

Puncturing Ability of Bat Canine Teeth: The Tip

PATRICIA W. FREEMAN AND WILLIAM N. WEINS

Abstract

Casts of upper canine teeth of 15 species of microchiropteran bats were fixed securely into a testing machine and made to puncture the bloom side of an apple. The force necessary to break through the surface of the apple was regressed against both shape of the tip of the canine and the size of the animal. Sharper tips require less force to puncture than blunt ones. Results were also verified using giant two-dimensional models of Plexiglas™ with sharp and blunt tips that were loaded onto a Plexiglas™ beam. Fringes, or stress lines, are more highly concentrated at the point of contact for a smaller radius than for a larger radius under the same loading conditions, indicating that stress in the beam is higher and more concentrated at the sharper tip. Ours is some of the first experimental work to quantify apical sharpness of a tooth and performance as it punctures a substance.

INTRODUCTION

That small heels of high-heeled shoes exert more force on the floor than do the broader heels of an Oxford is a well-known phenomenon called the principle of stress concentration. For all the description and emphasis placed on the form of mammalian teeth, little experimentation has been done to correlate dental form with dental function. Our objective in this study was simply to demonstrate that the principle of stress concentration could be applied to sharp and blunt canine teeth of some of the smallest mammals—bats. Seeing the actual patterns of stress at the tooth–food interface, however, was impossible given the small size of the teeth and the simple nature of our equipment. We verified that patterns of stress at the tooth-food interface are different by using oversized plastic models with sharp and blunt tips and photoelasticity, a technique borrowed from engineering.

Canine teeth are among the simplest teeth in the mammalian toothrow and are ideal for detailed study precisely because they are relatively simple. Methods devised to study function in canines could be used ultimately to study function in more complex teeth. Cross-sectional shapes of microchiropteran upper canine teeth are diverse (Freeman, 1992). The shanks of the canines have a variety of acute and obtuse edges that run the length of the tooth that are different from the round or oval shanks found in carnivorans. How swiftly and effectively a tooth penetrates a substance and how cracks radiate or propagate from that penetration through the substance could be important to animals that fly and capture prey. Forelimbs of bats are occupied with flying, and although the wings and tail membrane may be used as temporary scoops to bring food to the mouth, canines and other teeth must do the initial immobilizing and processing of food. Here we present data on how easily the tip of the canine of a bat punctures a surface, which is the first step in examining what is happening at the interface between tooth and food.

Quantification of canines as tools involves at least two aspects: first, the morphology of the tip and the force that the tip exerts to penetrate a substance; and second, the morphology of the shank, the force of continued penetration, and the potential pattern of food breakup. It is the tip that is the subject of this study. The force necessary for puncturing a food item with

Life Among the Muses: Papers in Honor of James S. Findley. Edited by T.L. Yates, W.L. Gannon & D.E. Wilson.

the tip can be precisely described and measured with a uniaxial compression tester.

Much has been written about the mechanical properties of teeth including hardness, biomechanics, and structure (Halstead, 1974; Waters, 1980). Still more has been written about the functioning of the jaws, mastication, occlusion, and wear facets (Maynard Smith and Savage, 1959; Mills, 1967; Hiiemae, 1967; Crompton and Hiiemae, 1969, 1970; Turnbull, 1970; Kay, 1975; Hiiemae, 1978). Little has been written about the shapes of teeth and how that shape may affect chewing efficiency or food breakup. Studies by Lucas (1979, 1980, 1982) are notable exceptions where he first describes the theoretical framework for how shapes of molars may be breaking up different foods, then examines the cusps on premolars and molars of humans experimentally for bluntness and ability to fracture food items. Another exception is a study by Osborn et al. (1988), who examined the spatulate incisors of higher primates and concluded the incisors are tilted forward for maximum efficiency for cutting food rather than for any other purpose. This paper is one of the few actually to test teeth experimentally. Cinefilm was used to analyze upper teeth as subjects bit into apples. Van Valkenburgh and Ruff (1987) describe the strength of canines of large carnivorous mammals and how changes in shape among different kinds of carnivorans are probably correlated with strength and the animal's biting behavior. No experimental work was done. Most studies concerning mechanical properties of teeth or patterns of movements of teeth are studies of premolars and molars. Canines are only cursorily considered, if at all, in the above studies, and variation in shape of canines has not been systematically examined.

Microchiropteran bats are ideal for study because they are a monophyletic group with remarkable dietary diversity. In addition, 1) there is high diversity in canine cross-section among species; 2) their teeth are the primary tools for gathering and consuming prey; and 3) the jaws of bats exhibit a tightly constrained system of centric occlusion and isognathy or near isognathy so that differences in canine shape may have substantial functional significance (Freeman, 1992).

The New World family Phyllostomidae exhibits the greatest diversity in diet and contains species that are insectivorous, carnivorous, frugivorous, nectarivorous, and omnivorous. The families Molossidae, Mystacinidae, Vespertilionidae (the most speciose family), and Natalidae are primarily insectivorous as are Rhinopomatidae, Emballonuridae, Mormoopidae, Rhinolophidae, and Nycteridae. Noctilionids exhibit both insectivory and piscivory, desmodontids are sanguivorous, and megadermatids are insectivorous and carnivorous. However, regardless of food habit, bat teeth are small—teeth in this study range in size from 2.0 to 8.5 mm above the gum line—and prohibit anything but simple analysis without more sophisticated techniques or equipment.

To mimic activity at the actual tooth–food interface of bats, we conducted similar compression tests with oversized two-dimensional Plexigas™ models that were wedge-shaped but had tips of different radii. Not only did large models help us see interactions better, we could also test the efficacy of using plastic to quantify differences made by different tips when loaded into a plastic material. The birefringent qualities of plastic when stressed produce patterns that can be viewed and quantified under a polariscope using an engineering technique called photoelasticity.

MATERIALS AND METHODS

We cast replicas of upper canine teeth of single male specimens of each of 15 species from six microchiropteran families (*Megaderma lyra, Hipposideros commersoni, Hipposideros pratti, Noctilio leporinus, Vampyrum spectrum, Chrotopterus auritus, Phyllostomus hastatus, Trachops cirrhosus, Artibeus lituratus, Centurio senex, Cheiromeles torquatus, Eumops perotis, Scotophilus heathi, Lasiurus cinereus*, and *Eptesicus fuscus*). Plastic-resin casts of the teeth were produced following procedures from vertebrate paleontology, but modified for our purposes (Goodwin and Chaney, 1994; G.W. Brown, pers. comm.). Casts were made with the following materials: 1) molds: General Electric RTV 700 Silastic with Beta 1 catalyst, a silicon rubber; 2) casts: standard casting epoxy-resin of medium hardness, TAPOX 4 : 1 formula.

We glued the replica of the tooth to a 6.5 mm diameter wooden pedestal about 50 mm long and inserted it into a chuck on an Instron uniaxial compression tester (Model TTCL, Instron, USA). Right and left replicate canine of each cast were separated and tested separately to get a cleaner puncture without obstruction from the other canine. Three punctures were made with each replica.

After experimentation with artificial materials, we found that apples were an excellent test-surface for several reasons. There is a large literature of tests on apples (Hamann, 1969; Hanna and Mohsenin, 1972; Mohsenin, 1977; Osborn et al., 1988), and a spherical object like an apple can be turned for better orientation and angle of attack. Punctures to the surface were made with as perpendicular an angle to the tip as possible. Finally, the surface of an apple is covered with a skin through which the tooth has to break. It is this penetration of the skin that gives clearcut results from the Instron compression testing. Artificial surfaces without a "skin" gave indistinct results (e.g., urethane and styrofoam). There is a bloom side (facing the sun) and non-bloom side to the apple that may affect results. Care was taken to stay within a small test area on the bloom side of a single apple for all punctures in this study. The loading rate was 20 mm min^{-1} with a maximum load scale of 4.91 N. The depth of deformation can also be recorded from the Instron; this is the distance traveled by the tooth from the point of contact to the point of breaking through the skin on the apple.

We quantified morphology of the tip of the canines using Hertzian theory, a materials science approach, that utilizes two orthogonal radii and produces a sharpness index (Mohsenin, 1986; Fig. 1). These radii were measured by mounting the skull of the bat on a right-angle stand under a Wild M5 dissecting microscope with camera lucida and drawing the tip from anterior and lateral (90° turn) views. Quantification was done by fitting circles from an engineering template. The sharpness-quantity is a reciprocal so that smaller numbers indicate sharper tips (Fig. 1). The size characteristic (SIZE) in Fig. 2 is one that the first author has used in several previous papers and is the sum of the natural logs of condylocanine length, zygomatic breadth, and temporal height (Freeman, 1984, 1988, 1992). Using a polariscope, we quantified stress patterns (fringes) created by models with different radii as they were loaded onto a photoelastic beam (a beam of Plexiglas™). The oversized "teeth" were two-dimensional Plexiglas™ points with radii of 12.7 mm, 5.1 mm, and 1.6 mm, and were loaded to 13.34 N and 22.27 N. Three trials were performed with each load using each point. The beam itself was 6.22 mm thick by 171.45 mm by 25.4 mm, and after calibration, three locations within the beam were selected for quantification. These were the point of contact, 25.4 mm from the side of contact, and the point opposite contact on the inferior surface of the beam. The fringes at all three points were counted by observation through a polariscope, and the stress represented by the fringes were quantified by a null balance compensator.

SHARPNESS

of tip

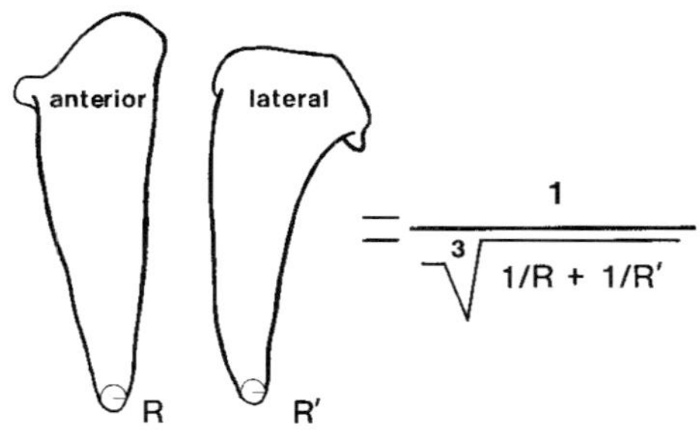

Fig. 1. Sharpness of tip depends on quantifying orthogonal radii of the tip and is used as an index. The smaller the number, the sharper the tip. Details are found in Materials and Methods.

RESULTS

Uniaxial compression tests performed on the Instron produced highly repeatable measurements of the force needed for a canine to puncture a material. Repeated measures of the force needed for a single tooth to penetrate a surface produced a coefficient of variation of only 9%. In an ANOVA, the variance in force was partitioned among replicate punctures of a single

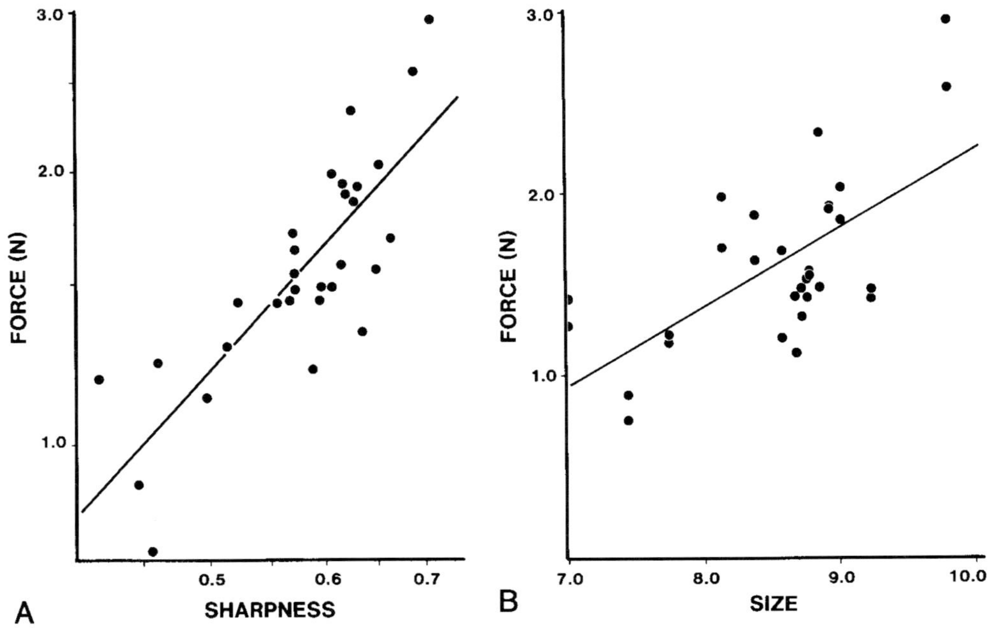

Fig. 2. Uniaxial compression tests showing the amount of force in Newtons (N) necessary for tips of varying sharpness (A, r = 0.79, b = 1.802, a = 1.446; b and a are logged values) and for animals of different size (B, r = 0.70, b = 0.441, a = -2.126) to penetrate an apple. Both are significant ($p < 0.0001$). Sharpness of tooth and size of animal are defined in Materials and Methods. Less force is required for sharp teeth, which have the smallest numbers and are found in small bats. In A, the vertical axis is in Newtons, but has been plotted on a logarithmic scale. Sharpness on the horizontal axis is also on a logarithmic scale.

tooth, different canines of the same species, and canines of different species. The significant variance in force among teeth of the same species was 13 times greater than the variance among replicate punctures of a single canine. There was also a highly significant effect by different species. The variance in force among species was 107 times greater than the variance among replicates of a single canine. Using a correlation analysis, we found that the sharpness of a canine (as measured by orthogonal radii of a canine apex, Hertz theory; Mohsenin, 1986; Fig. 2) explained 63% ($p < .0001$) of the variation in force required to puncture the material.

In general, the sharper the tip, the smaller the force required for penetration, and smaller bats use less force to penetrate an item than larger bats (Fig. 2). Big bats have absolutely duller teeth (but not necessarily relatively, i.e., within its size class) that take greater amounts of force to break through a surface. But size is not the reason more force is necessary. More force is necessary because the big bat lacks a sharp tip. Sharpness of the tip explains most of the variation in force. In Fig. 2, a regression line has been drawn to indicate how canines of differently-sized individuals vary. Actual heights of canine range from 2.0 mm to 8.5 mm. The depth at which the tooth punctures the surface of the apple, depth of deformation, occurs at about 10% of the length of the canine (a range from 8 to 13%). Results for the model "teeth" show that stress at points away from the contact point is constant regardless of the size of the radius (Fig. 3). As expected, stresses at the point of contact are higher for smaller radii. However, even though stress at the point of contact is less for larger radii, the stresses present at points that are a distance away from the tip (> 2.54 cm) are equal. That is, fringes are more spread out and less concentrated for larger radii.

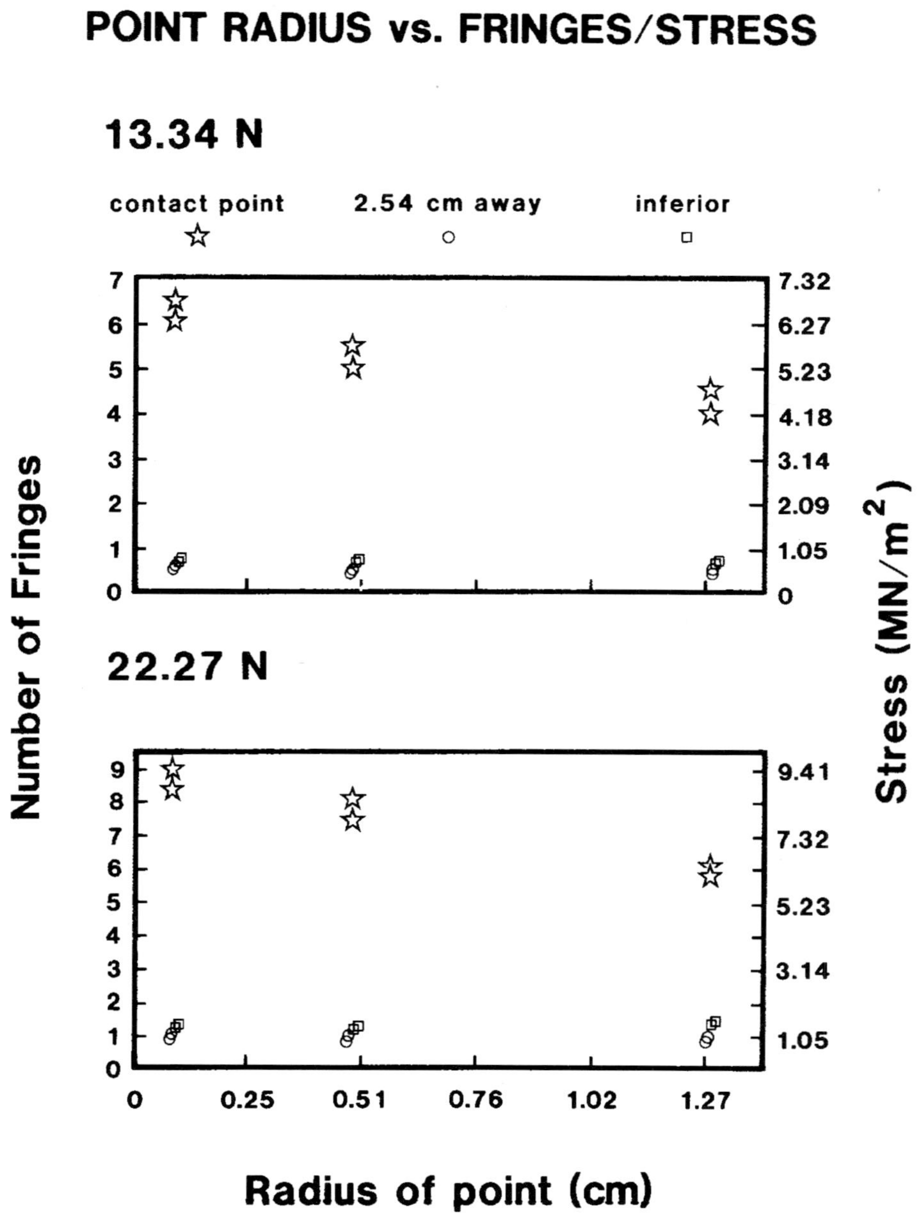

Fig. 3. Giant two-dimensional Plexiglas™ model "teeth" of different radii at the tip loaded onto a photoelastic beam under a 13.34 N and 22.27 N load regime. The points with the smallest radii produced the greatest number of fringes, or stress lines, at the point of contact, indicating higher stress and greater stress concentration.

DISCUSSION

This work is some of the first experimental work actually to quantify apical sharpness of a tooth and performance as the tooth punctures a substance. Lucas (1980, 1982) experimented with pressing items against cusps of molars and, because of the bluntness of the cusps, suggested that cheek teeth in humans were adapted to a diet of hard, brittle foods. In an earlier paper, Lucas (1979) proposed that shapes and configurations of mammalian teeth are largely an evolutionary response to kinds and textures of foods eaten.

Smaller teeth can puncture with greater ease because they concentrate the force in a smaller area (creating greater stress in that area). It is not surprising that the principle of stress concentration applies to small teeth, but it is a first step in matching canine form with canine function. Quantifying force and stress produced by different points under different loads makes possible future experiments that examine stresses created by the shank of the tooth rather than just the tip. Indeed, recent work has proven this as a viable technique (Freeman and Weins, 1994; Freeman, In Press).

What occurs at the interface between tooth and food is difficult information to obtain considering variation in both teeth and food. Some bats are not physically able to take certain hard prey items (Goldman and Henson, 1977), canine teeth can be worn or broken, and some insects can be tougher than others, depending on the amount of chitin in the exoskeleton. There must also be compromises between toughness and nutritional value of different foods, not to mention size of prey. Canines are on the "front line" of action for a bat and may be the most readily changed for dietary requirements. Maier (1984:304) suggests the same thing for primates, and states "that anterior teeth are more directly correlated with the gross structure of food and they may more directly reflect ecological adaptations of a species."

The final shape of the canine in flying predators like bats must be a compromise between sharpness, the ability to penetrate foods swiftly, and wear and breakage, which are blunting features that would impede swift penetration. Wear and breakage is a problem among terrestrial predators, even in hyaenas that have strong, blunt, and conical teeth (Van Valkenburgh and Ruff, 1987). Freeman (1992) thinks bats can afford to have longer, more slender canines because their prey lack hard, brittle, and potentially tooth-breaking substances. Once food is penetrated, how the shank may direct cracks in the food to other teeth could have significant functional implications for an animal that captures prey while flying. Indeed, there is now evidence that enlarged geometric models, mimicking the edged and non-edged cross-sectional nature of canine teeth in bats, produce different patterns of stress in photoelastic plastic (Freeman and Weins, 1994; Freeman, In Press). There is a concentration of stress in the substance being penetrated at the edge of a pyramid-shaped "tooth" that is not present in the conical, non-edged "tooth" that should make penetration easier. However, the initiation of the puncture and the resulting cracks that may radiate from it, no matter what the food may be, is the function of the tips of the canine teeth in bats. Sharp or blunt canines are features we can actually test.

ACKNOWLEDGMENTS

Our thanks to G.W. Brown for casting teeth, R.L. Jochum for running photoelastic tests, and C.A. Lemen for statistical help. We are grateful to curators of mammal collections at the University of Kansas and Field Museum of Natural History for use of their specimens for casting, and to M.A. Hanna for his expertise and use of his lab and equipment during the early stages of this study.

Jim Findley played a pivotal role in the scientific life of the first author. Findley asked (asks) thoughtful questions that go to the heart of the matter. It was those questions, his endless curiosity, and his encouragement of new ideas that made him an excellent advisor. The roots of the ideas in this paper were stimulated by time spent under Jim Findley's tutelage. For these reasons and others, the first author is grateful.

LITERATURE CITED

Crompton, A.W., and K. Hiiemae. 1969. How mammalian molar teeth work. Discovery, 5:23-24.

Crompton, A.W., and K. Hiiemae. 1970. Molar occlusion and mandibular movements during occlusion in the American opossum, *Didelphis marsupialis*. Zoological Journal of the Linnean Society, 49:21-47.

Freeman, P.W. 1984. Functional cranial analysis of large animalivorous bats (Microchiroptera). Biological Journal of the Linnean Society, 21:387-408.

Freeman, P.W. 1988. Frugivorous and animalivorous bats (Microchiroptera): Dental and cranial adaptations. Biological Journal of the Linnean Society, 33:249-272.

Freeman, P.W. 1992. Canine teeth of bats (Microchiroptera): Size, shape, and role in crack propagation. Biological Journal of the Linnean Society, 45:97-115.

Freeman, P.W., In Press. Form, function, and evolution in the skulls and teeth of bats. *In* Bats: Phylogeny, morphology, echolocation, and conservation biology (T.H. Kunz and P.A. Racey, eds.). Smithsonian Institution Press.

Freeman, P.W., and W.N. Weins. 1994. Quantifying the tooth-food interface with finite-element analysis and photoelasticity. Journal of Morphology, 220:345.

Goldman, L.J. and O.W. Henson, Jr. 1977. Prey recognition and selection by the constant frequency bat, *Pteronotus p. parnelli*. Behavioral Ecology and Sociobiology, 2:411-419.

Goodwin, M.B., and D.S. Chaney. 1994. Molding and casting: Techniques and materials. Pp. 235-271, *in* Vertebrate paleontological techniques, Vol. 1 (P. Leiggi and P. May, eds.). Cambridge University Press, New York, xvii + 344 pp.

Halstead, L.B. 1974. Vertebrate hard tissues. Wykeham Publications, London, 179 pp.

Hamann, D.D. 1969. Some dynamic mechanical properties of apple fruit flesh. Transactions of the American Society of Agricultural Engineering, 12:170-174.

Hanna, M.A., and N.N. Mohsenin. 1972. Pack handling of apples. Journal of Agricultural Engineering Research, 17:154-167.

Hiiemae, K.M. 1967. Masticatory function in the mammals. Journal of Dental Research, 46:883-893.

Hiiemae, K.M. 1978. Mammalian mastication: A review of the activity of the jaw muscles and the movements they produce in chewing. Pp. 359-398, *in* Development, function, and evolution of teeth (P.M. Butler and K.A. Joysey, eds.) Academic Press, London, xix + 523 pp.

Kay, R.F. 1975. The functional adaptations of primate molar teeth. American Journal of Physical Anthropology, 43:195-216.

Lucas, P.W. 1979. The dental-dietary adaptations of mammals. Neues Jahrbuch Geologie und Palontologie Monatshefte, 8:486-512.

Lucas, P.W. 1980. Biomaterials as food. Pp. 463-464, *in* The mechanical properties of biological materials (J.F.V. Vincent and J.D. Currey, eds.). 34th Symposia of the Society for Experimental Biology. Cambridge University Press, Cambridge, ix + 513 pp.

Lucas, P.W. 1982. Basic principles of tooth design. Pp. 154-162, *in* Teeth: Form, function, and evolution (B. Kurten, ed.). Columbia University Press, New York, ix + 393 pp.

Maier, W. 1984. Tooth morphology and dietary specialization. Pp. 303-330, *in* Food acquisition and processing in primates (D.J. Chivers, B.A. Wood, and A. Bilsborough, eds.). Plenum Press, New York, xi + 576 pp.

Maynard Smith, J., and R.J.G. Savage. 1959. The mechanics of mammalian jaws. School Science Review, 141:289-301.

Mills, J.R.E. 1967. A comparison of lateral jaw movements in some mammals from wear facets on the teeth. Archives of Oral Biology, 12:645-661.

Mohsenin, N.N. 1977. Characterization and failure in solid foods with particular reference to fruits and vegetables. Journal of Texture Studies, 8:169-193.

Mohsenin, N.N. 1986. Physical properties of plant and animal materials. Second ed. Gordon Breach, New York, xvii + 891 pp.

Osborn, J.W., F.A. Baragar, and P.E. Grey. 1988. The functional advantage of proclined incisors in man. Pp. 445-458, *in* Teeth revisited: Proceedings of the VIIth International Symposium on Dental Morphology, Paris, 1986 (D.E. Russell, J.-P. Santoro and D. Sigogneau-Russell, eds. Memoires du Museum National d'Histoire Naturelle, 53:1-462.

Turnbull, W.D. 1970. Mammalian masticatory apparatus. Fieldiana: Geology, 18:149-356.

Van Valkenburgh, B., and C.B. Ruff. 1987. Canine tooth strength and killing behavior in large carnivores. Journal of Zoology, London, 212:379-397.

Waters, N.E. 1980. Some mechanical and physical properties of teeth. Pp. 99-136, *in* The mechanical properties of biological materials, (J.F.V. Vincent and J.D. Currey, eds.). 34th Symposia of the Society for Experimental Biology. Cambridge University Press, Cambridge, ix + 513 pp.

Patricia W. Freeman, University of Nebraska State Museum, and William N. Weins, Department of Mechanical Engineering, University of Nebraska, Lincoln, Nebraska 68588-0514

The Effects of Daily and Seasonal Temperature Variation on a Model of Competing Lizard Species

J.S. Scheibe

Abstract

Simulation models were used to study the effects of age-structured population growth and thermal amplitude variance on competitive interactions between hypothetical model lizard species. The competing species were identical in all aspects except their position along a resource axis. The lizards grew on a physiological time scale, with fecundity, survival, and position on the resource axis dependent on age. Competition was modeled by allowing the younger age classes of species B to overlap the older age classes of species A. The model included the simulation of natural variance in daily and yearly temperature regimes. The presence of thermal amplitude variance and size-structured population growth produced patterns that were dissimilar from typical Lotka-Volterra patterns. In competitive encounters, it was not usually possible to predict whether competitive exclusion would occur, nor which species would be excluded. Also, correlations between competing species showed no consistent trend with increasing levels of competition. Both size-structured population growth and thermal amplitude variance produced patterns of population growth that obscured the effects of interspecific competition.

INTRODUCTION

Much theoretical development in community ecology over the last 30 years has been based on logistic population growth models (Pianka, 1982). There are two basic problems associated with this fact. First, the assumptions associated with these models are usually violated and often ignored. For example, it seems clear that populations of poikilotherms in particular have size-dependent fecundity and survival, and may exhibit stage-structured population growth. The Lotka–Volterra models assume implicitly that all size/age classes grow at the same rate, and that populations exist in a stable size/age distribution. Similarly, the models assume that competitive effects are symmetrical. Violations of these assumptions are usually considered to be minimal or to have no significant impact on the qualitative results of the study. Whether this is true or not has yet to be determined. The second problem associated with the use of Lotka–Volterra type growth models is that, as demonstrated by the work of May (1973, 1974) and Gonzales–Andujar and Perry (1993), even simple logistic population models can produce complex growth patterns. It is not clear what impact this behavior may have on our ability to discern patterns of competition in nature.

With respect to competition, field studies of community structure often produce equivocal results (Findley and Findley, 1985; Scheibe, 1987), and this is especially true in studies of lizards. The extensive work of Pacala and Roughgarden (1982), Pianka (1986), and Ricklefs et al. (1981) resulted in evidence for interspecific competition in lizards, while Dunham (1980, 1983), M'Closkey and Baia (1987), and Scheibe (1987) found weak or confusing patterns. This may be a consequence of epistemological problems (Conner and Simberloff, 1978, 1979; Strong et al., 1979; and Weins, 1977, 1984), but may also reflect the importance of climatic factors or phylogenetic history (Losos, 1990; but see also Losos, 1992; Roughgarden, 1992; Roughgarden and Pacala, 1989). Although reviews by Connell (1983), Schoener (1983), and Case and Bolger (1991)

Life Among the Muses: Papers in Honor of James S. Findley. Edited by T.L. Yates, W.L. Gannon & D.E. Wilson.

have provided strong support for the competition hypothesis, and in fact few would deny the existence of competition, there is some argument about the magnitude of its effects relative to other factors (Abrams, 1986; Den Boer, 1986; Giller, 1986; Roughgarden, 1986).

The patterns observed in lizard communities may be a consequence of population dynamics that are influenced by ectothermy. Unlike birds and mammals, lizards tend to exhibit indeterminate growth. Lizards, just as insects, exist in physiological time in the sense that their behavior, performance, and ecology is directly related to the thermal environment. In fact, variation in the thermal environment may influence lizard population growth and result in fluctuating populations. Schoener (1985) concluded that lizard populations tend to be temporally constant, although his review was not based on long-term data. Recent work by Andrews (1991) has demonstrated considerable fluctuations in lizard population sizes. The potential for fluctuating populations and size-dependent fecundity and survival suggest that lizards may deviate from the assumptions of the Lotka–Volterra models sufficiently to make the predictions of the models meaningless for many lizard communities.

In this paper, I demonstrate that variation in the thermal environment, together with size- and density-dependent fecundity and survival and asymmetrical competitive effects, produce population growth patterns that display persistent and apparently random fluctuations. The resultant patterns mask the effects of interspecific competition to the extent that detection of interspecific competition is usually impossible, even when it is known to exist.

METHODS

The growth of coexisting lizard populations was studied using a computer simulation model. The model was comprised of three basic components: thermal environment routines, population growth routines, and interspecific competition routines. The thermal environment routines were designed to simulate daily and seasonal thermal variation, and to incorporate daily and yearly variance in temperature amplitudes. The population growth routines modeled coexisting lizard species with age-dependent fecundity and survival using renewal equations (Charlesworth, 1980). The competition routines were used to model density-dependent fecundity and survival schedules in a pattern consistent with the expectations of competitive effects.

The thermal environment model was based on that of Taylor (1981) and used a Fourier series of the form:

$$(1)\quad T(t)=\overline{T}-\frac{a_2}{2}\cos(\frac{2\pi}{365}t+\theta)-\frac{a_1}{2}\cos(2\pi t)$$

where temperature at time t is a function of the mean yearly temperature and both seasonal and daily temperature fluctuations. The parameter a_2 represents the yearly temperature amplitude, a_1 represents the daily temperature amplitude, and theta provides a lag time. The equation produces a curve in which temperatures rise and fall daily and yearly. Parameters a_1 and a_2 were treated as normal random variables with means mu_1 and mu_2 and variances $sigma_1$ and $sigma_2$, respectively. Thus, by specifying the means and variances of a_1 and a_2, it was possible to produce daily and seasonal temperature curves that reflected accurately those found in nature.

The temperature-dependent development rate of the lizards was modeled using a normal curve:

$$(2)\quad R(T) = R_m\ EXP[-.5(T-T_m/T_\sigma)^2]$$

The parameters defining the development rate were the variance of the development curve and the mean of the curve, which represented the optimal development temperature. Thus, at low temperatures, the lizards grew very slowly, and as temperatures increased, growth increased up to the maximum development rate. Temperatures beyond the optimal temperature resulted in slower growth.

No attempt was made to model the effects of behavioral or physiological thermoregulation on the development rates. Thermoregulation by the lizards would result in more moderate body temperatures at high ambient temperatures, and higher body temperatures at low ambient temperatures. Thus, thermoregulation should result in a flatter curve. Preliminary trials revealed the general results of the simulations to be robust with respect to the shape of the development rate curve.

The accumulation of developmental 'units' by the lizards was modeled by the equation:

$$(3) \quad D(t) = \int_{o}^{t} R(T(t))dt$$

Thus, the development of the lizards was based on a physiological time scale, in which the animals developed rapidly (in clock time) during warm parts of the year and slowly (in clock time) during cold parts of the year.

Population growth was modeled using a renewal equation of the form:

$$(4) \quad B_f(t) = g(t) + \sum_{x=1}^{t} B_f(t-x)\ l(x,t)\ m(x,t)$$

$$(5) \quad g(t) = \sum_{x=t+1}^{d} n(x-t,o)\ l(x,t)\ m(x,t) \qquad (t<d)$$

In these equations, time was measured as clock time while age was measured as developmental units using Equation (3) for the accumulation of development given above. $B_f(t)$ represents the births of females at time t. This equation expresses population growth as a function of the initial cohort g(t) plus births of females at time t-x times the age-specific survival l(x,t) and age-specific fecundity m(x,t) of female lizards aged x at time t. The contribution of the initial cohort to the female births at time t is simply the number of lizards in the initial cohort, n(x-t,0), times the age-specific fecundity and survival of that cohort.

The age-specific survival function was a negative exponential of the form:

$$(6) \quad l_x = EXP\ [-kx(\ln(10+N_x))]$$

where k is a constant and N_x is the number of individuals of both species in the age (size) class. The shape of the survival curve under conditions of increasing competition is shown in Fig. 1.

Age-specific fecundity was modeled as:

$$(7) \quad m_x = [(b_o + b_1x)\ln x]\ EXP(-n_x/1000)$$

The form of the fecundity function was based on regressions of clutch sizes on female sizes for *Cnemidophorous* spp. and is shown in Fig. 2 as a function of increasing levels of competition. Thus, as lizards grew larger, they produced larger clutches. This phenomenon has been noted for many lizards (e.g., Abts, 1987; Vitt and Breitenbach, 1993). As the number of lizards within a specific age class approached the resource limit, fecundity declined as a consequence of the negative exponential term.

The life span of each species was chosen to be three years and was divided into 30 age/size classes. Initially, each size class contained an equal number of lizards. As a simulation progressed, animals accumulated 'developmental units' following Equation 3. The accumulation of 10 developmental units resulted in the transition of an animal from one age class to the next. Following the accumulation of 120 developmental units, animals could reproduce. If reproduction occurred, reproduction could occur again only after the accumulation of an additional 10 developmental units. Thus, the system models species that can potentially live three years, can begin reproduction in their second year, and can reproduce more than once each season. However, variance in the thermal regime, delayed onset of reproduction, and early seasonal cessation of reproduction limits the reproductive opportunities for the modeled organisms.

Since reproduction in lizards tends to be seasonal, reproduction was limited to warm months of the simulations. This was done by computing the maximum daily temperature every other day over a period of 10 days, then computing the regression coefficient of maximum temperature against day for that period. If the coefficient was positive, reproduction was possible 28 days later. Following the onset of reproduction, a similar regression coefficient was computed for the period 180 days later. Once the coefficients were negative, reproduction ceased

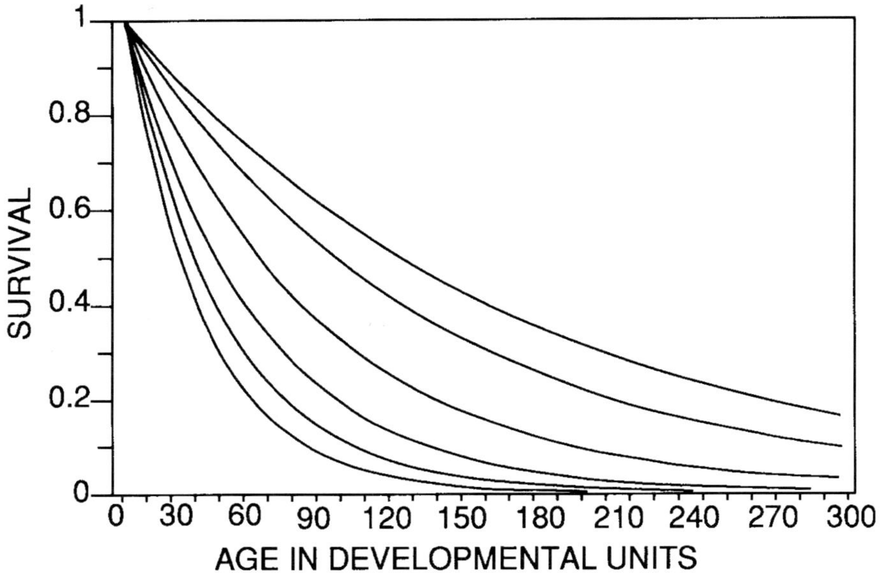

Fig. 1. Shape of the survival function in relation to increasing age (developmental units) and competition.

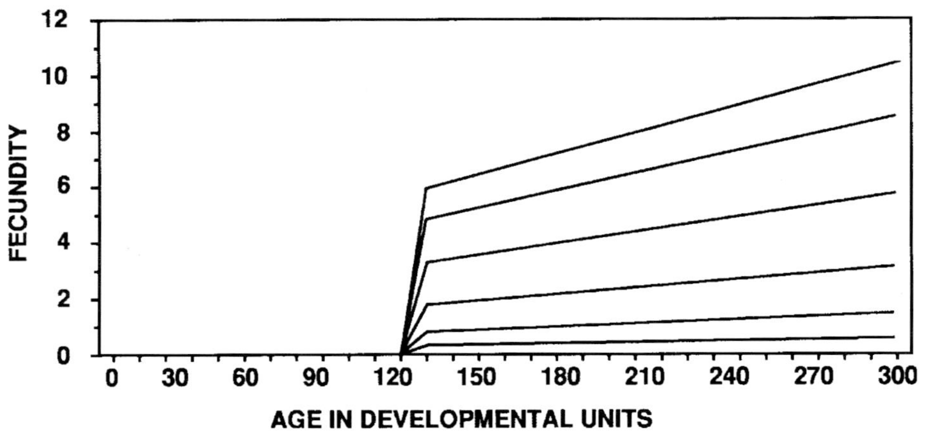

Fig. 2. Shape of the fecundity function in relation to increasing age (developmental units) and competition.

until the next season. This approach is not unrealistic, as indicated by the review of Duvall, Guillette, and Jones (1982). There is strong evidence that temperature plays a significant role in reproductive cycles of lizards. Although reproduction in lizards is more complex than a simple thermal relationship, the approach taken here provides a reasonable first step in the development of this modeling effort.

Each computer simulation modeled 100 years of population growth, one hour at a time. The simulations began with both species containing equal numbers of animals distributed evenly through all size classes. As the animals accumulated developmental units (determined by the thermal environment routines), they moved through the size classes of the population. For each hour, fecundity and survival were computed for each size class. Fecundity within reproductive size classes was dependent on the number of conspecific lizards in that size class and the number of lizards from the coexisting species that overlapped that size class. Survival in the first series of simulations was unaffected by competition. In the second series of simulations, survival of individuals in overlapping size classes decreased as per Equation (6). Thus, intraspecific competitive effects were manifested as fecundity effects. Interspecific competitive effects were manifested as fecundity effects for the first series of simulations and as both fecundity and survival effects for the second series of simulations. Interspecific overlap between species was modeled by allowing the older size classes of the first species to overlap with the younger size classes of the second species, under a uniform resource availability curve. That is, the carrying capacity for the overlapping size class was no larger than that for non-overlapping size classes. As population sizes of both species approached the carrying capacity for the overlapped size class, the fecundity and survival curves for both species were modified. Under these conditions, species B has a competitive advantage. Its most abundant size classes (newborns and young lizards) overlap initially with the least abundant and most fecund size classes of species A, thus significantly reducing the reproductive potential of species A.

The effects of interspecific competition were studied by performing independent replicates of simulations at each level of competitive overlap. Overlap was increased at 10% intervals from the initial series of simulations with no overlap to the final series with complete overlap. A total of 181 simulations were performed for the trials in which fecundity was affected by competition, and 113 simulations were performed for the trials in which fecundity and survival were affected. The fate of each species and the product moment correlations between species for each simulation were compared graphically with those obtained from initial trials in which there was no thermal amplitude variance.

RESULTS

The results of the competition trials were dependent on the extent of overlap between the model species, the nature of the competitive effect (whether it affected fecundity alone, or fecundity and survival), and the thermal amplitude variances. When amplitude variances were zero and only fecundity was affected by competition, the results of the competition trials were similar to those expected from systems using the Lotka–Volterra models. As the extent of overlap between species increased, the population of one species tended to increase while the population of the second species tended to decrease (Fig. 3). Unlike the expected Lotka–Volterra results, neither species went extinct, even with complete overlap. This was a consequence of the magnitude of the competitive effect on fecundity. Increasing the magnitude of the effect would have resulted in the extinction of the rarer species.

When competition affected both fecundity and survival in the absence of thermal amplitude variance (Fig. 4), the results were fundamentally different. Here, there was a marked reversal of fortune when competitive overlaps exceeded 60%. When overlap was less than 60%, the population size of species B exceeded that of species A, as in the previous trial. However, overlaps between 60 and 65% resulted in the reduction of species B relative to A and the extinction of A at 67% overlap. The extinction of species A was caused by competitive effects on the fecundity of all reproductive size classes. Relaxation of the competitive effect on survival and the ensuing increased number of fecund individuals tended to broaden this range. Even the extinction of species A, however, did not enable species B to grow significantly within the 100-yr simulation. Unlike a typical Lotka–Volterra system in which one species always has a competitive advantage, here that advantage was dependent on the extent of overlap.

The product moment correlations between yearly population sizes of the competing species in the absence of thermal amplitude variance are presented in Fig. 5. Typically, competition is expected to produce negative correlations between competing species, with the magnitude of the negative correlation an index of the intensity of competition. In this analysis, correlations were negative only when overlap was 67%. Furthermore, the correlations did not decrease monotonically with increasing competition. This suggests that correlations were a poor indicator of competition in this model. The fecundity and survival effects of competition resulted in changes in size structures of the populations, with subsequent changes in the intensity of competition between the species. Thus, in spite of extensive overlap between the species, changes in size structure alleviated the intensity of competition.

The incorporation of non-zero thermal amplitude variances in the model increased the variability in the results. Representative samples of population growth are shown in Fig. 6. Even in the absence of competition (Fig. 6a), there were considerable fluctuations in population size over the 100-yr interval. Since this sample trial models a population in the absence of interspecific competition, the fluctuations were a consequence of variable fecundity and survival, resulting from yearly and daily fluctuations in thermal amplitude. These fluctuations influenced the accumulation of developmental units and prevented the population from achieving a stable size distribution within the 100 yrs of the simulation. The sample trials for competing species, shown in Figs. 6b and 6c, illustrate the dynamic nature of the growth patterns and reveal that the competitive effects are not readily discernible from the underlying thermal, physiological, and size structure effects. These figures demonstrate that even careful multi-year field samples of compet-

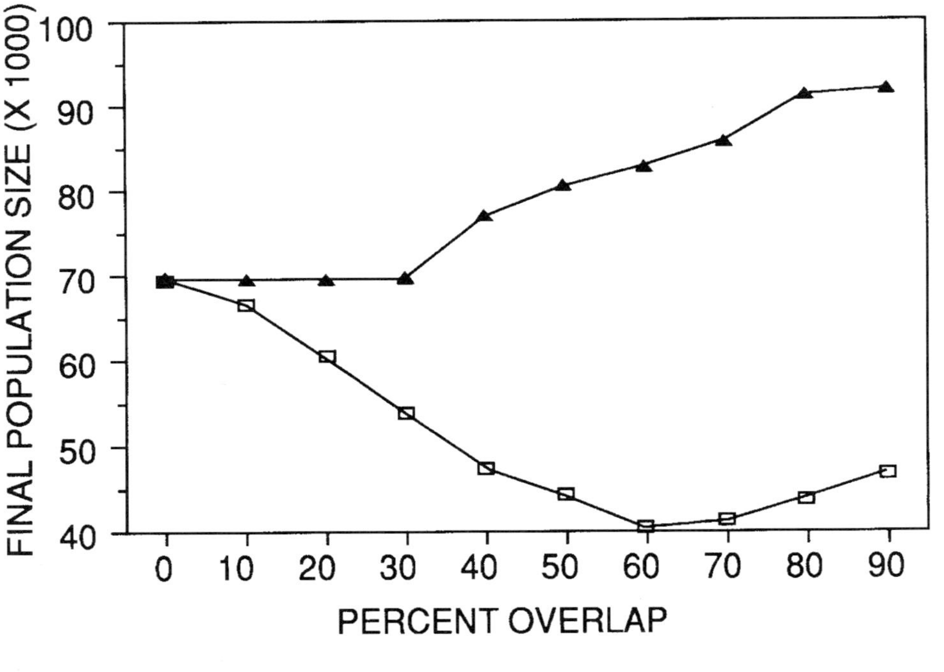

Fig. 3. Population sizes of species A (open rectangles) and species B (closed triangles) after 100 years of competition with no thermal amplitude variance. In these simulations, only fecundity was affected by competition.

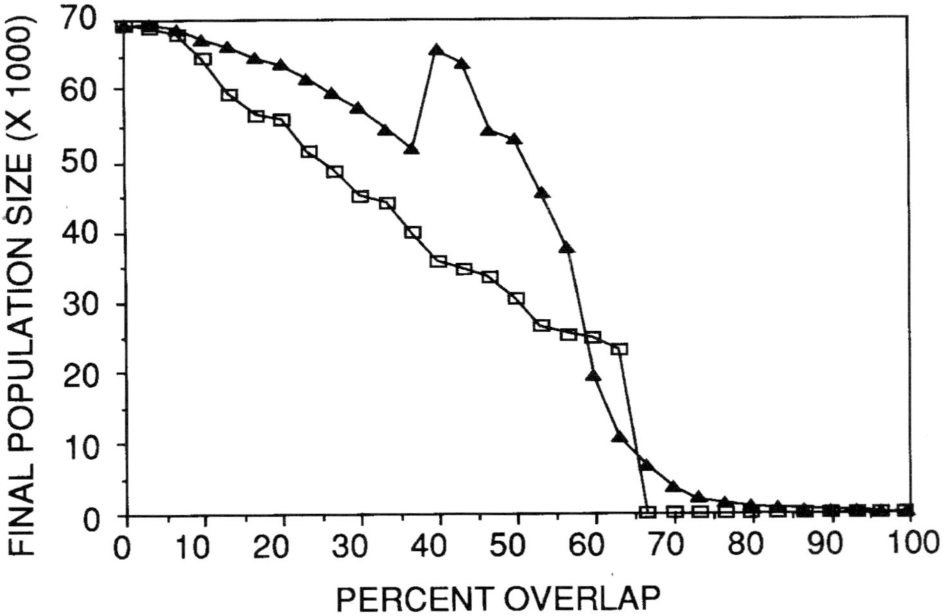

Fig. 4. Population sizes of species A (open rectangles) and species B (closed triangles) after 100 years of competition with no thermal amplitude variance. In these simulations, both the fecundity and survival functions were affected by competition.

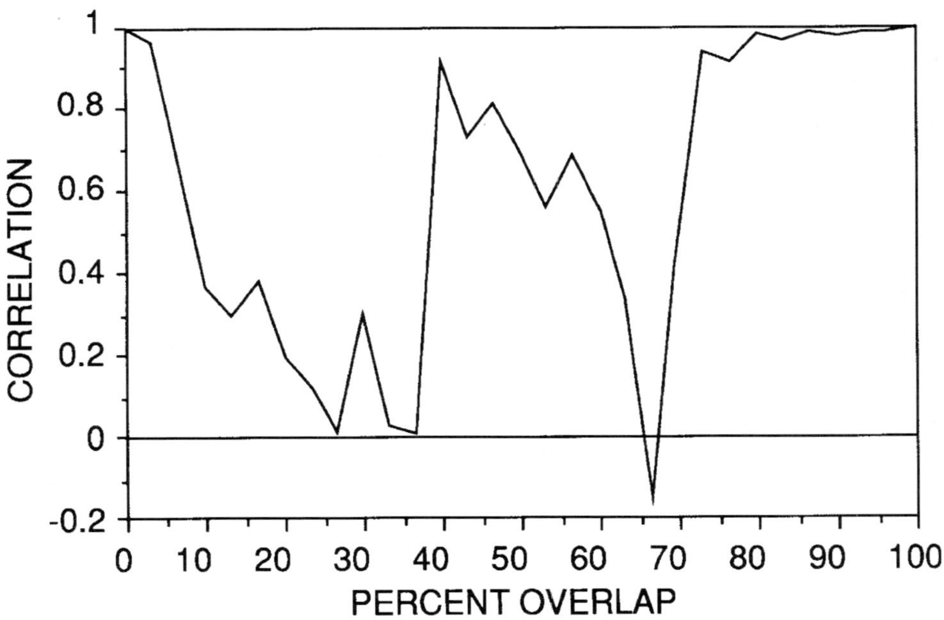

Fig. 5. Pearson product moment correlations between the population sizes of two competing species over 100 years, versus the level of overlap between competing species.

ing species may produce results that do not reveal interspecific competition.

The results of the correlation analyses for the trials with thermal amplitude variance are shown in Figs. 7a and 7b. In each figure, the upper, middle, and lower lines represent the upper quartile, median, and lower quartile of each distribution of correlation coefficients. It is clear that while there is some agreement between the expected and observed distributions, there is considerable variance in the correlation coefficients at most levels of overlap. The range of correlations for competing populations was smallest when overlaps were about 40 or 50%. It is within this range of overlaps that the youngest reproductive size classes of species B are first subjected to direct competitive effects. Below and above this range, variability in the correlations increased dramatically. Also, the median correlation coefficients were almost always positive, regardless of whether competition affected fecundity only, or survival and fecundity. There was no clear trend of increasingly negative correlations with increasing intensity of competition.

The fates of the two competing species in relation to their shared resource use are presented in Figs. 8 and 9. When fecundity was affected by competition, the abundance of species A relative to the final population of both species was usually greater than 50% at overlaps of less than 40% or greater than 70%. Between 40 and 70%, species A usually constituted less than 50% of the overall final population. The distribution of the quartiles over the range of overlaps indicates that species A generally dominated at low and high overlap values, but did poorly at intermediate overlaps. When both fecundity and survival were affected by competition, the abundance of species A relative to the overall final population was usually less than 50% at overlaps of 40 and 50% (Fig. 8b). At overlaps of 30 and 60%, species A generally predominated.

The extinction probabilities shown in Fig. 9a illustrate that species A is most likely to go extinct at intermediate levels of overlap, while species B is most likely to go extinct at low and high levels of overlap when competition affects fecundity. Species B almost always had a higher extinction rate than species A. The extinction probabilities, however, were not monotonically increasing with increasing level of competition. When fecundity and survival were affected by competition, the extinction probabilities for both species were low at intermediate overlap values and higher elsewhere. Species B almost always had a higher probability of extinction than species A. Again, the extinction probabilities were not monotonically increasing with increasing level of competition.

DISCUSSION

In general, fecundity and survival are size-dependent in lizards and many species exhibit some degree of indeterminant growth (Abts, 1987; Shine and Charnov, 1992). These generalizations, together with the periodic and variable nature of the temperate thermal environment, suggest that populations of temperate zone lizards are unlikely to reach a stable size distribution. Thus, the implicit assumption of the Lotka–Volterra models that populations exist in a stable size distribution may be violated by lizards, and predictions about competition based on those models may be inaccurate.

Size-structured population growth is sufficient to cause fluctuating population densities. Although Schoener's (1985) review of the literature found population constancy in lizards, the reviewed work entailed relatively short time periods. Recent work by Andrews (1991) on tropical *Anolis* lizards revealed 5- to 8-fold changes in density over a 19-year period. The density fluctuations she observed were statistically related to seasonal rainfall. She hypothesized the dynamic nature of the fluctuations to result from random perturbations alleviated by density-dependent interactions. It has been demonstrated by Den Boer (1991) that in some empirical systems fluctuations in population density can be mimicked with sufficient closeness by random walk models. Similarly, even simple model systems are capable of chaotic growth (Allen, 1990; Batterman, 1993; May, 1973, 1974; Schaffer and Kot, 1986). Thus, there is good reason to doubt the universality of population constancy in lizards. The randomness and unpredictability observed in the model presented here suggest that temperate zone lizard communities may be so-called "non-equilibrium communities"

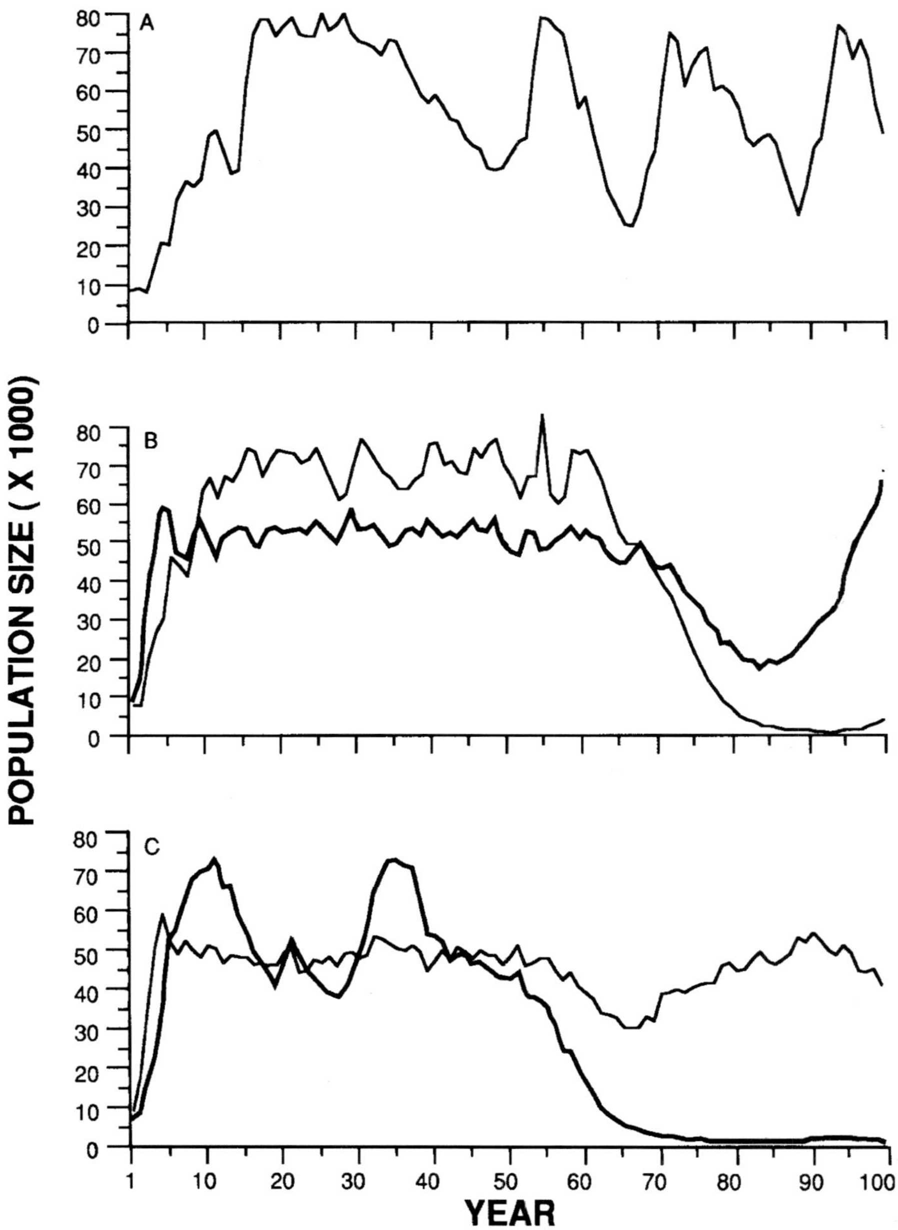

Fig. 6a-c. Representative growth curves for 1) a population without a competitor (6a), 2) a competitive system with 20 percent overlap (6b), and 3) a competitive system with 30 percent overlap (6c). In each simulation competition affected both fecundity and survival, and there was thermal amplitude variance.

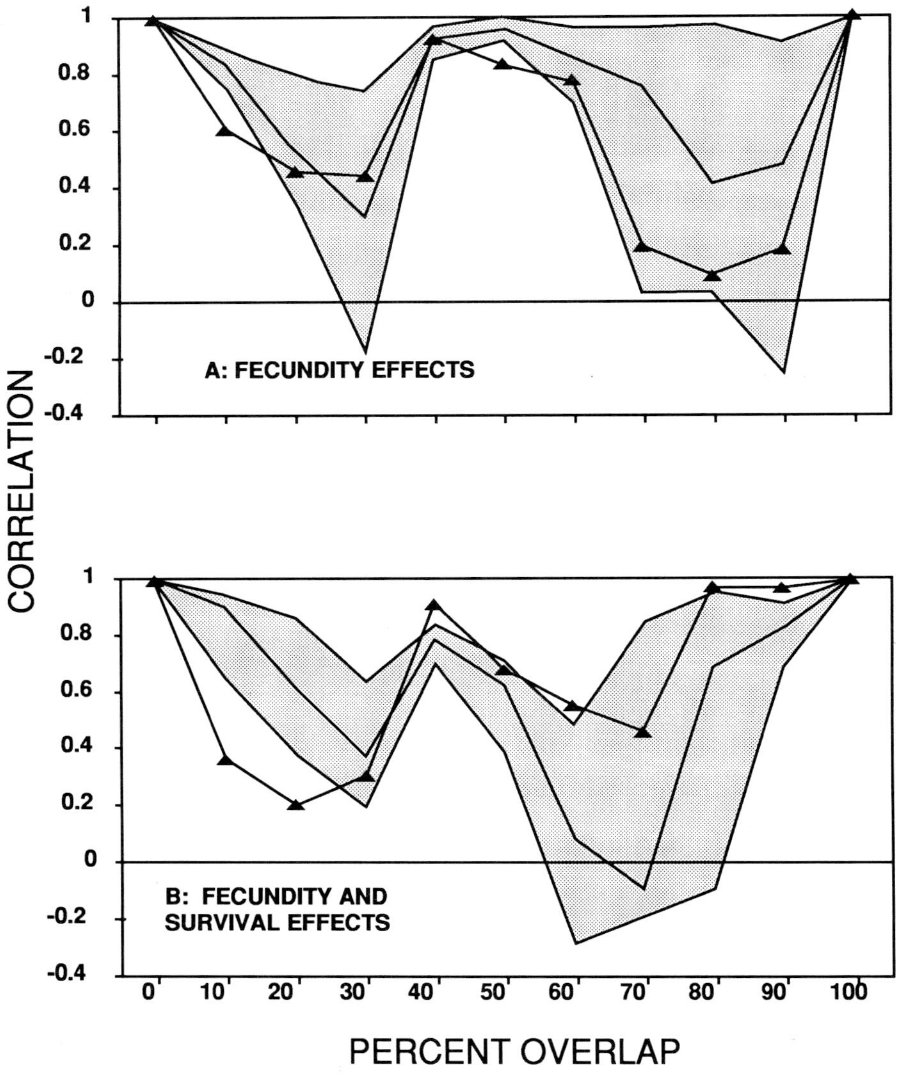

Fig. 7a-b. Correlation results for simulation trials in which competition affects fecundity (7a) and fecundity and survival (7b), relative to percent overlap. In each figure, the expected correlation (based on trials with no thermal amplitude variance) is indicated by closed triangles. The upper, middle, and lower lines in each figure represent the third quartile, median, and first quartile for each distribution of correlations.

(Weins, 1984). Attributes of such assemblages include biotic decoupling, abiotic limitation, and large stochastic effects.

Competition between the modeled lizard species was not symmetrical. Resource use was assumed to be size-dependent and the species existed on a size continuum. The larger, older, most fecund, and least abundant size classes of the smaller species overlapped the smaller, younger, and most abundant classes of the larger species. Although some work has been done on coexisting teiid lizards (Casas–Andreu and Gurrola–Hidalgo, 1993; Medica, 1967; Milstead, 1957a, 1957b, 1965; Mitchell, 1979), the life history and diet data necessary to validate assumptions of symmetrical competition are not available for lizards. The assumptions are not unreasonable, however, and certainly asymmetric interactions between species are not uncommon and may have im-

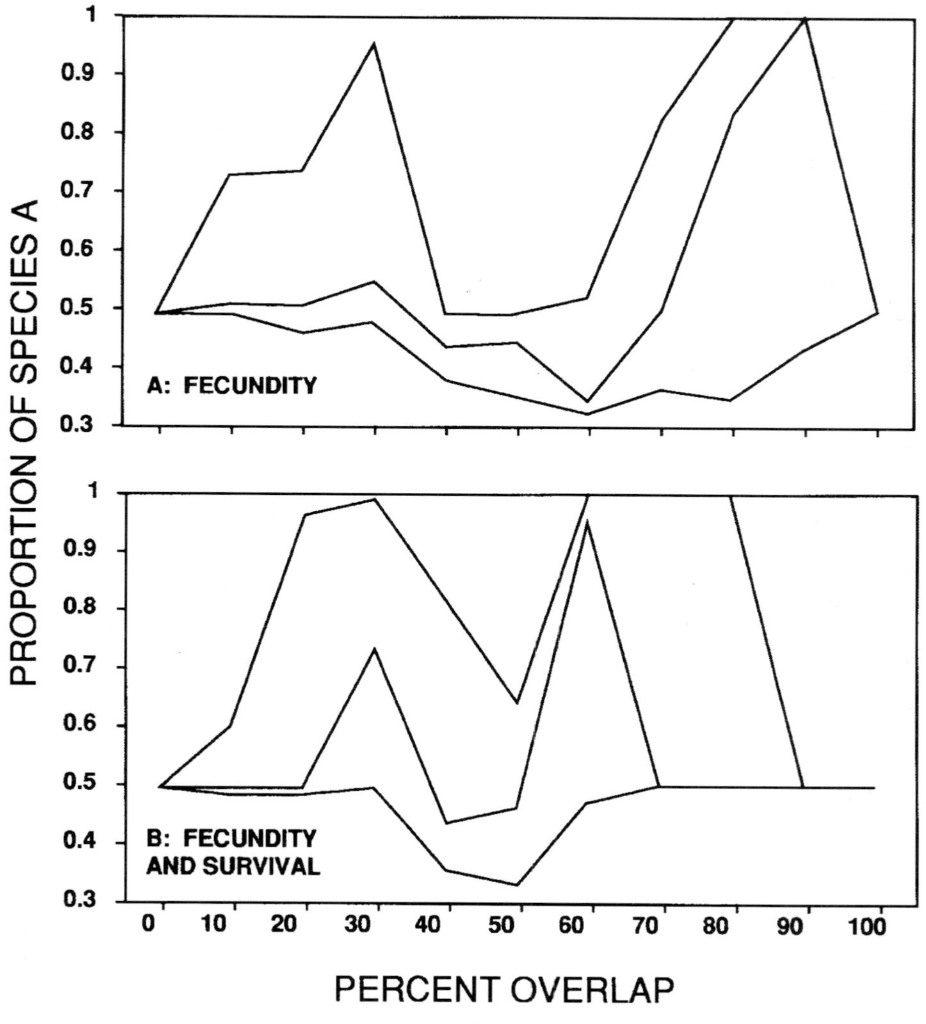

Fig. 8a-b. Proportion of species A in the two species population following 100 years of competition, for trials in which competition affected fecundity only (8a) and fecundity and survival (8b). In each figure, the upper, middle, and lower lines represent the first quartile, median, and third quartiles of the distributions respectively.

portant consequences for the behavior of the system, especially if the competitor populations are size structured.

Overlapping size structured species may experience variable intensities of competition as a consequence of fluctuating densities and size structures. Thus, it is not surprising that in this model increased levels of competition did not result in obvious density-dependent effects. As the intensity of competition increased, the expected interspecific correlations did not become increasingly negative, nor did they decrease monotonically. Furthermore, even when thermal amplitude variance was zero, increased competition did not produce monotonically decreasing correlations between species. Thus fluctuating size-structured populations with asymmetric overlaps did not respond to interspecific competition in the expected way. With the inclusion of non-zero thermal amplitude variance, interspecific correlations were generally positive and were not monotonically decreasing. This suggests that the size structure of the populations and fluctuating densities, both driven by the physiologically based development routines and influenced by the variable thermal environment,

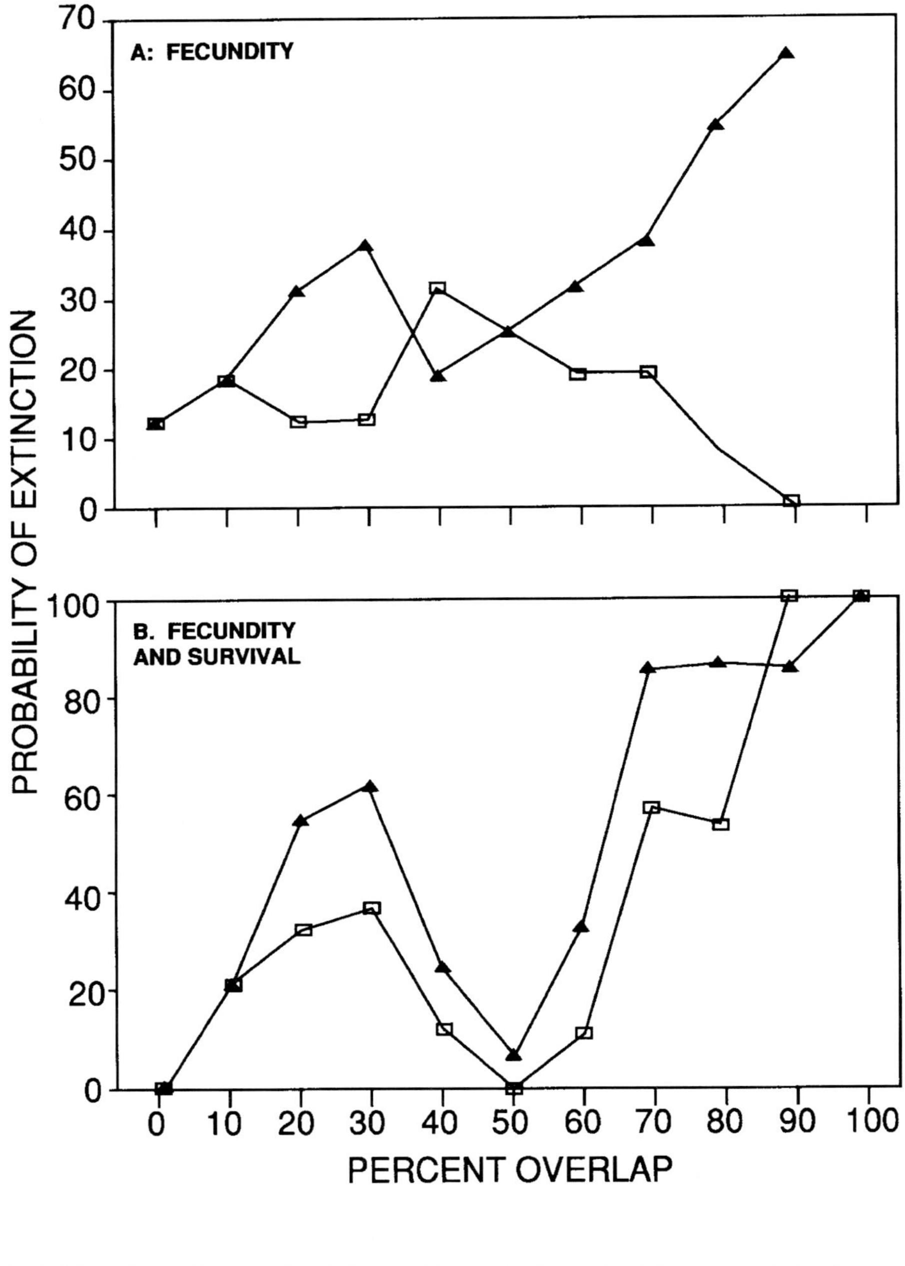

Fig. 9a-b. Probability of extinction versus level of competitive overlap for species A (open rectangles) and species B (closed triangles). In 9a, only fecundity is affected by competition while in 9b both fecundity and survival are affected.

may have ameliorated the effects of competition on the overall densities of the species.

The simulation results were dependent on competitive mechanisms and the presence of thermal amplitude variance. That is, when competition affected fecundity and survival, the results were different than when competition affected fecundity alone. Even though the results shown in Fig. 3 (no thermal amplitude variance; competition affects fecundity, but not survival) tend to be consistent with patterns expected from a Lotka–Volterra model, the introduction of competitive effects on survival produced an entirely different result. Here, increasing levels of competition did not result in a consistent increase in the probability that the larger species B would displace species A. In fact, species displacement was dependent on the amount of overlap. Even in a simple world (no thermal amplitude variance) it is difficult to predict if competitive exclusion will occur and which species will be excluded.

Numerous studies have demonstrated the effects of resource limitation or environmental uncertainty on lizard reproduction (Abts, 1987; Patterson, 1991; see also Shine and Harlow, 1993; and Shine and Schwarzkopf, 1992). Although survival effects are more difficult to demonstrate, they are also an expected consequence of resource limitation (Shine and Charnov, 1992). Detailed knowledge of how multiple species systems respond to resource limits and environmental fluctuations seems essential if we are to understand how competitive effects are manifested. As shown here, the presence of thermal amplitude variance produced results inconsistent with the predictions of competition theory. Repeated trials exhibited a great deal of variance with respect to long-term persistence in competitive encounters (Fig. 6). Although species B had a competitive advantage, with low levels of overlap species A usually did better than species B. Only when overlap was about 40 or 50% did species A follow the predicted trend. As overlap increased beyond 70%, species B again exhibited a competitive advantage most of the time. But, the variation from one trial to the next (Figs. 7 and 8) illustrates clearly how the system is unpredictable. This suggests that even a comprehensive knowledge of the competitive mechanism may be insufficient if the populations are size structured and responsive to environmental variance.

I used a model based on physiological time and size-structured population growth to observe the effects of interspecific competition when it did in fact exist. If the results reflect natural processes, then our difficulty in demonstrating competition in the field is understandable. The environmental and demographic forces impinging upon population size may overwhelm or mask the effects of competition.

LITERATURE CITED

Abrams, P.A. 1986. The competitive exclusion principle: Other views and a reply. Trends in Ecology and Evolution, 5:131-132.

Abts, M.L. 1987. Environment and variation in life history traits of the chuckwalla, *Sauromalus obesus*. Ecological Monographs, 57:215-232.

Allen, J.C. 1990. Factors contributing to chaos in population feedback systems. Ecological Modelling, 51:281-298.

Andrews, R.M. 1991. Population stability of a tropical lizard. Ecology, 72:1204-1217.

Batterman, R.W. 1993. Defining chaos. Philosophy of Science, 60:43-66.

Casas–Andreu, G., and M.A. Gurrola–Hidalgo. 1993. Comparative ecology of two species of Cnemidophorus in coastal Jalisco, Mexico. Pp. 133-150, *in* Biology of Whiptail Lizards (Genus *Cnemidophorus*) (J.W. Wright and L.J. Vitt, eds.). Oklahoma Museum of Natural History, Norman, Oklahoma, 417 pp.

Case, T.J., and D.T. Bolger. 1991. The role of interspecific competition in the biogeography of island lizards. Trends in Ecology and Evolution, 6:135-139.

Charlesworth, B. 1980. Evolution in age-structured populations. Cambridge studies in Mathematical biology: 1 (C. Cannings and F. Hoppensteadt, eds.), Cambridge University Press, Cambridge, 300 pp.

Connell, J.H. 1983. On the prevalence and relative importance of interspecific competition: Evidence from field experiments. American Naturalist, 122:661-696.

Conner, E.F., and D. Simberloff. 1978. Species number and compositional similarity of the Galapagos flora and avifauna. Ecological Monographs, 48:219-248.

Conner, E.F., and D. Simberloff. 1979. Assembly of species communities: Chance or competition? Ecology, 60:1132-1140.

Den Boer, P.J. 1986. The competitive exclusion principle: Other views and a reply. Trends in Ecology and Evolution, 5:132-133.

Den Boer, P.J. 1991. Seeing the trees for the wood: Random walks or bounded fluctuations of population size? Oecologia, 86:484-491.

Dunham, A.E. 1980. An experimental study of interspecific competition between the iguanid lizards *Sceloporus merriami* and *Urosaurus ornatus*. Ecological Monographs, 50:309-330.

Dunham, A.E. 1983. Realized niche overlap, resource abundance, and intensity of interspecific competition. Pp. 261-280, *in* Lizard Ecology; studies of a model organism (R.B. Huey, E.R. Pianka, and T.W. Schoener, eds.). Harvard University Press, Cambridge, Massachusetts, USA.

Duvall, D., L.J. Guillette, Jr., and R.E. Jones. 1982. Environmental control of reptilian reproductive cycles. Pp. 201-231, *in* Biology of the Reptilia, Vol. 13, Physiology D, Physiological Ecology (C. Gans and F. H. Pough, eds.), 345 pp.

Findley, J.S., and M.T. Findley. 1985. A search for pattern in butterfly fish communities. American Naturalist, 126:800-816.

Giller, P. 1986. The competitive exclusion principle: Other views and a reply. Trends in Ecology and Evolution, 5:132.

Gonzalez–Andujar, J.L., and J.N. Perry. 1993. Chaos, metapopulations, and dispersal. Ecological Modeling, 65:255-263.

Losos, J.B. 1990. Ecomorphology, performance capability, and scaling of West Indian Anolislizards: An evolutionary analysis. Ecological Monographs, 60:369-388.

Losos, J.B. 1992. A critical comparison of the taxon-cycle and character-displacement models for size evolution of *Anolis* lizards in the Lesser Antilles. Copeia, 1992:279-288.

May, R.M. 1973. Stability and complexity in model ecosystems. Monographs in Population Biology 6, Princeton University Press, 265 pp.

May, R.M. 1974. Biological populations with non-overlapping generations: Stable points, stable cycles, and chaos. Science, 186:645-647.

M'Closkey, R.T., and K.A. Baia. 1987. Assessment of competitive interactions between two iguanid lizard species. Oikos, 48:206-210.

Medica, P.A. 1967. Food habits, habitat preference, reproduction, and diurnal activity in four sympatric species of whiptail lizards (*Cnemidophorus*) in south central New Mexico. Bulletin of the Southern California Academy of Science, 66:251-276.

Milstead, W.W. 1957a. Observations on the natural history of four species of whiptail lizards, *Cnemidophorus* (Sauria: Teiidae), in Trans-Pecos Texas. Southwestern Naturalist, 2:105-121.

Milstead, W.W. 1956b. Some aspects of competition in natural populations of whiptail lizards (genus *Cnemidophorus*). Texas Journal of Science, 9:410-447.

Milstead, W.W. 1965. Changes in competing populations of whiptail lizards (*Cnemidophorus*) in southern Texas. American Midland Naturalist, 73:75-80.

Mitchell, J.C. 1979. Ecology of Southeastern Arizona whiptail lizards (*Cnemidophorus*: Teiidae) population densities, resource partitioning, and niche overlap. Canadian Journal of Zoology, 57:1487-1499.

Pacala, S.W., and J. Roughgarden. 1982. Resource partitioning and interspecific competition in two-species insular *Anolis* lizard communities. Science, 217:444-446.

Patterson, J.W. 1991. Rainfall and reproduction in females of the tropical lizard *Mabuya striata striata*. Oecologia, 86:419-423.

Pianka, E.R. 1983. Evolutionary ecology. University of Texas, Austin. 416 pp.

Pianka, E.R. 1986. Ecology and natural history of desert lizards. Princeton University Press, Princeton, NJ, 208 pp.

Ricklefs, R.E., D. Cochran, and E.R. Pianka. 1981. A morphological analysis of the structure of communities of lizards in desert habitats. Ecology 62:1474-1483.

Roughgarden, J. 1986. The competitive exclusion principle: Other views and a reply. Trends in Ecology and Evolution, 5:132.

Roughgarden, J. 1992. Comments on the paper by Losos: Character displacement versus taxon loop. Copeia, 1992:288-295.

Roughgarden, J., and S. Pacala. 1989. Taxon cycle among Anolis lizard populations: Review of evidence. Pp. 403-432, *in* Speciation and its consequences (D. Otte and J. Endler, eds.). Academy of Natural Sciences, Philadelphia, PA, 679 pp.

Scheibe, J.S. 1987. Climate, competition, and the structure of temperate zone lizard communities. Ecology, 68:1424-1436.

Schaffer, W.M., and M. Kot. 1986. Chaos in ecological systems: The coals that Newcastle forgot. Trends in Ecology and Evolution, 1:58-63.

Schoener, T.W. 1983. Field experiments on interspecific competition. American Naturalist, 122:240-285.

Schoener, T.W. 1985. Are lizard population

sizes unusually constant through time? American Naturalist, 126:633-641.

Shine, R., and E.L. Charnov. 1992. Patterns of survival, growth, and maturation in snakes and lizards. American Naturalist, 139:1257-1269.

Shine, R., and P. Harlow. 1993. Maternal thermoregulation influences viability in a viviparous lizard. Oecologia, 96:122-127.

Shine, R., and L. Schwarzkopf. 1992. The evolution of reproductive effort in lizards and snakes. Evolution, 46:62-75.

Strong, D.R. Jr., L.A. Szyska, and D.S. Simberloff. 1979. Tests of community-wide character displacement against null hypotheses. Evolution, 33:897-913.

Taylor, F. 1981. Ecology and evolution of physiological time in insects. American Naturalist, 117:1-23.

Vitt, L.J., and G.L. Breitenbach. 1993. Life histories and reproductive tactics among lizards in the genus *Cnemidophorus* (Sauria: Teiidae). Pp. 211-243, *in* Biology of Whiptail Lizards (Genus *Cnemidophorus*) (J.W. Wright and L.J. Vitt, eds.). Oklahoma Museum of Natural History, Norman, OK, 417 pp.

Weins, J.A. 1977. On competition and variable environments. American Scientist, 65:590-597.

Weins, J.A. 1984. On understanding a non-equilibrium world: Myth and reality in community patterns and processes. Pp. 439-457, *in* Ecological Communities: Conceptual Issues and the Evidence (D.R. Strong Jr., D. Simberloff, L.G. Abele, and A.B. Thistle, eds.). Princeton University Press, Princeton, NJ, 613 pp.

John S. Scheibe, Department of Biology, Southeast Missouri State University, Cape Girardeau, Missouri 63701

A Comparison of Morphometric Techniques to Distinguish Sympatric Mussel Species (Family Unionidae) with Similar Shell Morphology

PATRICIA MEHLHOP AND RICHARD L. CIFELLI

Abstract

New techniques for measuring and analyzing morphology were tested to distinguish similar shells of two sympatric species of freshwater mussels of the family Unionidae. Using the more traditional approach of principal components analysis on 10 measures reflecting valve size and shape, the species were not clearly distinguished from one another. Rotational fit methods demonstrated differences in the ways that 19 shell landmarks varied between the species, but little of the variation represents uniform shape change. In a principal components analysis based on coefficients derived from elliptic Fourier analysis using three harmonics, 73.9% of the variation was explained by the first three components, while in the same analysis using 10 harmonics, only 41.8% of the variation was explained. Because outlines based on 10 harmonics follow the original outlines closely, we infer that mathematical modeling of small irregularities along individual valve margins entered random variation into the approximated measures. These results support the application of curve-fitting models to morphological problems when few landmarks can be measured reliably, and suggest that the source of variation (systematic or random) should be considered when developing such models.

INTRODUCTION

Freshwater mussels of the family Unionidae are diverse in the Atlantic drainage of the United States, but have suffered dramatic declines in this century because of habitat destruction and introduction of exotic species (Mehlhop and Vaughn, in press; Neves, 1992). The distribution of species within a river is strongly influenced by gradient and substrate (Way et al., 1989), and species that occupy similar habitats often have similar shell size and shape. When such species co-occur, they are difficult, if not impossible, to distinguish by their shells. In a comparison of approaches to assessing relationships among North American Unionidae, Davis (1983) concluded that conchology was less effective than comparative anatomy of the soft parts, which was, in turn, less effective than allozyme analyses. Similar conclusions have been drawn elsewhere (Davis and Fuller 1981; Davis et al., 1981). The latter two approaches, however, require that live individuals be collected when often they are difficult to find, especially in adequate sample sizes. This is particularly problematic for rare species that occur in such low abundance that not only is sample size potentially low, but collecting live individuals could impact populations or subpopulations (Mehlhop and Vaughn, in press). This was the case in the Ouachita Mountain drainage of western Arkansas and eastern Oklahoma where two unionid species with similar shell morphology co-occur. *Obovaria jacksoniana* is relatively widespread in the lower Mississippi River drainage, occurring from Alabama to Texas, while *Villosa arkansasensis* has a narrow distribution associated with the southern Ouachita Mountain drainage. We wanted to know more precisely the distribution and rarity status of *V. arkansasensis*, but had difficulty in visually distinguishing its shells from those

Life Among the Muses: Papers in Honor of James S. Findley. Edited by T.L. Yates, W.L. Gannon & D.E. Wilson.

of the more common *O. jacksoniana*, and we were unable to find many live individuals. Study of shell morphology in these bivalves is difficult due to the paucity of distinct landmarks for measurement. We undertook this study to determine whether new morphometric techniques in measurement and analysis that have emerged over the past decade (Ferson et al., 1985; Rohlf 1990a,b,c; Rohlf and Slice 1990a,b) would facilitate discrimination between bivalves with similar shell morphology.

METHODS

Analysis of Shell Measurements

One hundred twenty-four identified specimens (105 belonging to *O. jacksoniana* and 19 to *V. arkansasensis*) were available for study; these derive from the known distributional ranges of the two species in the lower Mississippi River drainage. Taxonomic identity was determined by comparative anatomy of soft parts or, when outside the range of sympatry, by shell identification. Twenty-three landmarks were defined on the internal surface of the left mussel valve. These landmarks represent boundary points, local geometric features, and endpoints of greatest diameter, as discussed by Bookstein (1990; Bookstein et al., 1985). Of these, four proved to be unreliable because of measurement error, leaving 19 points (Fig. 1). Two-dimensional (X,Y) coordinates of the landmarks were captured using a Reflex Microscope (cf. MacLarnon, 1989). Nine measurements (Fig. 1), reflecting valve length, width, and other shape factors, were selected for analysis, with absolute distances between landmarks being determined from the coordinates. Valve thickness, measured with digital calipers at landmark 17, was added to the data set, which therefore included 10 mensural variables. The data were standardized by dividing the maximum value for each variable into all values for the respective variable. Values were converted to their natural logarithms for bivariate study, but were left unadjusted for the purposes of multivariate analyses. All analyses were performed with the SYSTAT 5.0 programs (Wilkinson, 1992). A principal components analysis, using a varimax rotation and the correlation matrix, was performed on the data.

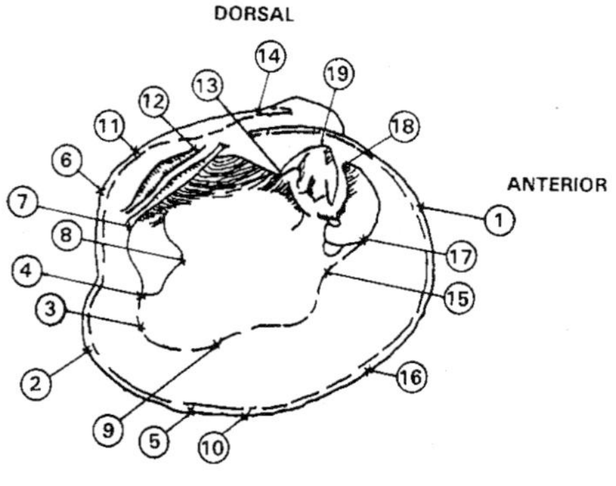

Fig. 1. Diagram of a unionid mussel showing the 19 landmarks used in this study. Nine linear variables for use in multivariate analyses were generated as the distance between the following landmarks: 1-2, 1-11, 1-17, 5-11, 8-17, 9-11, 9-12, 9-14, and 15-19. Shell thickness was added as a tenth variable.

Analysis Using Rotational Fit Methods

Of the available sample, 42 specimens (24 belonging to *O. jacksoniana*; 18 to *V. arkansasensis*), representing individuals collected from the range of sympatric occurrence in adjacent areas of Oklahoma, Arkansas, and Texas, were selected for analysis using rotational fit methods. The landmark data mentioned above, including 19 points for each specimen, were analyzed using the Generalized Rotational Fit (GRF) program (Rohlf and Slice, 1990a). Various trials were run, employing generalized and ungeneralized resistant-fit and least-squares methods, both with and without affine transformations (cf. Rohlf, 1990a). To interpret the results, the first and second principal component axes of fitted landmark locations with respect to a consensus configuration were plotted with a magnitude of one standard deviation in the respective direction; constant frequency ellipses, with a scale of two standard deviations, were also plotted (Rohlf and Slice, 1990b).

Analysis of Outline Shape

For a pilot analysis of outline data, we chose to employ Elliptic Fourier Analysis (EFA), which was used in a similar study (Ferson et al., 1985; see also Rohlf, 1990b). Twelve specimens of each species, taken from the range of sympatry, were used in this analysis. Sample sizes were admittedly small; however, they allowed us to evaluate the applicability of the techniques. The outline of each specimen was traced, reversed where necessary, and xerographically enlarged 200%. Point coordinates along the margin of each specimen were captured using a digitizing tablet; each specimen was represented by 110 to 120 points, concentrated in zones of tight curvature (cf. Bookstein et al., 1985). Elliptic Fourier Analyses, using a program written by Rohlf (1990c), were undertaken, varying the number of harmonics employed (and, therefore, preciseness of fit of reconstructed to original outline). The results were made invariant to specimen starting point, size and orientation. Each analysis resulted in 4N-3 coefficients, where N is the number of harmonics employed; of these, the first three are uninformative (Ferson et al., 1985) and were not subjected to further analysis. Principal components (using a varimax rotation and the correlation matrix) and discriminant functions analyses were performed on the remaining coefficients derived from the EFA, again using the SYSTAT software package.

RESULTS

Analysis of Shell Measurements

Principal components analysis of ten shell measurements resulted in explanation of 97.2% of variance by the first three rotated components. The first axis, which we interpret as size, accounted for 49.6% of the variance. Seven of the 10 variables, all representing length and width measurements, had high loadings (> 0.60) on the first axis. Specimen scores for both species are widely distributed along this first component (Fig. 2, top), reflecting the size variability of each in the original sample. One variable had a high loading (0.85) on the second axis, which explained 26.5% of the variance. This variable was a measurement of the short distance from the anterior valve margin to the ventral surface of the anterior adductor muscle (Fig. 1), more a measure of shape. Although specimen scores for both species overlap on the second axis (Fig. 2, top), *O. jacksoniana* is broadly ranging, while distribution of specimens of *V. arkansasensis* is restricted to lower values, generally below 1.0. We interpret this to indicate that, for this character at least, variability of the former species is much greater. Rotated component 3 explained 21.1% of the variance. Only one variable, shell thickness, had a relatively high loading (0.79) on this axis. Some separation of the two species is apparent from the distribution of specimen scores on this third axis (Fig. 2, bottom). Specimens of *V. arkansasensis* tend to have lower scores than those of *O. jacksoniana*. As with the case for adductor margin distance (component 2), shell thickness is notably less variable in *V. arkansasensis* than in *O. jacksoniana*, with specimen scores for the latter species being broadly distributed along axis 3. These observations are supported by the bivariate distribution of logged shell length (which had a high loading on the first principal component and was therefore taken as an approximation of size) against thickness (Fig. 3). For *O. jacksoniana*, these variables are highly correlated (Pearson correlation = 0.887, probability of non-linear relationship < 0.0001; N = 105), whereas there is no significant linear relationship of shell thickness to length in *V. arkansasensis* (Pearson correlation = 0.298, probability of non-linear relationship < 0.215; N = 19).

Analysis Using Rotational Fit Methods

Analysis using resistant-fit and least squares procedures produced generally similar results; because resistant-fit methods are more robust to localized variation in landmark position (Rohlf and Slice, 1990b), we confine our discussion to the results of analysis based on this method. The results of generalized resistant-fit analyses, both with and without affine transformation, were similar; we interpret this to indicate that little of the variation represents uniform shape change. The configurations of consensus centroids for the two species are highly similar (Fig. 4), with the most important differences occurring at landmarks 1, 15. and 16. As indicated by the general size of the ellipses, *O. jacksoniana* appears to be somewhat more variable at landmarks 1, 17, 18, and 19, whereas *V. arkansasensis* may be more variable at landmarks 12, 4, and 9. In addition, some differences in pattern of within-group variation is apparent, as reflected in divergence, between species, of the direction (landmarks 1, 2, 3, 7,

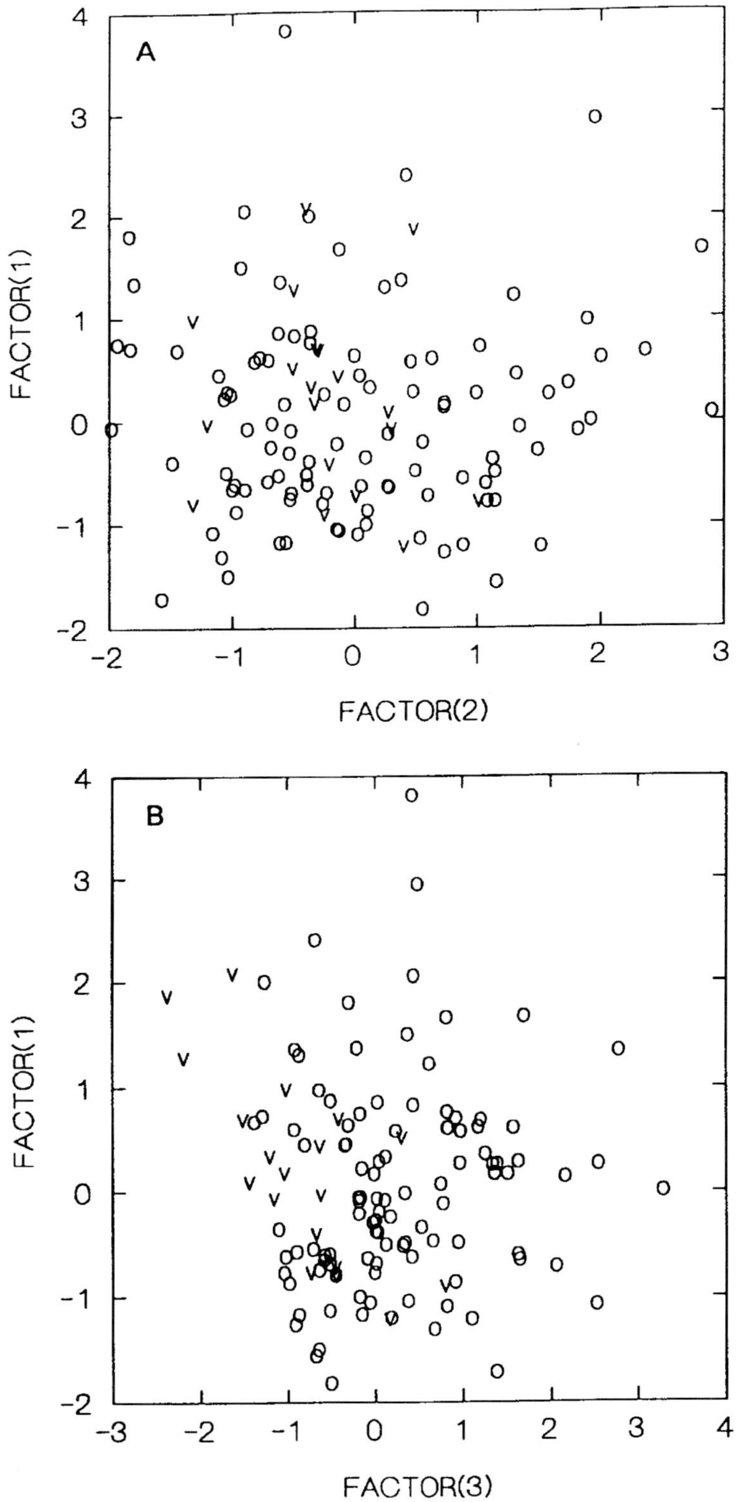

Fig. 2. Plots of the distribution of *Obovaria jacksoniana* (O) and *Villosa arkansasensis* (V) on the first three axes in a principal factor analysis. See Fig. 1 for variables.

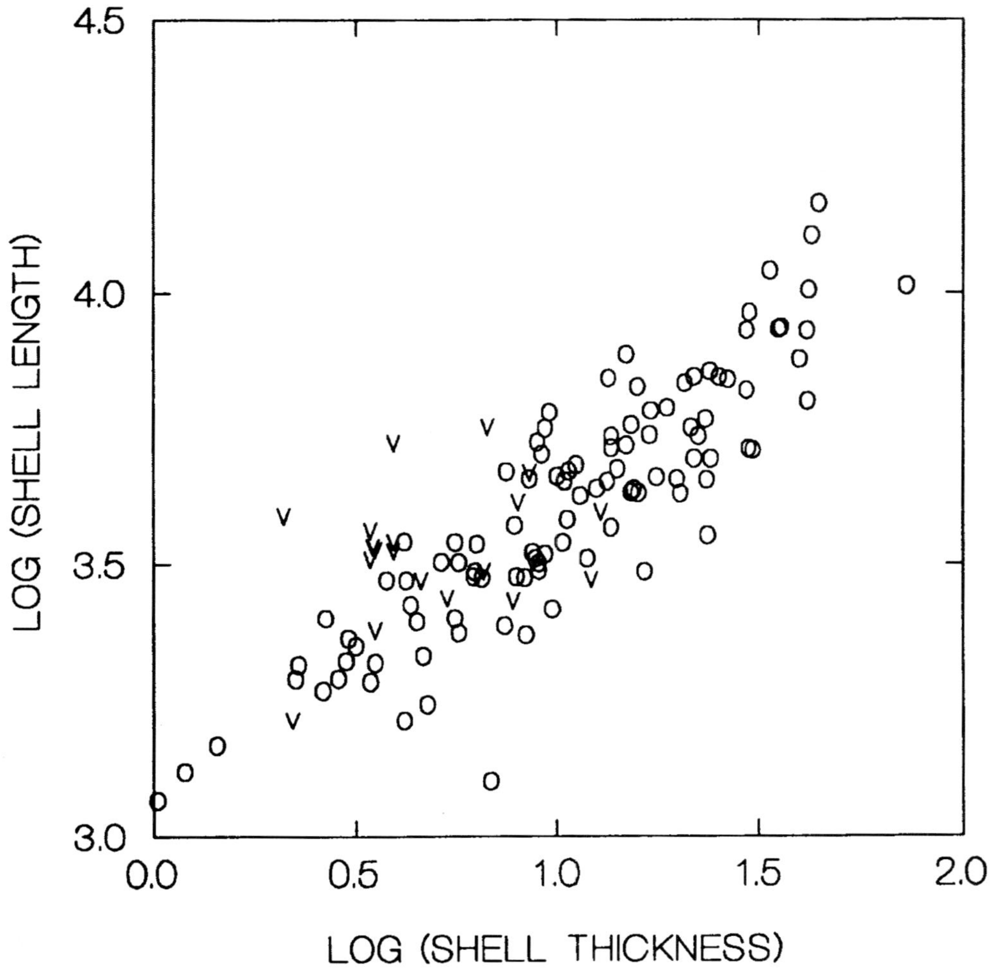

Fig. 3. Plot of the correlation between shell length and thickness for *Obovaria jacksoniana* (O) and *Villosa arkansasensis* (V), r = 0.887, n = 45.

16) and relative magnitude (landmarks 1, 9, 12, 17, 18, 19) of the principal axis.

Elliptic Fourier Analysis

An example of the digitized outline of a specimen of *V. arkansasensis* (Oklahoma Museum of Natural History specimen 1850.4) is shown in Fig. 5. The reconstructed outline based on EFA using 10 harmonics follows the original outline very closely (Fig. 5, top). Principal components analysis based on the 37 coefficients derived from the 10-harmonic EFA produced ambiguous results, with the first three rotated components accounting for only 41.8% of the total variance. EFA based on three harmonics also produced a reconstructed outline that closely mirrors the original data (Fig. 5, bottom). Principal components analysis of the nine coefficients so obtained resulted in explanation of 73.9% of the total variance by the first three rotated components, a considerable improvement. Factor 1, which explained 29.9% of the variance, included high loadings (> 0.75) for four of the variables (coefficients); the second component accounted for 22.6% of the variance and had high loadings (> 0.75) for two variables; and 21.3% of the variance was explained by factor 3, for which two other variables had high loadings (> 0.75). One variable did not have a loading in excess of 0.5 on any of the

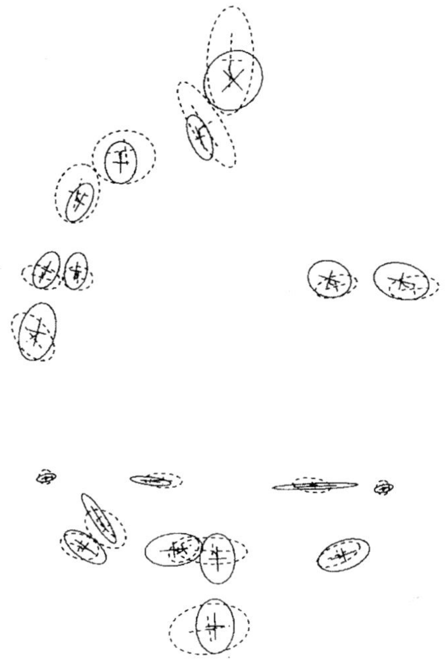

Fig. 4. Results of generalized resistant-fit analysis of valves of *Obovaria jacksoniana* (solid lines) and *Villosa arkansasensis* (dashed lines). Axes are the first two principal components of fitted landmark localities with respect to a reference landmark, scaled to have lengths of one standard deviation. Constant frequency ellipses show two standard deviations.

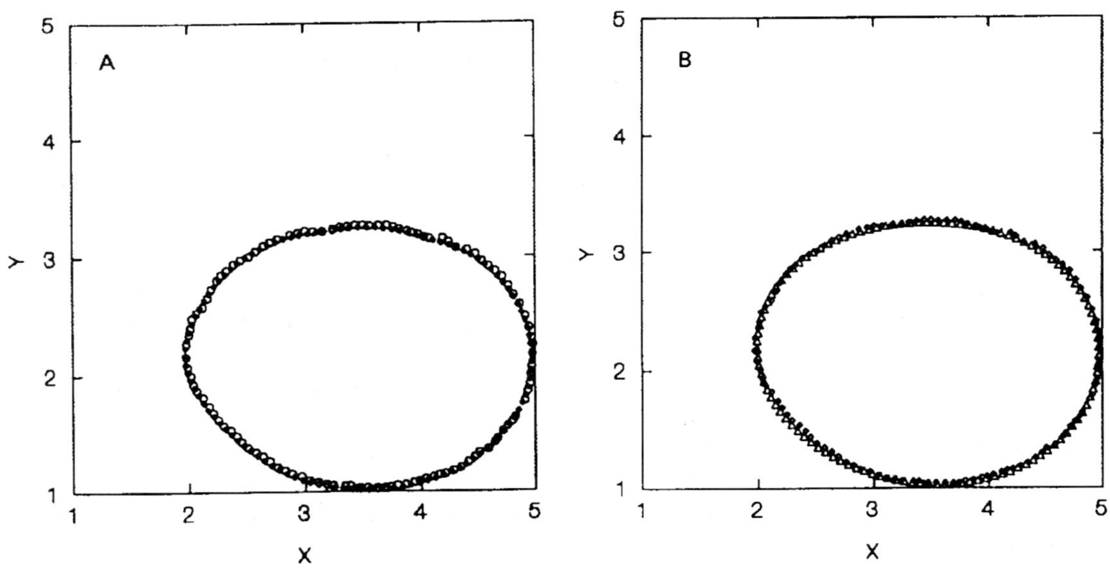

Fig. 5. Reconstructed X,Y coordinates based on Elliptic Fourier Analysis using: (A) ten harmonics (circles) and (B) three harmonics (squares) of original data (triangles).

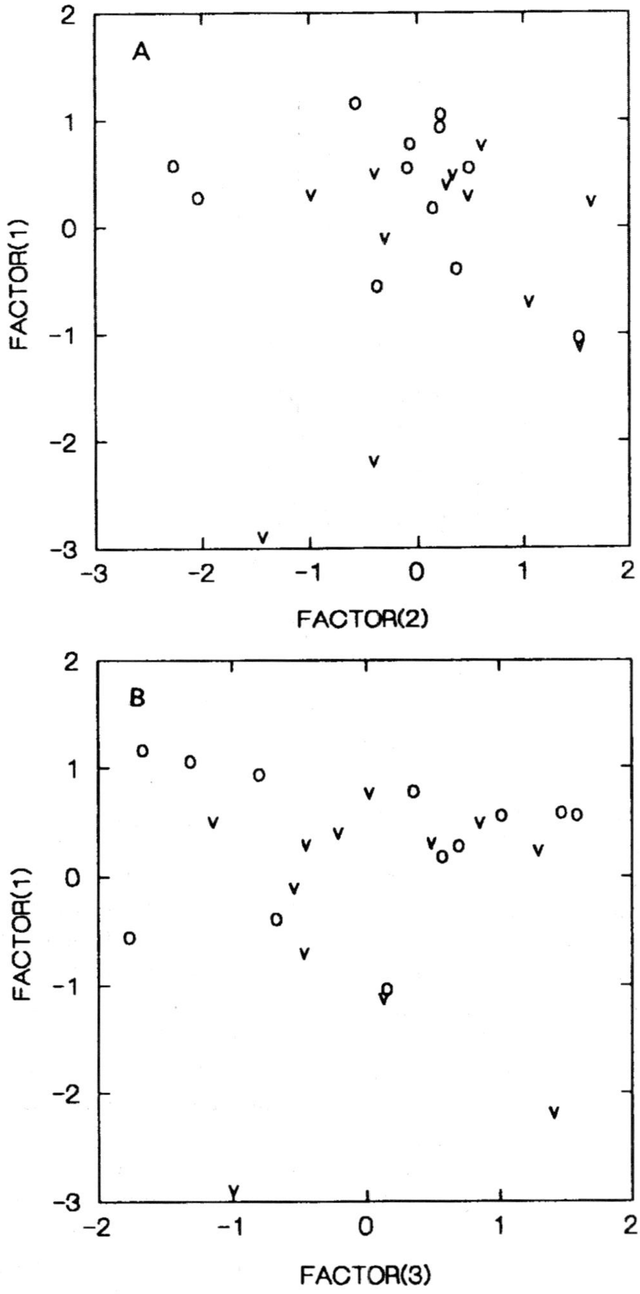

Fig. 6. Plot of the first three factors from principal components analysis based on nine coefficients produced from Elliptic Fourier Analysis using three harmonics. 6 + O = *Obovaria jacksoniana*; V = *Villosa arkansasensis*.

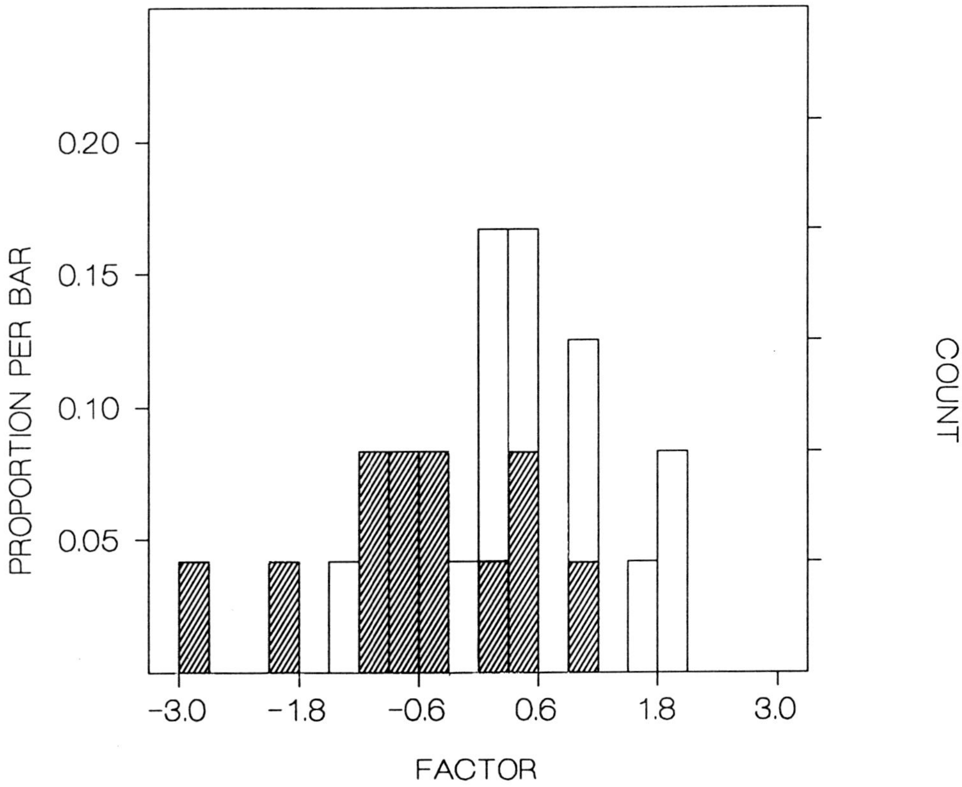

Fig. 7. Plot of *Obovaria jacksoniana* (open bars) and *Villosa arkansasensis* (hatched bars) in discriminant space.

first three components. Specimen scores overlap broadly on all three axes (Fig. 6), but *V. arkansasensis* appears to be more widely ranging on the first axis, whereas *O. jacksoniana* is more broadly distributed on axis 2.

The density distribution of specimens on the first factor resulting from a discriminant functions analysis is shown in Fig. 7. Both univariate and multivariate test statistics for the discriminant functions null hypothesis contrasting the two species were insignificant, which we interpret as a result of the facts that they are closely similar and the sample size of this pilot was small (N = 24). For this reason, we urge caution in interpreting the results and consider them to be suggestive rather than indicative. In terms of classification prediction, however, individuals were correctly assigned to species in 75% of cases (Pearson chi-square test statistic, $P < 0.01$, N = 24); 2 specimens of 12 *O. jacksoniana* specimens and 4 of 12 *V. arkansasensis* were incorrectly classified.

DISCUSSION

This preliminary exploration of differences between two sympatric mussels based on shell morphology has demonstrated the usefulness of both landmark and elliptical Fourier techniques when few discrete morphometric characters are present.

Multivariate analyses based on measurements failed to discriminate these species. It did show, however, that species differ in variability of certain characters, particularly the distance from the anterior margin to the ventral surface of the anterior adductor muscle and in valve thickness. For the latter character, at least, there is a strong correlation with size in *O. jacksoniana*, but not in *V. arkansasensis*. This suggests that shell thickness varies ontogenetically in the former species, but that other factors, perhaps ecomorphologic in nature, may control its expression.

The differences in variability between the two species is more clearly explained by results of the GRF, in which the direction and magnitude of variation at individual landmarks can be appreciated. The difference in relative position in the anterior adductor muscle, for example, is expressed as differing length and orientation of principal axis 1 for landmark 17. There was no evidence for uniform shape changes in either species. The EFA fits a function to a sequence of points such that the function can be made to mathematically model an original outline with arbitrarily defined precision (Rohlf, 1990b). The resulting parameters of the fitted functions, which are used in subsequent multivariate analyses, cannot be directly related to morphologic attributes.

One of the most interesting results of the EFA is methodologic in nature. Because we know of no practical quantitative means of assessing the accuracy of an outline reconstructed by EFA, it is difficult to determine how many harmonics are appropriate; results can be compared visually or analyzed in a trial and error fashion. We initially followed Ferson et al. (1985) in using 10 harmonics because those authors were successful in distinguishing different genetic populations of the marine clam *Mytilus edulis* with that number. However, the fact that less than half of the total variance was explained on the first three axes suggests that the data include a large component of random shape variation. Because the outlines based on ten harmonics followed the originals closely, we interpret the randomness as mathematical modeling of small irregularities along individual valve margins. This conjecture is supported by the fact that considerably better results were obtained by principal components analysis on EFA with only three harmonics, in which the first three components accounted for 73.9% of the variation. The difference in results between this study and that of Ferson et al. (1985) probably reflects morphologic differences among bivalves; the valves of *M. edulis* are thicker than those of the species we studied and, thus, would be less prone to nicks and other physical damage. The results obtained here indicate that the potential for random variation should be considered when mathematical modeling of outlines using EFA or similar techniques is undertaken, and that general pattern can be masked by too closely approximating outlines.

Although statistical discrimination between species in this study was precluded by small sample sizes combined with very similar shell morphology, tendencies toward morphological differences were achieved and prediction of species classifications with discriminate function analysis was moderately successful. These results support the application of curve fitting models to morphological problems when few landmarks are available.

ACKNOWLEDGMENTS

We thank F.J. Rohlf for fruitful discussions that led us to conduct this study, and C.M. Mather for enlightening us on the ecology and distribution of the two mussel species. E.K. Miller assisted with field work. We thank the Mississippi Museum of Natural Science, Oklahoma Museum of Natural History, University of Science and Arts of Oklahoma, and U.S. National Museum for kindly providing specimens for this study.

LITERATURE CITED

Bookstein, F.L. 1990. Introduction to methods of landmark data. Pp. 215-226, *in* Proceedings of the Michigan Morphometrics Workshop (F.J. Rohlf and F.L. Bookstein, eds.). Special Publication 2, University of Michigan Museum of Zoology.

Bookstein, F.L., B. Chernoff, R. Elder, J. Humphries, G. Smith, and R. Strauss. 1985. Morphometrics in evolutionary biology. The geometry of size and shape change, with examples from fishes. The Academy of Natural Science of Philadelphia, Special Publication No. 15, 227 pp.

Davis, G.M. 1983. Relative roles of molecular genetics, anatomy, morphometrics, and ecology in assessing relationships among North American Unionidae (Bivalvia). Pp. 193-222, *in* Protein Polymorphism: Adaptive and Taxonomic Significance (G.S. Oxford and D. Rollinson, eds.). Systematics Association Special Volume No. 24, Academic Press, New York.

Davis, G.M., and S.L.H. Fuller. 1981. Genetic relationships among Recent Unionidae (Bivalvia) of North America. Malacologia, 20:217-253.

Davis, G.M., W.H. Heard, S.L.H. Fuller, and C. Hesterman. 1981. Molluscan genetics and speciation in *Elliptio* and its relationships to other taxa of North American Unionidae (Bivalvia). Biological Journal of the Lynnean Society, 15:131-150.

Ferson, S., F.J. Rohlf, and R.K. Koehn. 1985. Measuring shape variation of two-dimensional outlines. Systematic Zoology, 34:59-68.

MacLarnon, A.M. 1989. Applications of the Reflex instruments in quantitative morphology. Folia Primatologia, 53:33-49.

Mehlhop, P., and C.A. Vaughn. In press. Threats to and sustainability of ecosystems for freshwater molluscs in the United States. In Conference on Sustainable Ecological Systems. USFS Rocky Mountain Range and Forest Experiment Station, Technical Report, Fort Collins, CO.

Neves, R.J. 1992. A state-of-the-unionids address. Symposium on the Conservation and Management of Freshwater Mussels: 9 (Abstract).

Rohlf, F.J. 1990a. Rotational fit (Procrustes) methods. Pp. 227-236, *in* Proceedings of the Michigan Morphometrics Workshop (F.J. Rohlf and F.L. Bookstein, eds.). Special Publication 2, University of Michigan Museum of Zoology.

Rohlf, F.J. 1990b. Fitting curves to outlines. Pp. 167-177, *in* Proceedings of the Michigan Morphometrics Workshop (F.J. Rohlf and F.L. Bookstein, eds.). Special Publication 2, University of Michigan Museum of Zoology.

Rohlf, F.J. 1990c. A simple program to compute elliptic Fourier coefficients given an outline represented by a sequence of x,y-coordinates of points. Software distributed with Proceedings of the Michigan Morphometrics Workshop (F.J. Rohlf and F. L. Bookstein, eds.). Special Publication 2, University of Michigan Museum of Zoology.

Rohlf, F.J., and D. Slice. 1990a. Program for generalized rotational fitting (least-squares and resistant fit of two or more objects represented by x,x-coordinates). Software distributed with Proceedings of the Michigan Morphometrics Workshop (F.J. Rohlf and F. L. Bookstein, eds.). Special Publication 2, University of Michigan Museum of Zoology.

Rohlf, F.J., and D. Slice. 1990b. Extensions of the Procrustes method for the optimal superimposition of landmarks. Systematic Zoology, 39:40-59.

Way, C.M., A.C. Miller, and B.S. Payne. 1989. The influence of physical factors on the distribution and abundance of freshwater mussels (Bivalvia: Unionidae) in the lower Tennessee River. Nautilus, 103:96-98.

Wilkinson, L. 1992. SYSTAT for Windows: Version 5 ed. Systat Inc., Evanston, IL.

Patricia Mehlhop, New Mexico Natural Heritage Program, Department of Biology, The University of New Mexico, Albuquerque, New Mexico 87131-1091, and Richard L. Cifelli, Oklahoma Museum of Natural History and Department of Zoology, University of Oklahoma, Norman, Oklahoma 73019

Evaluation of Methods for Permanently Marking Kangaroo Rats (*Dipodomys*: Heteromyidae)

DANIEL F. WILLIAMS, WALTER TORDOFF III, AND DAVID J. GERMANO

Abstract

We analyzed the loss and failure rates, misreading errors, mortality during applying and reading marks, and relative costs of marking kangaroo rats, *Dipodomys heermanni*, *D. ingens*, and *D. nitratoides*, during an 8-year study. Tattooing was considered to be an unacceptable method because of the high rates of misread marks (3.9%) and mortality during handling (0.5%). Cheek-pouch tags showed unacceptable problems with infections and injuries from the tags. Ear tags were the least expensive, but had the highest rates of loss (a range of 1-58% for time intervals of 6 days to > 1 year) and a relatively high rate of reading errors (0.5%). Infections often were associated with tags after they had been attached for several weeks or months. Passive integrated transponders (PIT tags) had the highest equipment and supply costs, but required the least amount of time to identify marked animals. They had lower loss rates (1-7%) than ear tags and a very low misreading rate (0.005%). Readers that store numbers can eliminate all misreading and transcription errors. No infections or other pathology were noted for PIT tags. The rate of mortality during handling for applying and reading ear and PIT tags (most were marked by both methods) was 0.04%; all mortalities involved the smallest-sized species, *D. nitratoides*.

INTRODUCTION

Most long-term population studies require that individual animals have a unique, permanent identifying mark. Statistical models for estimating population size assume that marks are read correctly and the animals do not lose their marks during a study (Pollock et al., 1990). Different marking methods are prone to different rates of loss and reading error. Further, some methods of marking may risk injury to the animal during handling and cause infection, reducing fitness of marked individuals for these or unforeseen reasons.

Methods of marking small mammals have been analyzed and reviewed by several researchers. These include reports by Fullagar and Jewell (1965), Twigg (1975), Bazhenov et al. (1984), and Genoud (1984). Currently, the most common methods of marking small mammals are toe-clipping, attaching ear tags, and injecting passive integrated transponders (PIT tags). The *ad hoc* Committee on Acceptable Field Methods in Mammalogy (ASM; 1987) included tattooing, ear tagging, and toe clipping among the acceptable methods. Despite the ASM endorsement, there is a virtual lack of information on the potential problems and efficacy of tattooing small mammals (Twigg, 1975). The California Department of Fish and Game does not permit toe-clipping of small mammals with protected status, so this method was unavailable to us. Though we do not review toe clipping in detail, comparative information can be found in Stoddart (1970), Fairley (1982), Pavone and Boonstra (1985), Korn (1987), Wood and Slade (1990), and Bias et al. (1993).

Ear tags are lost at rates that may be unacceptably high, depending upon the species, duration, and objectives of the

Life Among the Muses: Papers in Honor of James S. Findley. Edited by T.L. Yates, W.L. Gannon & D.E. Wilson.

study (Twigg, 1975), but are inexpensive and easily applied by one person in the field. Metal tags also have been attached to the edge of the opening of cheek pouches of pocket gophers (Geomyidae; J.L. Patton, pers. comm.), but the effects of the tag on manipulation and transport of food and rate of tag loss are unknown. For pocket gophers and other species such as pocket mice (*Chaetodipus, Perognathus*), tags cannot be secured to the ears because they are too small.

Plastic- and glass-encased PIT tags, implanted under the skin or intra-abdominally, have been used to mark a variety of vertebrates. Transponders are programmed with a unique code during manufacture. The battery-free device is activated by a low-frequency electromagnetic signal with an external reader. The signal energizes the transponder and causes it to transmit its code (Fagerstone and Johns, 1987). Since its first use for marking wild vertebrates, the technology has been modified several times, and there are different transponder frequencies and designs available. Yet data on field use and failure are mostly unavailable (Fagerstone and Johns, 1987; Camper and Dixon, 1988; Rao and Edmondson, 1990; Ball et al., 1991; Germano and Williams, 1993; Schooley et al. 1993).

Also of concern, but seldom discussed or analyzed, are the personnel time to apply and read marks, equipment costs, and the number of errors made in reading marks. Error in reading codes from toe-clipped animals was mentioned by Le Boulenge–Nguyen and Le Boulenge (1986). Their and our (D.F. Williams, unpubl. observ.) experiences are that both novices and veteran field workers may make errors in marking and reading toe-clip codes at rates that could compromise the objectives of the study. Also, natural amputations occur at unknown rates in different populations of small mammals, which introduces possible additional error in identifying animals.

Rates of tag loss and reading error, effects on fitness, ethical issues such as minimizing pain to the animal, and costs should be considered when selecting a marking method. Yet, surprisingly little information is available to evaluate marking methods for most groups of small mammals. Without adequate information, researchers and agency personnel who issue permits are likely to select traditional or low-cost marking methods. Cost often is the overriding factor.

During a long-term demographic study of endangered giant kangaroo rats (*Dipodomys ingens*), we were initially required by permit-issuing agencies to mark by tattooing, but we changed methods of marking as accumulating experience showed unacceptable problems with tattooing and tagging cheek pouches. Because we lacked information on the failure rates for ear- and PIT-tagging, we simultaneously used two tagging methods to reduce the number of individuals with lost identity, and to gather information on tag losses for both methods. In a subsequent demographic study of Tipton (*D. nitratoides nitratoides*) and Heermann's (*D. heermanni*) kangaroo rats, we continued to double tag. While this paper was in review, we continued to collect data and updated values from two sites to better compare loss rates, but did not update other statistics. Herein we report and discuss data and observations on four methods of marking kangaroo rats: tattooing, attaching metal tags to cheek pouches and ears, and injecting passive integrated transponders.

MATERIALS AND METHODS

Field studies were conducted on the Carrizo Plain Natural Area, San Luis Obispo Co., California (*D. ingens, D. nitratoides*), and Pixley National Wildlife Refuge, Tulare Co., California (*D. heermanni, D. nitratoides*). Animals were captured between 6 July 1987 and 25 February 1995 on 12 100- to 196-trap grids and from trap stations at burrow systems spread over an area of about 40 ha. A total of 2,884 kangaroo rats were marked; they were captured a total of 26,707 times. Some animals were marked by all four methods. Cost of equipment and supplies are 1993 dollars and do not include shipping or taxes. Manufacturers and suppliers are listed where appropriate, but this does not denote endorsement of a particular supplier or product. Marking methods changed over the course of the studies and are described below in the sequence of their adoption.

Tattooing

In the field, 148 kangaroo rats (97 *D. ingens* and 51 *D. nitratoides*) were tattooed between July and October 1987. The tattooing machine was custom-made. Power to operate the device in the field was generated by a 110-V portable, gasoline-powered generator. A 25-V transformer with an in-line rheostat controlled power to the magnets. Two carbon-based black inks, India and Higgins Eternal, and Higgins Waterproof Red, No. 4085 (Faber Castell Corp., Newark, NJ 07107 USA), were used. One hundred animals were tattooed with India, 34 with Eternal, and 50 with Waterproof Red ink.

Two *D. heermanni* and one *D. nitratoides* were tattooed in the laboratory to practice techniques and find the best site for tattooing. These three laboratory animals and 14 field-collected *D. nitratoides* first were anesthetized with halothane (2-bromo-chloro-1, 1,1-trifluoroethane; Halocarbon Laboratories, Inc.), a fast-acting respiratory anesthetic. During the period when we were tattooing animals, all animals captured in the field were brought to a central point for weighing and reading or applying tattoos.

Efficient handling of both anesthetized and unanesthetized animals for tattooing required that one person hold the animal while a second tattooed the ear pinna. Because animals recovered quickly from effects of halothane, firm, two-handed restraint was necessary—one hand holding the animal by the skin on the neck, the other holding the legs and thighs. Unanesthetized animals were difficult to restrain for tattooing without causing them injury.

Tattooed marks were the letters A-Z, applied singly to the left or right ear, then letters on both ears. A few whose tattoos became illegible due to damaged ears were retattooed on the pads at the base of the digits of the manus. Toe pads were marked with a large dot, using a numbering sequence commonly used for toe clipping (DeBlase and Martin, 1981).

Cheek-pouch Tags

Size-1, monel, numbered, self-piercing tags (style 4-1005-1, National Band and Tag Co., Newport KY 41072 USA) were attached to the anterior edge of the cheek pouch with applicator pliers (style 4-1005S-1). Tags were applied in the field at the site of capture.

Ear Tags

Size-1, self-piercing tags were applied to ear pinnae using applicator pliers. Initially, a single tag was applied to one ear in the field at the site of capture. Subsequently, from approximately January 1989 through September 1992, an identically numbered tag was applied to each ear. After September 1992, only a single ear was tagged.

PIT Tags

Passive integrated transponders were injected subdermally using a 2.54-cm long, 12-gauge, Luer-Lok stainless-steel needle (Jorgensen Laboratories, Inc., 2198 W. 15th Street, Loveland, CO 80538 USA) in a modified, plastic, disposable, Luer–Lok, 3-cc syringe (Fig. 1). Modifications of the syringe included removing the flexible gasket at the tip of the plunger from its stalk and reinserting the gasket in the barrel of the syringe; cutting a straight slot through one side of the barrel of the syringe and inserting a plunger formed from brass welding rod approximately 90 mm long. The plunger was made by heating and bending it as shown in Fig. 1. The rod was inserted into the syringe through the slot cut in the barrel. Its tip was forced through the gasket and out the opening at the tip of the syringe.

Transponders were glass-encapsulated, approximately 11 mm long and 2.1 mm in diameter and activated at 400 kHz. Transponders and readers were manufactured by Destron Identi-

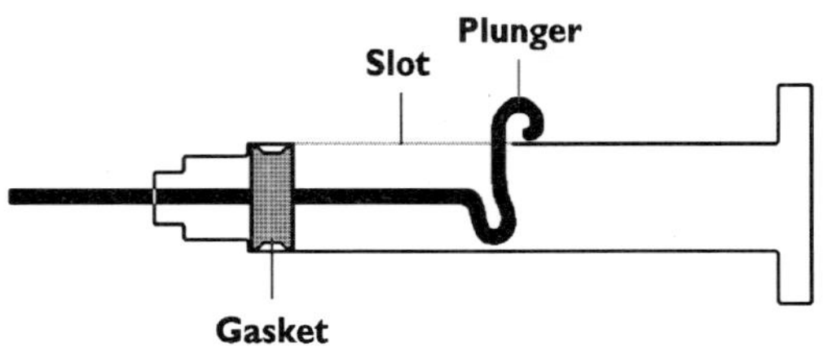

Fig. 1. Diagram of 3-cc, Luer-Lok plastic syringe, modified to inject passive integrated transponders (PIT tags). A slot about 2 mm wide and 20 mm long was cut in the barrel of the syringe from the gasket to behind the position of the plunger. The plunger was made by heating and bending a brass welding rod 1.62 mm in diameter.

fication Devices, Inc. and obtained from Biosonics, Inc. (3670 Stone Way N., Seattle, WA 98103 USA).

PIT tags and needles were sterilized by soaking in 70-100% isopropyl alcohol for several minutes. Tags were then coated with triple-antibiotic ointment and inserted into the tip of the needle. The coating of ointment kept the PIT tag from falling out of the needle. Loaded needles were capped with a segment of tygon tubing a few mm longer than the shaft and stored in sterilized plastic 35-mm film canisters. Capping and storing needles kept them clean and prevented their points from being dulled. Needles were sterilized and reused repeatedly until they became dull.

Three *D. heermanni* and four *D. nitratoides* were injected with PIT tags in the laboratory. They were sacrificed and autopsied one month later to determine effects of tagging.

A single worker injected animals with PIT tags in the field. Animals were tagged at their site of capture. They were removed from traps into a cloth bag, then restrained by grasping the loose skin on their back with one hand so that the skin of the animal's rump area was held between the thumb and forefinger (the reverse of the normal grip). The needle was slipped through the skin being held taut by the fingers of the gripping hand. As the PIT tag was injected, it was gripped under the skin and pushed forward and to the side of the injection site. The needle was then withdrawn as the fingers of the gripping hand held the animal's skin against the shaft of the needle. The skin at the site of injection then was squeezed tightly for a few seconds while the PIT tag was being read.

Error Detection, Failures and Statistical Analysis

Tattoos that became indecipherable because of damage to ears or disappearance of ink were scored as marking failures. Animals with a torn cheek pouch but no tag were scored as having lost a tag. Some of these animals also had been tattooed, so the lost tag was verifiable and the animal's identity was retained. Animals with a torn ear were scored as having lost a tag if the damage was consistent with having been tagged. Ears with tags that had been in place for a few weeks before loss typically had a rounded hole at the medial edge of the tear. Since most ear-tagged individuals initially had a tag on each ear and a PIT tag, almost all losses were verifiable and the identity of the animal was retained. During the first year of use, we attempted to determine if the PIT was lost or inoperable by carefully feeling over the animal's body for a tag. Because almost all cases of "failure" within about 30 days of injection were from lost tags, we ceased spending much time feeling for an inoperable tag before retagging the animal. All lost cheek, PIT, and ear tags were scored as marking failures. Inoperable PIT tags also were scored as marking failures, though the nature of failure was recorded whenever it was determined.

Misread tattoo marks, cheek, or ear tags were identified by comparing data on field forms with permanent records for an animal. When discrepancies in data for a marked animal were found, they were reviewed to find the erroneous datum. Often, tattooed animals had not yet been released, so the source of error could be identified. Because tagged animals were released after examination at the site of capture, there was no opportunity to immediately identify the source of error. For animals with both ear and PIT tags, misread or mistranscribed ear or PIT tags were identified by discrepancy between the two. Because the PIT readers electronically stored the numbers read, these records were uploaded to a computer, printed, and checked against field forms to resolve the source and type of error.

Loss rates of PIT and ear tags were compared by species overall and temporally: recaptures within six days of marking (most census sessions lasted five or six days); recapture between 6 and 42 days ("monthly" censuses varied from 4-6 weeks apart depending on weather); recaptures between six weeks and one year of marking and recapture more than one year from marking. Animals that were not recaptured within an interval were not included in the tallies. Because most animals were double or triple-tagged (one or two ear tags, one PIT tag), few animals with a lost tag were unidentifiable. Error and loss rates were compared statistically using the Row by Column G-test for independence (Pimentel and Smith, 1990). Significant values were $p \leq 0.05$.

RESULTS AND DISCUSSION

Tattooing

Animals were captured and handled a total of 781 times (*D. ingens* 542; *D. nitratoides* 229) between 6 July 1987 and 10 October 1987, when we stopped tattooing. Some problems associated with tattooing and handling animals apparent in the first months after adopting this marking method are summarized in Table 1.

The inner surface of the ear was sparsely covered with small hairs, had relatively little dark pigment, and the skin was backed by a layer of cartilage, making this the best site for a tattoo. Tattooing the inner surface of the ears of the three captive individuals with a single letter on each ear was easy when the animals were anesthetized first and held by a second person. The marks, using suspended carbon-particle inks, were permanent and easily read four years after tattooing. After tattoo wounds had healed, reading tattoos was easier when the ear first was swabbed with water or dilute ethanol.

Applying a tattoo that did not damage ear tissue, yet was deep enough to be permanent, was more difficult. Penetration of the needles through the dermis into the cartilage layer resulted in wounding of the ear and, often, subsequent loss of the pigment when the scab fell off. These tattoos were usually legible as scars, however. On a few animals, ink moved into cracks and channels in the cartilage creating unintentional marks and causing difficulty reading tattoos. For *D. nitratoides*, wounding often was more serious than for *D. ingens*, resulting in loss of portions of the pinna and illegible marks. The ears of many kangaroo rats were also cut and torn by fighting. Kangaroo rats with extensive damage to ears that precluded retattooing the ear were tattooed on the front foot pads (the soles and pads of the hind feet are too densely furred for tattooing, so only a few individuals could be uniquely marked in this way).

The carbon-based inks were viscous and dried too quickly in field use. They also were messy and obscured the position of tattooed marks during application. If an animal moved and interrupted tattooing, the ear had to be cleaned before resuming. The red ink was less opaque and tattoos could be distinguished from surrounding ink, so cleaning the ear was not required when a subject moved during tattooing. Both carbon-based inks were permanent, but tattoos of animals marked with Higgins Waterproof Red ink began to disappear after about a week. These animals had to be retattooed with black ink when they were recaptured. Not all red marks remained legible and some letters had to be deciphered from scars. Tattoos of three individuals could not be read and they had to be assigned new letters. Because red ink was

Table 1. Summary of problems associated with marking *Dipodomys* spp. with tattoos. Individuals with damaged ears had tears or scars that occurred after marking and include injuries due to fighting and tattooing that rendered the marks illegible or difficult to read. Values in parentheses are number of times animals were handled and tattoos read and numbers of individuals. Individuals were retattooed if original marks were partly or wholly illegible, or if first tattooed with red ink. All data are from field-marked animals.

Problem	Species					
	D. ingens		*D. nitratoides*		Pooled Species	
Illegible marks leading to lost data	0.5%	(2 of 443)	1.7%	(3 of 176)	0.8%	(5 of 619)
Misread marks	2.9%	(13 of 443)	6.3%	(11 of 176)	3.9%	(24 of 619)
Individuals with damaged ears	9.3%	(9 of 97)	24.5%	(12 of 49)	14.4%	(21 of 146)
Individuals retattooed	33.0%	(32 of 97)	36.7%	(18 of 49)	34.2%	(50 of 146)
Mortality from anesthetic	—		14.3%	(2 of 14)	14.3%	(2 of 14)
Mortality during handling	0.4%	(2 of 542)	0.9%	(2 of 227)	0.5%	(4 of 769)

used relatively late in the field studies in 1987, 28 of the 50 animals marked with red ink were not recaptured in time to decipher their marks. For the next trapping sessions in February and March 1988, only 9 of the 28 marked individuals were identified. Probably less than half the other individuals were still present on plots, though, if the proportion of recaptures of these individuals was similar to the proportion of animals with black, carbon-particle ink tattoos.

Halothane proved to be too toxic to use in the field on kangaroo rats. The first two *D. nitratoides* died of exposures of less than 20 s. Subsequent administration of Halothane to 13 *D. nitratoides* was successful, but the animals typically recovered prior to completion of tattooing. Due to these problems, use of Halothane was terminated and all other kangaroo rats were marked without benefit of anesthetic.

Mortality during tattooing was relatively high (Table 1). Handling time was relatively long for previously marked animals because many tattoos were difficult to read and records had to be checked to verify identity of animals with ambiguous marks. Efficient processing of animals so that they could be released before daytime ambient temperatures approached a lethal level for *D. ingens* (about 35°C) required three or four workers to handle an average of 40 captives from two 100-trap grids.

The 0.8% loss from illegible tattoos (Table 1) was less than the 2-11% loss of ear tags reported in various studies of small mammals (Krebs et al., 1969; Le Boulenge–Nguyen and Le Boulenge, 1986; Stoddart, 1970; this study). However, because these data were based only on a 4-month period, additional losses and errors from illegible tattoos from damage to ears from fighting could be expected. The rate of errors from misreading tattoos exceeded errors for numbered tags. (see Tag Reading and Recording Errors).

Cheek-pouch Tags

A total of 112 kangaroo rats were tagged in a cheek pouch (107 *D. ingens*, 5 *D. nitratoides*) and 2.8% of them lost the tags within three months (three *D. ingens*, no *D. nitratoides*). Some animals retained cheek-pouch tags for about three to four years, when they disappeared from plots, but we discontinued attaching tags to cheek pouches after using this marking method for about three months in 1988. Problem tags (i.e., numbers turned inside; infections or tears in the skin) were removed and replaced by ear tags. Thus, we are unable to measure a loss rate over a longer period.

Problems associated with cheek-pouch tags were apparent soon after adopting this method. Though the tags were attached so that the numbered side was outside, many tags soon turned so that the numbered side was inside the pouch. Reading these tags required careful manipulation of the skin a few mm from the incisors. Of greater concern was massive infections in three animals at the site where the tag pierced the double layer of skin of the cheek pouch. We drained these abscesses and removed the tags with wire cutters, replacing them with ear tags. One animal tore a 2-cm opening in the skin on the side of the junction between head and neck, exposing its salivary glands and muscles. The tear probably resulted from catching a toe in the tag.

Ear Tags

Rates of ear tag loss differed significantly among the species (Table 2; $G = 44.35$, $p < 0.01$). The loss rates for *D. ingens* (10.6%) and *D. nitratoides* (8.0%; Table 3), captured at the same sites on the Elkhorn Plain, did not differ significantly ($G = 1.41$, $p = 0.23$), but the difference in their overall loss rates were highly significant ($G = 44.23$, $p < 0.01$). The loss rate for *D. nitratoides* was about twice that of *D. ingens*. The overall loss rate for *D. ingens* was significantly lower than that of *D. heermanni* ($G = 4.13$, $p = 0.04$; Table 2).

When loss rates were compared temporally, those of *D. nitratoides* were significantly greater than for the other species for all time intervals for which comparisons were appropriate (Table 4). About 38% of all individuals of *D. nitratoides* lost ear tags between 7 and 42 days of marking, and almost 58% lost ear tags between 42 days and one year from marking. All rates of loss between *D. nitratoides* and *D. ingens* were highly significant ($p < 0.01$), as were the loss rates for *D. nitratoides* and *D. heermanni*. In stark contrast, none of the loss rates differed significantly for *D. ingens* and *D. heermanni*.

Table 2. Numbers of tags applied and lost for three species of *Dipodomys* over all studies. Total Applied is the total of all tags applied, including retagged animals.

	Ear Tags					PIT Tags		
Species	Single Tagged	Double Tagged	Individuals Tagged	Lost[1]	Total Applied	Tagged	Lost	Total Applied
D. ingens	537	920	1,439	191 (9.1%)	2,532	1,278	38 (2.9%)	1,316
D. heermanni	301	5	306	39 (11.1%)	350	306	8 (2.6%)	314
D. nitratoides	767	137	904	225 (15.0%)	1,497	865	63 (8.7%)	928

[1] Animals with two ear tags were not retagged when one was lost.

Table 3. Numbers of tags applied and lost during specific studies of *Dipodomys* spp.

	Ear Tags			PIT tags		
Species	Applied	Lost	Loss	Applied	Lost	Loss
	Elkhorn Plain 1 Jan. 1988–3 Sep. 1992					
D. ingens	1,043	111	10.64%	532	16	2.84%
D. nitratoides	263	21	7.98%	117	0	0%
	Soda Lake 14 July 1989–25 Feb. 1995					
D. heermanni	26	0	0%	25	0	0%
D. ingens	1,089	73	6.70%	665	18	2.71%
	Pixley National Wildlife Refuge 3 Dec. 1992–7 Nov. 1994					
D. heermanni	340	39	11.47%	309	8	2.59%
D. nitratoides	1,155	435	37.66%	783	63	8.05%

D. nitratoides is the smallest kangaroo rat—adults weigh about 32-45 g; *D. heermanni* is intermediate in size (mean of about 72 g), and *D. ingens* is the largest (mean weight about 135-140 g). The much higher loss rate for *D. nitratoides* compared to *D. heermanni* and *D. ingens* (Table 3) suggests that the small, thin pinnae of *D. nitratoides* are not suited to retaining ear tags. The small ears of *D. nitratoides* make it difficult to properly attach a tag to them, especially if the tips of the applicator pliers are not ground off so that they can fit properly behind the ear. Yet, even with this modification and despite careful application, the number of lost tags was unacceptably high for this species.

For animals that were properly tagged (Fig. 2), most losses were caused by necrosis of tissue around the area where the tag pierced the

Table 4. Tag loss of *Dipodomys* spp. during four time periods. Values in parentheses are the number of animals that lost a tag and the total number of animals captured during that period. Values are pooled for all studies of > than one year duration.

Period	*D. heermanni* Tag		*D. nitratoides* Tag		*D. ingens* Tag	
(Days)	Ear	PIT	Ear	PIT	Ear	PIT
1-6	0% (0/132)	3.03% (4/132)	5.92% (17/287)	6.97% (20/287)	0.78% (2/255)	1.18% (3/255)
7-42	3.11% (5/161)	0.62% (1/161)	38.43% (191/497)	6.64% (33/497)	3.64% (12/330)	1.82% (6/330)
42-365	18.97% (33/174)	1.15% (2/174)	57.61% (227/394)	2.54% (10/394)	15.99% (51/319)	2.19% (7/319)
>365	6.25% (1/16)	6.25% (1/16)	0% (0/26)	0% (0/26)	11.43% (8/70)	2.86% (2/70)

pinna. The movement of the tag at this site sometimes caused a necrotic callous to build. This callous apparently occluded normal blood circulation, causing edema, infection, and death of additional tissue. Eventually, a large plug of tissue was lost from the site, freeing the tag to be easily ripped from the ear. The larger, thicker ears of *D. ingens* appeared to be more prone to this syndrome than those of *D. nitratoides* and *D. heermanni*, but its rate of lost tags was significantly lower than either of the latter species, suggesting their smaller, thinner ears were more prone to having loose tags ripped out.

Overall rates of loss of ear tags for the two larger kangaroo rats (Table 2) were within the range reported for other small mammals, but on the high side. These other studies were of shorter duration and some did not exclude animals marked but not recaptured, lowering the apparent loss rates. Reported loss rates were 2.2% and 5.1% for *Microtus ochrogaster* and *M. pennsylvanicus*, respectively (Krebs et al., 1969), 5% for *Arvicola terrestris* (Stoddart, 1970), 9.5% for *Mus musculus* (Bias et al., 1993), and 16% for *M. ochrogaster* (Wood and Slade, 1990). All these studies used the Style-1 tags manufactured by National Tag and Brand Co. Le Boulenge–Nguyen and Le Boulenge (1986) found losses of 4.1% for *Apodemus sylvaticus* and 8.1% for *Clethrionomys glareolus* for a tag made from metal surgical wound clips.

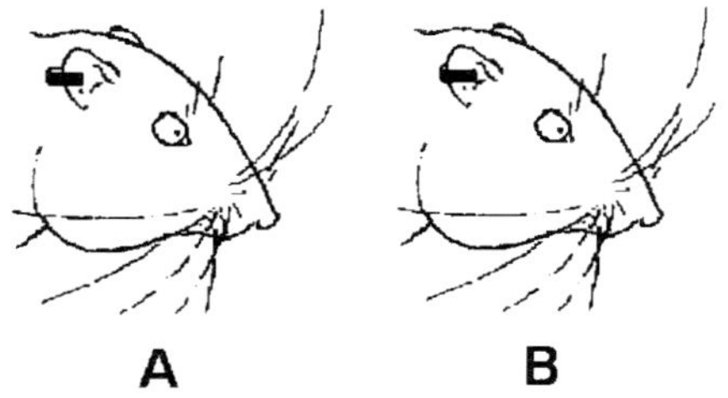

Fig. 2. Heads of kangaroo rats (Dipodomys), illustrating: A—improper; and B—proper attachment of ear tags.

These studies also did not partition loss rates temporally. When this is done, kangaroo rats have significantly higher loss rates for individuals that persist on plots for more than a single 5-6 day census, and rates increase significantly for animals that are present six weeks after marking (Table 4). Whether these rates are inordinately high compared to other small mammals cannot be determined, but these values at least strongly suggest that some objectives of long-term studies of kangaroo rats that rely solely on ear tags for marking individuals risk being compromised by tag loss.

PIT Tags

On autopsy of animals tagged in captivity, the tag of one *D. nitratoides* was found projecting partly out of the opening made by the needle and adhering to a scab that had formed around the edges. Tags of three of five other kangaroo rats had become loosely bound in connective tissue 30 days after injection, while tags of two still were freely mobile. There was no evidence of infection or inflammation either at the injection sites or in the subdermal region where the tags were located. Ball et al. (1991) and Rao and Edmondson (1990) also found no adverse tissue reactions to implanted transponders for laboratory rats (*Rattus norvegicus*) and mice (*Mus musculus*), respectively. Ball et al. (1991) did not detect effects on normal body weight gain nor food consumption. For field-tagged animals, no infection or other problems were noted in our study.

The overall 4.3% rate of loss of PIT tags from kangaroo rats (Table 2) was much less than the 30.4% failure rate for PIT tags reported by Fagerstone and Johns (1987) for captive domestic ferrets (*Mustela putorius furo*). PIT tags used in their study were encased in plastic. Failures were from moisture leaking through the plastic. Rao and Edmondson (1990) reported 5% (three lost and four failed tags) of 140 implanted in laboratory mice, though the tags used were from a different supplier and the glass casing was capped with polypropylene to elicit a mild tissue reaction so that it would be immobilized by connective tissue at the site of implantation.

The total failure rate (from losses and failures of all kangaroo rats combined) of PIT tags (4.3%) was significantly lower than that of ear tags (20.8%; Table 2; $G = 76.67$, $p < 0.01$; Table 2) and generally similar to within-year rates for *Spermophilus townsendii* (Schooley et al., 1993). The loss rates of the two larger kangaroo rats, however, were lower than the pooled loss rate (3.4%) for *S. townsendii*, while that of *D. nitratoides* was almost twice as great. The rates of loss of PIT tags between kangaroo rat species were significantly different when data for all sites were pooled ($G = 19.81$, $p < 0.01$). The rates of loss for *D. ingens* and *D. heermanni* were lower than that of *D. nitratoides* ($p < 0.01$ for both comparisons), but the rates of the two larger species did not differ ($G = 0.10$, $p = 0.75$). Yet, loss rates did not differ for *D. ingens* and *D. nitratoides* in the Elkhorn Plain study ($G = 0.235$, $p = 0.633$; Table 3).

Only three PIT tags were known to have become inoperative in animals at the Elkhorn Plain Ecological Reserve, though we did not attempt to find the source of "failure" of about half the tags. In contrast, later batches of tags, made by a different unit of the supplier, were much more prone to failure. Eighteen of the latter were known to have failed, though we suspect many more failures were not distinguished from lost tags. Inoperative tags probably failed by leaking moisture, either by the glass cover being cracked or because of improper sealing of the ends after inserting the transponder during manufacture.

Losses can be minimized by taking care to inject the tag sufficiently posterior so that it can be pushed far forward and to the side of the injection site. We believe that firmly squeezing the injection site for several seconds after withdrawing the needle also reduces losses. Clipping the hair close to the skin at the injection site and applying liquid suture to the wound left by the needle probably would further reduce losses (Patrick A. Kelly, pers. comm.). This technique, however, greatly increases handling and the time to apply a tag, rendering it impractical when large numbers must be marked by each field worker. Soaking the tags for several days in an alcohol solution, then reading before use should identify bad transponders.

When battery power to readers became weak, the reader would fail to detect the tag on the first few passes. Thus, 10-12 animals (two were not recaptured for verification) were tagged more than once. The simultaneous signals from two or more PIT tags caused the reader to fail to respond. In at least five cases, this was interpreted as a lost or failed PIT, leading to injection of yet another tag. The two animals that were not recaptured for verification probably had three and four tags injected, but this was not noticed until data were being transcribed to permanent records. The extra PIT tags were located and removed from the other individuals.

Tag Reading and Recording Errors

The size-1 tags applied to ears and edges of cheek pouches were prone to reading errors. For 1988 and 1989, when cheek-pouch and ear tags were read every time the animal was captured, the misreading/recording error rate was

0.5% (23 of 4,771). This rate was significantly lower than the 3.9% misread tattoos (Table 1; G = 43.742, $p < 0.01$). The small size, font, and the deep stamping of numbers combined to make it difficult to distinguish readily some numbers (i.e., 4, 7, and 9), especially when the tag had a veneer of dirt. The greatest number of mistakes was due to tags being read upside down, resulting in errors for some number combinations (e.g., 66-99, 68-89, 606-909, 106-901). Other reading errors were caused by failing to record the thousand digit, which was on the folded edge of the tag, in a different plane from the last three digits. Having all personnel attach tags to the ear on the same side and in the same orientation should minimize the former type of error.

Reading and transcription errors were few for PIT tags, despite the complex, 10-digit hexadecimal numbers. Only five errors were detected out of 10,208 tags read (0.005%), a rate two orders of magnitude lower than for ear tags (G = 30.114, $p < 0.01$). Errors were readily identified and corrected when the ear tag number also was recorded. By ensuring the PIT tag readers are set to scan and store numbers and regularly uploading stored files to a computer, all such errors should be easily corrected. Having both ear and PIT tags provides the greatest margin of safety against loss of identity of marked animals and the need to repeat a study or trapping session because of the occasional failure of a reader's power supply.

Mortalities While Tagging and Reading Tags

Mortalities during tagging and reading tags were few—five animals died while tagging or reading tags out of 12,905 times animals with ear or PIT tags were handled. All involved *D. nitratoides*, which was handled 2,714 times—only 21% of the total for all kangaroo rats. Only the difference in mortality rates for *D. ingens* and *D. nitratoides* was significant in species-pair comparisons (G = 15.792, $p < 0.01$). The rate of mortality during applying and reading of tattoos (Table 1) was significantly greater than that for ear and PIT tags for all species (G = 11.229, $p < 0.01$).

Costs

Though we did not precisely measure time for each method, applying and reading tattoos and returning animals to the site of capture required three-four people and took the longest amount of time (typically 4-5 h) to operate two grids with a combined daily capture of about 20-60 animals. Two people could operate two grids with a combined daily capture of 120-160 animals in less time (2-4 h) using PIT and ear tags. The tattoo device cost $200, but this price would vary widely, as would a power supply to tattoo in the field. The transformer for converting electricity from AC to DC was made from spare parts and was estimated to cost about $30 if the parts were purchased new. The cost of ink was negligible.

When animals only had to be handled to read their mark, reading PIT tags required less than half the time needed for reading ear tags, reducing personnel costs. The newer model PIT tag reader (HS5105L118K110) could search stored records and immediately signal (by a double beep from the scanner) recaptures, reducing time to process animals to about half that when using the old style readers. Ear tags were $0.08 each in lots of 500 and applicator pliers cost $13.50; no other costs were associated with ear tagging. PIT tags were $5.00 each in lots of 500. Reader prices varied: the newer model cost $1,130; the larger, bulkier model (HS5101) cost $1,888. Operating costs included electricity to recharge batteries (unmeasured) and periodic replacement of batteries. Batteries for model HS5101 (two lead-acid batteries) cost $60.00; batteries had to be replaced every 2-3 years. The newer model's battery life varied from about 1 to 18 months and averaged about 12 months. The small lead-acid battery was welded with epoxy to the electronic circuit board. Readers had to be returned to the manufacturer for replacement, which cost $150 and took from 10 days to more than three months. Needles cost $0.45 each and had to be replaced after 20-50 uses. Syringes cost $0.11 in lots of 100; a syringe was used hundreds of times. Other minor costs associated with PIT-tagging included costs of alcohol, tygon tubing, and containers for storing and sterilizing tags and equipment.

CONCLUSIONS

Tattooing required excessive handling of animals, was slow, resulted in unacceptable errors in reading marks, and required more time and a larger field crew to process animals and read their marks, compared to other marking methods. Mortality from handling during applying and reading marks was relatively high compared to handling during applying and reading ear and PIT tags. We consider tattooing to be unacceptable for marking kangaroo rats.

Cheek-pouch tags had a low short-term rate of loss, but this method also is unacceptable because of infections and injuries associated with the attached tag. Error rates for cheek-pouch tags were not distinguished from rates for the identical tags attached to ears.

Infections at the site of application of ear tags often were seen after the tags had been in place for several weeks or months. Infection sometimes led to loss of the tag. We have no data on other consequences of infection, but are concerned by the relatively common occurrence in *D. ingens*. No infections or other problems were found to be caused by PIT tagging.

Ear tags had a lower rate of misreading errors than tattoos, but much higher loss and misreading rates than PIT tags. Reading errors for PIT tags were correctable when the readers were set to store scanned numbers (not all models do). Loss of ear tags varied from about 7.5 to 15.0% without regard to duration. Time-dependent rates, however, showed increasing loss rates between the first six days and one year, such that about 58% of the smallest kangaroo rat (*D. nitratoides*) lost ear tags between six weeks and a year from marking. Though short-term loss rates for ear and PIT tags did not differ significantly, greater loss rates for ear tags were highly significant compared to PIT tags for periods longer than six days (Table 4).

For one time, short-duration population censuses, marking with a single ear tag is probably sufficient, especially for all but the smallest species of kangaroo rats (*D. nitratoides* and its relatives). Attaching a tag to each ear is advisable for all longer studies that do not simultaneously use two marking methods. The rates of ear tag loss for all three species for periods over six weeks may be unacceptably high for long-term demographic studies. For studies longer than about a week, we recommend double-marking with an ear and a PIT tag. If an ear tag is lost, a second tag should be applied to the opposite ear. If the second tag is lost and the PIT tag still is functional, applying a third ear tag is not advisable because both ears probably already have been torn and the third tag likely will be lost more quickly.

The cost of PIT tags and readers may make this marking method too expensive for many studies, though other manufacturers and models with lower costs are available now. The occasional failure of reader batteries requires that a backup reader always be available or that ear tags also be applied, else information can be lost or studies may have to be repeated. Because PIT-tags can be read without holding and inspecting the animal, however, the time required to operate a trapping plot is significantly less for studies where the animal does not have to be physically inspected each time it is captured. Savings in personnel costs can more than offset the expense of PIT tags and readers for large and long-term studies. Less handling also is desirable because it minimizes adverse effects to the animals and lessens the potential for biasing results.

ACKNOWLEDGMENTS

These studies were supported by funds from: California Department of Fish and Game, Nongame-Heritage Program, Endangered Species Tax-check-off funds, California Tobacco Tax funds, and Section-6 funds; The Nature Conservancy; USDI Bureau of Land Management; and USDI Bureau of Reclamation. Susan D. Williams helped with various phases of the field work and compiled, entered and verified computerized data bases. Dena Bos and Rebecca Roth also helped with data management. Mike Allenger, Karla Allred, Richard Anderson, Steve Bailey, Larry Berkoben, Ellen Cypher, Mary Davis, Laurissa Hamilton, Cheryl Johnson, Joshua Lerner, Tony Mann, Mark McFall, Stephen Meyer, Suzanne Nelson, Nancy Nicolai, Larry Saslaw, Ronald Schlorff, Bill Silva, Donna Snell, Linda Spiegel, Terrence Trassatti, and Cynthia Wirth contributed to field work, data management, or both. Allison Banks,

John Brode, Wendy Duron, Rosalie Faubion, Bill Lehman, Jim Lidberg, Gail Kobetich, Ted Rado, Richard Reiner, Andy Robinson, Larry Saslaw, Ron Schlorff, and George Sheppard contributed logistic support in obtaining equipment, funding, or permits. The PIT-tag injector was modeled after a design described by Galen Rathbun (pers. comm.). Work with the endangered *D. ingens* and *D. nitratoides* was authorized by USDI Fish and Wildlife Service permit number WILLDF-9 and a Memorandum of Understanding between the California Department of Fish and Game and California State University, Stanislaus.

LITERATURE CITED

ad hoc Committee on Acceptable Methods in Mammalogy. 1987. Acceptable field methods in mammalogy: Preliminary guidelines approved by the American Society of Mammalogists. Journal of Mammalogy, 68(4), supplement:1-18.

Ball, D.J., G. Argentieri, R. Krause, M. Lipinski, R.L. Robinson, R.E. Stoll, and G.E. Visscher. 1991. Evaluation of a microchip implant system used for animal identification in rats. Laboratory Animal Science, 41:185-186.

Bazhenov, A.V., V.N. Bol'shakov, and O.F. Sadykov. 1984. New method for marking small mammals. Soviet Journal of Ecology, 15:96-98 [English translation of Russian journal].

Bias, M.A., N.L. Breuner, and M. L. Morrison. 1993. House mice as indicators of potential marking effects on salt marsh harvest mice. 1992 Transactions of the Western Section of the Wildlife Society, 28:34-37.

Camper, J.D., and J.R. Dixon. 1988. Evaluation of a microchip marking system for amphibians and reptiles. Texas Parks and Wildlife Department, Research Publication, 7100-159:1-22.

DeBlase, A.F., and R.E. Martin. 1981. A manual of mammalogy with keys to families of the world. Second ed. W.C. Brown Company Publishers, Dubuque, IA, 436 pp.

Fagerstone, K.A., and B.E. Johns. 1987. Transponders as permanent identification markers for domestic ferrets, black-footed ferrets and other wildlife. Journal of Wildlife Management, 51:294-297.

Fairley. 1982. Short-term effects of ringing and toe-clipping on the recapture of wood mice (*Apodemus sylvaticus*). Journal of Zoology (London), 197:295-297.

Fullagar, P.J., and P.A. Jewell. 1965. Marking small rodents and the difficulties of using leg rings. Journal of Zoology (London), 147:224-228.

Genoud, M. 1984. Activity of *Sorex coronatus* (Insectivora, Soricidae) in the field. Zeitschrift für Säugetierekunde, 49:74-78.

Germano, D.J., and D.F. Williams. 1993. Field evaluation of using passive integrated transponder (PIT) tags to permanently mark lizards. Herpetological Review, 24: 54-56.

Korn, H. 1987. Effects of live-trapping and toe clipping on body weight of European and African rodent species. Oecologica, 71:597-600.

Krebs, C.J., B.L. Keller, and R.H. Tamarin. 1969. *Microtus* population biology: Demographic changes in fluctuating populations of *Microtus ochrogaster* and *M. pennsylvanicus* in southern Indiana. Ecology, 50:587-607.

Le Boulenge–Nguyen, P.Y., and E. Le Boulenge. 1986. A new ear-tag for small mammals. Journal of Zoology (London), 209:302-304.

Pavone, L.V., and R. Boonstra. 1985. The effect of toe-clipping on the survival of meadow voles (*Microtus pennsylvanicus*). Canadian Journal of Zoology, 63:499-501.

Pimentel, R.A., and J.D. Smith. 1990. BioΣtat I. A univariate statistical toolbox version 2 tutorial manual. Sigma Soft, Placentia, CA, 392 pp.

Pollock, K.H., J.D. Nichols, C. Brownie, and J.E. Hines. 1990. Statistical inference for capture-recapture experiments. Wildlife Monographs, 107:1-97.

Rao, G.N., and J. Edmondson. 1990. Tissue reaction to an implantable identification device in mice. Toxicologic Pathology, 18:412-416.

Schooley, R.L., B. Van Horne, and K.P. Burnham. 1993. Passive integrated transponders for marking free-ranging Townsend's ground squirrels. Journal of Mammalogy, 74:480-484.

Stoddart, D.M. 1970. Individual range, dispersion and dispersal in a population of water voles (*Arvicola terrestris* (L.)). Journal of Animal Ecology, 39:403-425.

Twigg, G.I. 1975. Marking mammals. Mammal Review, 5:101-116.

Wood, M.D., and N.A. Slade. 1990. Comparison of ear-tagging and toe-clipping in prairie voles, *Microtus ochrogaster*. Journal of Mammalogy, 71:252-255.

Daniel F. Williams and Walter Tordoff III, Department of Biological Sciences, California State University, Stanislaus, Turlock, California 95382; David J. Germano, Department of Biology, California State University, Bakersfield, California

Influence of Proximity to Rivers on Chipmunk Vocalization Patterns

William L. Gannon

Abstract

Analysis of geographic variation in characteristics of chip calls of three species of chipmunks revealed that characteristics of the calls differed more between species and varied less within each species at riverine boundaries separating the species than elsewhere in their geographic ranges. Moreover, variation in characteristics of calls of each species increased with distance from the species' zones of potential contact along rivers. Call characters respond ecologically to the physical nature of a riverine boundary, which acts to slow mixing of chipmunk species and encourage species-specific call types. Compression of call character variation at these riverine edges is compatible with the concept of character displacement, although no attempt was made in this study to document competition or other processes traditionally required to claim this concept.

INTRODUCTION

Lineages undergoing evolutionary expansion into different niches exhibit accompanying differences in character states during adaptive radiation (Futuyma, 1986). Furthermore, parapatric populations can either adapt similarly to an environment (introgression) or evolve differing strategies to co-exist. Western North American chipmunks (subgenus *Neotamias*) have undergone adaptive radiation into ca. 21 species and 80 subspecies west of 105° W. latitude (Jones et al., 1992; Sutton, 1992; Wilson and Reeder, 1993). These forms exhibit many, but often subtle, phenotypic differences (e.g., Blankenship and Brand, 1987; Brand, 1976; Gannon and Lawlor, 1989; Patterson and Heaney, 1987). Divergence of taxa is thought to result from several processes, including natural selection for phenotypes best suited to a resource type, reproductive isolation, and interspecific competition.

The role of competition in structuring communities and fostering adaptive radiation is, however, controversial (Dayan et al., 1989; Findley, 1993; Schluter and McPhail, 1992; Schluter, 1993) and the subject of extensive modeling studies (Kelt et al., 1995; Taper and Case, 1992; Scheibe, this volume). Character displacement is a result of an ecological process by which there is a differential morphological response by two taxa of similar phenotype (usually when only a few taxa are present) for resources (Grant, 1972:44).

North American chipmunks, which occur from the Arctic (64° N latitude) into northern Mexico (29° N latitude) and from sea level to about 4250 m, may be good subjects for studies of interspecific competition and character displacement. Taxonomy of western chipmunks (Wilson and Reeder, 1993) has been controversial at both the generic (e.g., Nadler et al., 1977; Patterson and Heaney, 1987; Sutton, 1987) and specific levels because of strong phenotypic similarity among some taxonomic groups (Gannon and Stanley, 1991; Sullivan, 1985). Populations of different species of chipmunks overlap both in ecological and in geographic distribution (e.g., Brown, 1971; Heller, 1971). Some species of chipmunks occur parapatrically or sympatrically with as few as one to as many as five other species (J. Brown, pers. comm.; Chappell, 1978; Hall, 1981; Johnson, 1943). In the Sierra Nevada mountains of California, many of these contiguously allopatric, parapatric, or sympatric species coexist only in nar-

Life Among the Muses: Papers in Honor of James S. Findley. Edited by T.L. Yates, W.L. Gannon & D.E. Wilson.

row strips or border each other at the junction of major life zones that, in turn, correspond to differences in elevation and slope aspect (Chappell, 1978; Heller, 1971; Sharples, 1983) and the realized niche of each species may be maintained by competitive exclusion (Brown, 1971; Meredith, 1977). Thus, habitat and resource partitioning may have allowed chipmunks to undergo the adaptive radiation that has occurred. For instance, Sharples (1983) found that four overlapping species of *Tamias* from the same elevation in the Sierra Nevada of California exhibited preferences for different habitats; *T. speciosus* was associated with scattered trees, *T. amoenus* preferred brushy areas, *T. quadrimaculatus* preferred transition/disturbed habitats, and *T. senex* preferred old-growth closed-canopy forest. The behavioral influence of one population on another may determine which species co-occur in an area. In his study of *T. dorsalis* and *T. umbrinus* in the Sheep Mountains of southwestern Nevada, Brown (1971) found that the more terrestrial and aggressive *T. dorsalis* excluded the more arboreal *T. umbrinus* from rocky, piñon-juniper habitats at lower elevations. At higher elevations, however, the more arboreal *T. umbrinus* excluded *T. dorsalis* from forested habitats (primarily *Pinus* and *Cercocarpus*). Since 1990, however, *T. umbrinus* appears to have become extinct in these mountains (unpubl. data;W. Stanley, pers. comm.). *T. dorsalis* now occurs to the peaks of the Sheep Mountains, presumably expanding its range in the absence of competition from *T. umbrinus*. Similarly, *T. townsendii* (a closed-canopy chipmunk) dominates *T. amoenus* in sympatry, but during habitat modifications such as logging, *T. amoenus* becomes more abundant and successful than *T. townsendii* (Trombulak, 1985). Speciation may be promoted by behavioral changes superseding other morphological alterations in a species' phenotype.

How species may divide their niche can be elucidated by looking closely at the emerging differences of two sister-related chipmunk species such as those in the Townsend chipmunk species group. I report here a pattern that reflects the divergence of character states between two parapatric pairs of species of chipmunks—*T. senex* and *T. siskiyou*, and *T. siskiyou* and *T. townsendii*. Call morphology of one species has become more stereotyped when in proximity of another, neighboring species. The systematic and taxonomic background of Townsend-group chipmunks is provided for a more complete understanding of the results.

Taxonomic Background

Five subspecies of *Tamias townsendii* from Northern California, Oregon, and Washington were recognized by Johnson (1943). Three of these subspecies were subsequently elevated to species with three rivers as their respective borders. *T. ochrogenys* has a southern range to Marin County, California, and is bounded on the north by the Eel River. *T. senex* has the Eel River at its southern border and the Klamath River its northern edge. *T. siskiyou* is bordered on the southern portion of its range by the Klamath River and Rogue River as its northern edge. Finally, *T. townsendii* ranges from the Rogue River northward into southern British Columbia (Hall, 1981; Sutton and Nadler, 1974). Sutton and Nadler (1974) and Williams (1980) performed discriminant function analyses on morphometric and bacular data and found significant differences between these taxa. The Eel River was determined to be important as a species boundary, between *T. ochrogenys* and *T. senex*, by Kain (1985; bacular and cranial characters) and by Gannon and Lawlor (1989; unique call patterns).

Chipmunk Call Character Variation

In this study, I analyzed the differences in vocal characters between populations of *T. siskiyou*, *T. senex*, and *T. townsendii* along the Rogue River and Klamath River to examine variation of characters describing call patterns of the chip vocalization (Fig. 1). Call characters were recorded from a transect moving away from these river boundaries and along the riverine shores ("T-pattern" analysis; Fig 1). I predicted that means of call characters should be equal and that the range in variation in call characters ought to be constant among populations within each species if distance of a population from a riverine boundary has no effect on call characters.

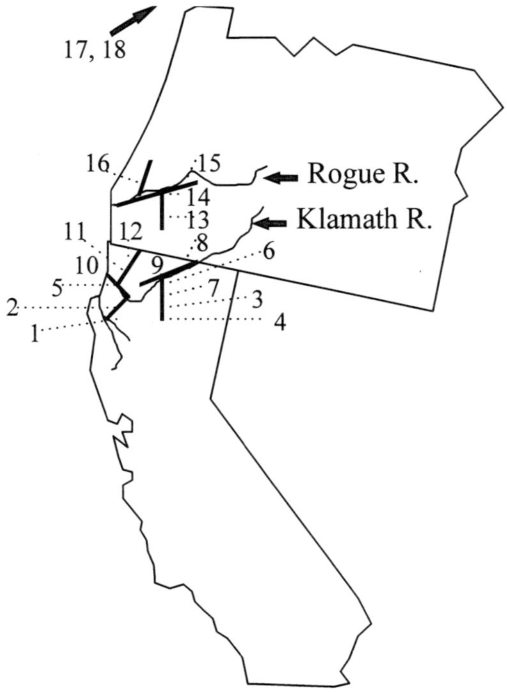

Fig. 1. Map of collection localities in California and Oregon, pertinent river systems, and species ranges of *Tamias senex*, *T. siskiyou*, and *T. townsendii* (after Hall 1981). "T"s along rivers form transects away from and parallel to river strips. Localities are: (1) Carlotta, (2) Snow Camp, (3) Hawkins Bar, (4) Friday Ridge, (5) South Klamath, (6) Orleans, (7) Gridder Ridge, (8) Oak Flat, (9) Fish Lake, (10) Klamath Glen, (11) Mill Creek, (12) Big Flat, (13) Brookings, (14) Squirrel Peak, (15) Rogue, (16) Illahe, (17) Mt. Hood, (18) Lake Wenatchee.

METHODS AND ANALYSES

Calls were recorded from chipmunks in their natural habitat (mostly coniferous forest) using a Sony Walkman WM-D6 cassette recorder and a Sennheiser ME-80 directional microphone along three transects (two parallel and one perpendicular) of the lower drainages (last 200 km) of the Klamath River and Rogue River, following methods suggested by Gannon and Foster (1996). The perpendicular transect that ran from the river's edge toward the center of each species' distribution formed a T-shaped sample pattern for each of the two sets of species (Fig 1). Each syllable is defined as a single sound pulse and constitutes a component of a call. Six characteristics were analyzed for 1,393 "chip" calls, including 1,134 calls from the vicinity of the Klamath River (from *T. senex* and *T. siskiyou*) and 259 calls from the vicinity of the Rogue River (*T. siskiyou* and *T. townsendii*).

Call structure was examined by analyzing variables in frequency (in kHz) and time (in s) using a Kay sound spectragraph (Sonagraph 7029A) and with Canary spectrographic analysis software (Cornell Laboratory of Bio-Acoustics) on a MacIntosh Quadra 700. Time and frequency were measured to the nearest 0.01 s and 0.05 kHz, respectively. Six variables describing call morphology were used based on Gannon and Lawlor (1989): maximum frequency (MAXF), the most discernible maximum frequency of each syllable (in kHz); minimum frequency (MINF), the most discernible minimum frequency of the upsweep of each syllable (in kHz); harmonic number (HARN), including the fundamental, the number of clearly identifiable harmonics in a syllable; syllable-peak to syllable-peak (SPSP), the time from the frequency maximum (peak) of the first syllable to the frequency peak of the next syllable (in s); syllable number per call (SNUC), the total number of syllable in a call sequence; and total call duration (TOLC), the total duration of a call sequence (in s).

Spectrographic data were analyzed statistically at species and population levels, as well as at distances from rivers using general statistics (means and standard deviations) and a nested analysis of variance procedure (SAS Institute, 1985). Data meeting assumptions of ANOVA were tested using a comparison of means within and between species. Multiple comparison procedures (i.e., Fisher's least significant distance method, Duncan's Multiple Range Test, and Bonferoni comparisons) was performed. A correlation matrix and regression analysis were then performed to ascertain relative distances between species and populations in contact with another. Finally, calls were grouped as being recorded from the edge of the river to 5 km ("near"), from more than 5 km but less that 20 km from the river ("medium"), and from points more than 20 km from the river but not infringing upon another species' riverine edge ("far").

RESULTS

Populations of each species at medium and far distances from riverine boundaries have character means not differing significantly from each other, but with high variance (CV). Populations adjacent to riverine boundaries parapatric to another species have means that differ significantly from each other and they have a lower CV (Table 1, Fig. 2).

The nested analysis of variance design tested the relationship of chip calls between species, among populations, and with distance from zones of potential contact. All species-wise comparisons showed significant differences for the six variables examined (Table 2). Comparisons between species for all call characters analyzed as to near, medium, and far distances from rivers are as follows:

The Klamath River: *T. senex* versus *T. siskiyou*

For "near" sites adjacent to the Klamath River, mean values for maximum frequency were 12.75 kHz (*T. senex*) and 14.89 kHz (*T. siskiyou*; Table 1). At "far" sites, the maximum frequency means were 14.65 kHz (*T. senex*) and 15.07 kHz (*T. siskiyou*). "Near" populations of the two species differed more from each other than did "far" populations (Fig. 2). Duncan's test ranked means for *T. senex* as increasing with distance from river from 12.17 kHz (near), to 12.80 kHz (medium), and 13.85 kHz (far; Table 2). Variance (CV) also increased with distance (2.62, 3.21, and 5.11, respectively). Conversely, for *T. siskiyou*, Duncan's test ranked mean values for maximum frequency as decreasing with increasing distance from the river from 15.07 kHz (near), to 14.39 kHz (medium), and 14.13 kHz (far). Similar patterns are found for maxi-

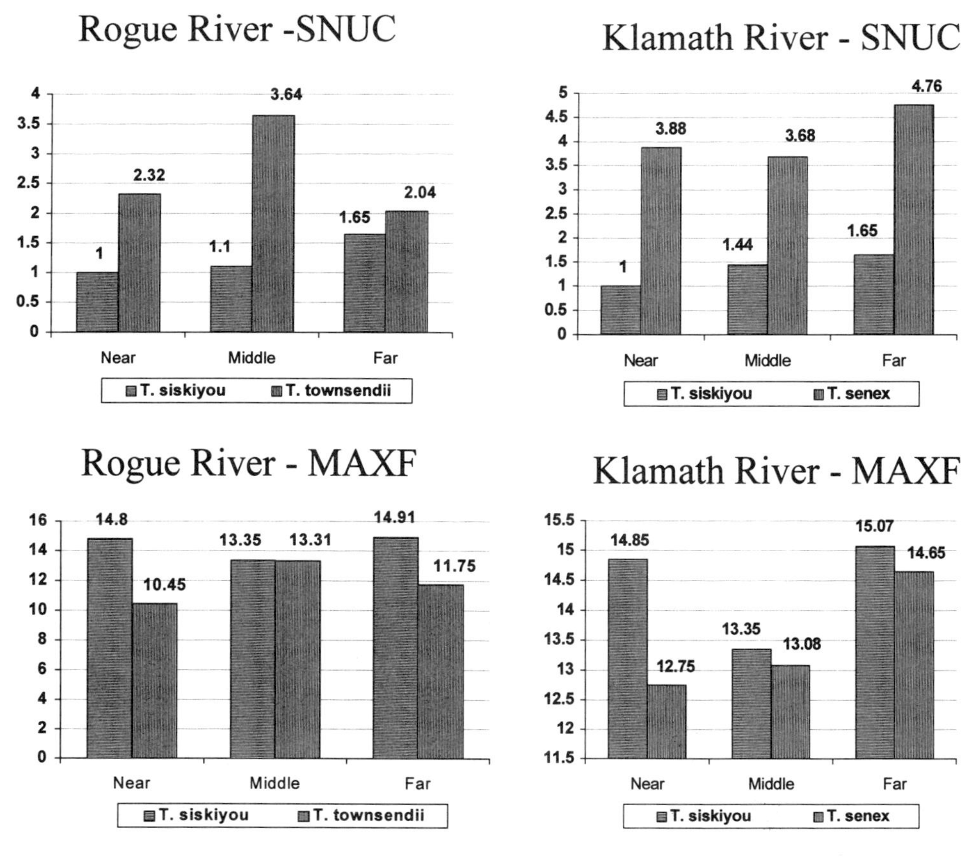

Fig. 2. Mean values for syllable number (SNUC) and maximum frequency (MAXF) plotted near, medium, and far distances from river boundaries of the Rogue (left) and Klamath (right) rivers, CA.

Table 1. **Means, standard deviations (in parentheses), and coefficients of variation (below means) of six variables by population. Numbers before locality name correspond to Fig. 1. Letters after locality name refer to locations that are near (N), medium (M), or far (F) distances to rivers. Refer to text for character abbreviations.**

Populations	No. of Calls	MAXF(kHz)	MINF(kHz)	SPSP(s)	SNUC	TOLC(s)	HARN
KLAMATH							
1) Carlotta (F)	164	12.65 (1.02)	4.47 (.63)	.33 (.06)	3.63 (.77)	.75 (.23)	3.21 (.85)
		4.43	8.22	3.71	11.32	8.76	13.22
2) Snow Camp (M)	226	12.78 (1.11)	5.01 (.58)	.39 (.02)	3.43 (.73)	.76 (.17)	3.45 (.87)
		4.74	7.99	3.33	10.84	9.35	10.32
3) Hawkins Bar (F)	74	14.65 (.66)	5.28 (.43)	.36 (.02)	4.76 (.89)	.94 (.18)	5.01 (.26)
		4.92	8.21	3.54	8.78	9.09	11.57
4) Friday Ridge (F)	348	12.66 (.93)	4.14 (.69)	.39 (.02)	4.01 (.77)	.97 (.26)	4.54 (.75)
		5.11	8.01	2.99	10.26	9.49	8.39
5) S. Klamath (N)	70	12.75 (.77)	4.82 (.63)	.38 (.01)	3.88 (.71)	.73 (.18)	3.60 (.80)
		3.76	5.47	1.06	9.97	13.46	14.31
6) Orleans (N)	131	12.89 (.69)	4.74 (.66)	.31 (.02)	3.96 (.55)	.70 (.09)	3.47 (.55)
		2.62	6.00	0.75	8.92	8.34	6.77
7) Gridder Ridge (M)	19	13.08 (.48)	4.32 (.34)	.28 (.03)	3.68 (.66)	.73 (.11)	2.42 (.50)
		3.21	5.97	0.93	9.28	9.06	9.50
8) Oak Flat (N)	16	14.66 (.87)	3.97 (.49)	—	1.00 (.01)	.28 (.04)	5.56 (.92)
		5.21	4.93		2.25	4.37	5.33
9) Fish Lake (N)	19	14.63 (1.13)	4.27 (.61)	—	1.04 (.02)	.27 (.07)	5.37 (.40)
		2.75	5.01		2.56	4.21	4.88
10) Klamath Glen (N)	67	14.89 (.62)	3.67 (.37)	—	1.00 (.01)	.21 (.02)	5.44 (.76)
		1.88	4.78		2.14	6.93	7.87
ROGUE							
11) Mill Creek (M)	36	13.35 (1.99)	3.96 (.49)	.22 (.03)	1.44 (.36)	.41 (.11)	6.11 (1.31)
		4.65	6.24	1.85	3.26	7.32	12.32
12) Big Flat (F)	67	15.07 (.72)	4.01 (.50)	.20 (.02)	1.10 (.40)	.36 (.09)	6.14 (1.34)
		5.11	13.68	0.22	1.77	2.45	2.37
13) Brookings (F)	69	14.91 (1.11)	3.62 (.99)	.20 (.02)	1.65 (.33)	.40 (.17)	5.96 (.79)
		3.88	9.32	0.07	1.96	2.03	3.01
14) Squirrel Peak (N)	34	14.80 (.27)	3.70 (.27)	—	1.00 (.01)	.22 (.04)	5.10 (.23)
		2.57	7.45		0.73	1.87	3.33
15) Rogue (N)	14	14.39 (1.31)	4.46 (.53)	—	1.00 (.01)	.21 (.03)	5.86 (.56)
		2.61	6.23		0.69	1.67	2.84
16) Illahe (N)	19	10.45 (.81)	4.58 (.48)	.34 (.02)	2.32 (.49)	.74 (.05)	3.74 (.45)
		3.27	6.88	1.11	1.11	3.44	3.27
17) Mt. Hood (M)	16	13.31 (.57)	4.34 (.63)	.56 (.02)	3.64 (.39)	.80 (.05)	4.19 (.75)
		6.32	9.73	2.05	8.49	9.34	9.79
18) L. Wenatchee (F)	4	11.75 (.65)	3.88 (.25)	.44 (.02)	2.04 (.43)	.71 (.18)	3.50 (.58)
		5.58	10.22	5.61	10.21	10.21	10.31
Total Calls 1,393							

The headers MAXF(kHz) through HARN are grouped under the spanning header "Characters".

Table 2. **Results of nested ANOVA and Duncan's Multiple Range test. Below are listed variables charted against results and means of each comparison. NS denotes non-significant differences (P > 0.01). All those cases marked with * denote significant differences at P < 0.0001. Duncan's means are listed as near to, medium distance from, and location far from river. See text for further explanation.**

	Significant Difference			Duncan's Mean	
Variable	Between Species	Within Populations	By Distance	Near, Medium, Far *T. senex, T. siskiyou*	Overall Mean
AT KLAMATH RIVER					
MAXF	*	*	*	12.17, 15.07 12.80, 14.39 13.85, 14.13	13.24
MINF	*	NS	*	4.82, 3.67 4.54, 3.96 4.47, 3.97	4.47
SPSP	*	NS	*	0.33, 0.44 0.36, 0.38 0.29, 0.39	0.36
HARN	*	*	*	3.44, 5.66 4.96, 4.92 3.74, 5.32	4.20
SNUC	*	NS	*	3.28, 1.00 3.34, 1.10 4.16, 1.65	2.88
TOLC	*	*	*	0.92, 0.20 0.81, 0.34 0.70, 0.28	0.55
AT ROGUE RIVER				for *T. townsendii* only	
MAXF	*	NS	*	10.45 13.31 11.75	12.43
MINF	*	NS	*	4.58 4.34 3.88	4.48
SPSP	*	NS	*	0.30 0.56 0.44	0.33
HARN	*	NS	*	3.74 4.19 3.50	4.87
SNUC	*	NS	*	2.99 3.64 2.04	2.02
TOLC	*	NS	*	0.73 0.80 0.71	0.55

mum frequency where population means significantly differed, and variance deceased (CV) in most call characters at near sites and increased with distance away from Klamath River. For SPSP, means for populations at near and medium distance from the river did not differ significantly from each other. However, both means were significantly different in populations that were far from the river ($P < 0.05$).

Syllable number (SNUC) was the most distinctive among the six call variables examined. At the level of species, *T. siskiyou* and *T. senex* differed significantly from each other (mean 3.42 syllables and 1.35 syllables, respectively). Significant differences in syllable number were also found between the near ($\overline{X} = 3.28$), medium ($\overline{X} = 3.34$), and far ($\overline{X} = 4.16$) samples as measured using Duncan's Multiple Range test. *T. siskiyou* populations nearest the river always had one syllable per call. Those at a medium-distance had a mean of 1.44 syllables per call, and far populations had a mean of 1.65 syllables per call (Fig. 2). Duncan's Multiple Range test revealed significant differences among means at all three distances (Table 2).

The Rogue River: *T. siskiyou* versus *T. townsendii*

The results of comparing *T. siskiyou* and *T. townsendii* parallel those of *T. siskiyou* and *T. senex* at the Klamath River, but due to the small sample size, they should be interpreted more cautiously. Similar patterns were found for maximum frequency, minimum frequency, and number of harmonics in that population means differed significantly with increased variance (CV) in call characters with distance away from the Rogue River. For *T. townsendii*, Duncan's test showed means of the variable time-between-syllables peaks as 0.30 s (near), 0.56 s (medium), and 0.44 s (far). Because there was only one syllable per call for *T. siskiyou*, time between syllable peaks was zero. Similarly, for syllable-number-per-call, Duncan's test showed means

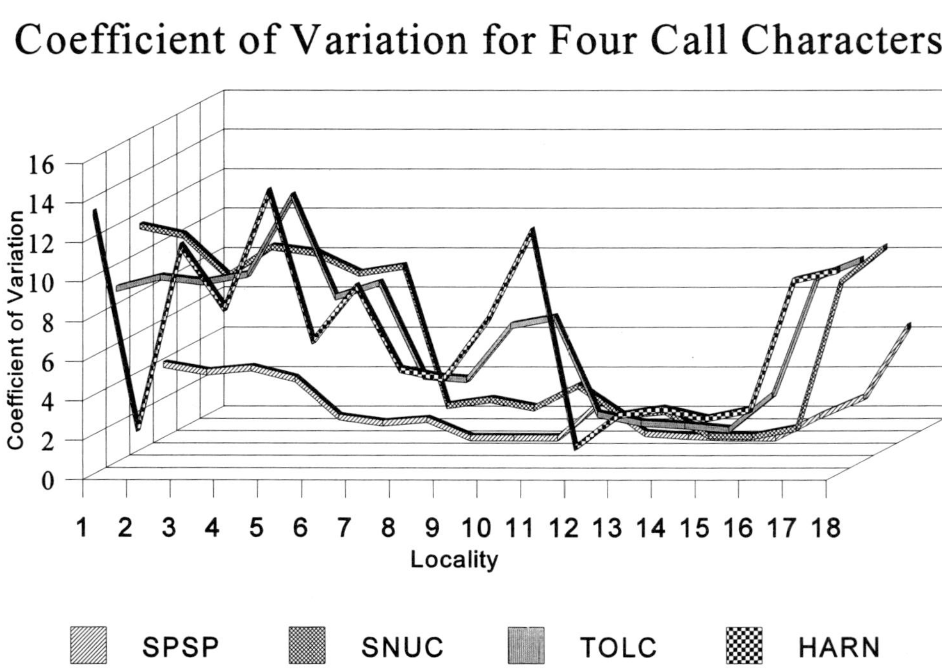

Fig. 3. Plot of the coefficient of variation (CV) for each population sampled for calls. The four variables represented here are: time between syllable peaks (SPSP), number of harmonics (HARN), number of syllables per call (SNUC), and total call length (TOLC). See text for definitions of these variables. Fig. 1 lists population localities.

of 2.32 (near), 3.64 (medium), and 2.04 (far) syllables per call for *T. townsendii*. For *T. siskiyou*, mean syllables per call was 1.0 (near), 1.10 (medium), and 1.65 (far). The means of total call length (TOLC) for these species was ranked by Duncan's test as 0.73 s (near), 0.80 s (medium), and 0.71 s (far). *T. siskiyou* had a call duration of 0.21 s (near), 0.22 s (medium), and 0.40 s (far). Chipmunks at a medium distance from the river had a greater number of syllables—hence, greater call length than either near or far locations. Fig. 3 shows a pattern whereby the CV decreased for four variables (SPSP, SNVC, TOLC, and HARN) at sites near the Klamath (localities 5, 6, 8, 9, 10) and the Rogue (localities 14, 15, 16) rivers. This is most dramatic at sites bordering the Rogue River.

DISCUSSION

The null hypotheses, that variation of call characters is constant among all populations without regard to species differences, is not supported. Furthermore, patterns of intra- and interspecific variation among call characters suggest a behavioral and perhaps an evolutionary response of one species to another. Means of call character states of one species differ more significantly where local populations meet those of another species at parapatry (rivers' edges) than at locations distant from parapatry. Call variation also appears to increase in most cases with distance from the area of direct parapatry. The level of variation for call character states does not remain constant. Parapatry at rivers' edges calls of *T. townsendii* occur in pairs or triplets of medium maximum frequency (10.45 kHz) for a long duration (0.74 s). Away from the Rogue River, calls are paired, higher in maximum frequency (11.75 kHz), and slightly shorter in call duration (0.71 s). Compared to *T. townsendii*, *T. siskiyou* emits one syllable of short duration (0.21 s) with a high maximum frequency (14.59 kHz) at the Rogue River. At a distance from that river, but not in the proximity of the Klamath River, *T. siskiyou* emits 1-2 syllables (mean 1.65 syllables per call) at a maximum frequency of 14.9 kHz for a call of longer duration (0.4 s). At the Klamath River, *T. siskiyou* gives a one syllable "chip" call of high maximum call frequency (14.89 kHz), a low minimum call frequency (3.97 kHz), and lasts for a brief 0.21 s. Furthest away from the Klamath River, syllable number and call duration increases (mean of 1.65 syllables per call for 0.41 s), but both maximum and minimum call frequencies remain about the same as near-river samples. Bordering, but across the Klamath River from *T. siskiyou*, *T. senex* has four syllables per call (mean of 3.96) of a narrower frequency range (maximum frequency 12.89 kHz, minimum frequency 4.74 kHz), and that lasts an average 0.73 s. At a medium distance from its riverine border with *T. siskiyou*, *T. senex* calls contain 3-5 syllables of narrow frequency range but increased call length (0.97 s). Coefficients of variation increase in all cases with distance away from both Klamath River and Rogue River. Weakness of these data are in the unbalanced number of cases among comparisons for the Rogue River. However, sample sizes and the patterns they reflect are robust for the Klamath River comparisons.

For these chipmunk species it appears that character compression (or where call characters exhibit less variance about a populational mean) is occurring between the pairings of *T. senex* with *T. siskiyou* and of *T. siskiyou* with *T. townsendii* that occur parapatrically. Furthermore, call character variation increases with distance from the edges of each species' range, the edge being a parapatric strip bordered by another chipmunk species. Patterns occur among call characters that suggest a reaction of one species to the presence of another. Fig. 2 and Tables 1 and 2 show that variation is "compressed" at areas of parapatry, then released with distance from this strip, a pattern not unlike that describing character displacement by Grant (1972:43). I use the term "character compression" because character variance is reduced and calls are more precise where species come into contact. Why is this not clearly character displacement? From this study, I have not been able to infer an underlying process of competition for resources—only a character displacement-like pattern based on behavioral data.

Character Compression In Parapatry? Leaky Barriers

There were marked differences in both means and variances (CV) with distance from rivers for each species generally for all six variables. CVs are lower in areas of parapatry but call variation is dynamic with increasing distance from rivers (Fig. 3).

As originally defined, character displacement refers to character variation patterns of two *overlapping* sympatric populations (Brown and Wilson, 1956). Since Brown and Wilson (1956) first presented the case for character displacement, argument has raged regarding the influence of competition on this phenomenon (Slatkin, 1980), whether character displacement occurs at all, and if so, is it convergent or divergent (Grant, 1972; Strong, 1984; Strong et al., 1979). Interactions to document interspecific competition or to detect differences in resource utilization, habitat quality, or differences between species level of aggressive behavior (*sensu* Brown, 1971) could further classify these call patterns or some other character set (e.g., morphology) as character displacement. In order to produce a pattern of character displacement, there also must be some form of selection. In this case, selection may be occurring by the rivers' action of separating species' populations and by the likelihood that this barrier is "leaky" (Fig. 4). Although flow rates for these coastal rivers are among the highest (100,000 mps; maximum values for the Klamath and Rogue rivers) of *any* continental U.S. river, (Norris and Webb, 1990:364, U.S. Geological Survey, 1996) a chipmunk from *T. senex* can traverse the Klamath River during summer when river water levels are low to mix with individuals of *T. siskiyou* on the other side. This occasional mixing of diverging sibling species acts to enhance emerging differences in characters (such as with calls or *os genitalia*) between taxa. The case discussed in Gannon and Lawlor (1989) in which chipmunks at a site near the mouth of the Eel River near Rio Dell, California, displayed call characters that were intermediate, probably reflect greater mixing or

Fig. 4. Wide gravel and sediment-laden river bar of the Klamath River ca. Klamath Glen, Del Norte Co., California. Sound travels differently along this noisy, relatively sparse corridor compared with nearby coniferous forest. Sound space is of a different nature here than in nearby coniferous forest. Along this gravel bar, emitted calls may be greatly altered by wind or temperature changes, and noise generated by water flow.

hybridizing between emergent forms (*T. ochrogenys* and *T. senex*).

Environmental Selection of Calls

Morton (1975) found that for neotropical bird species in forest habitat, attenuation of sound was reduced for calls between 1.6 and 2.5 kHz. Edge and grassland habitats had a lower sound "window" of optimal sound transmission for calls between 1.0 to 2.0 kHz. Open areas such as grasslands or river gravel bars are poor for sound transmission because of gradients in wind, temperature and turbulence (Fig. 4). Chipmunks calling along a riverine zone may simply have to be more distinctive in their calling in order to overcome their noisey environment; that is, calling may be better "understood" in the forest away from the noise and sound disturbance of a riverine environment. It remains to be seen whether chipmunk calls correlate with differences in acoustical traits of habitats (*sensu* Morton, 1975). If no sound "window" effect is present in edge habitats (environmental selection), evolution of chipmunk calls at edges may be more influenced by selection pressures that favor species identity or distinctiveness over species competing with a neighboring, parapatric species for resources (Paterson, 1982).

Chipmunks, Calls, and Variability

Call characters have relatively higher variance than most morphometric and behavioral characters and may be an early indicator of incipient diversification. For instance, total call lengths (TOLC) may have a CV of 10 or higher, whereas greatest length of skull have a CV of less than two (this study; Sullivan and Petersen, 1988). Calls are crucial for species identity (Bolles, 1988). Because chip calls are frequently used by chipmunks when confronted by a conspecific or for warning of an intruder (Blakenship and Brand, 1987; Brand, 1976), these calls have clearcut species-specific characteristics that are taxonomically useful (Bolles, 1988; Dunford and Davis, 1975). They may even be used for predator alarm and evolved by kin selection maintained by reciprocity (Smith, 1978). Like soricid communities (Kirkland, 1985) and some insectivorous bat communities (Findley, 1993), chipmunks have immense, widely dispersed, renewable, and dependable resources (mostly seeds and fungi). Therefore, overall competition may be reduced and less limiting. However, chipmunks are unique in that rarely does more than one "type" occur in one microhabitat. Altitudinal zonation (e.g., Heller, 1971) and other exclusionary patterns (*sensu* Brown, 1971) suggest that chipmunks of any given species do not tolerate resource sharing with another species. This fits well with the pattern I have found. Rivers slow dispersal and isolate chipmunk populations enough for differentiation, but not exclusion, to ensue. The character compression of calls may act to minimize wasted efforts at defense, territoriality, or reproduction and serve to enforce species identity (Paterson, 1982). Sufficient variation exists in chipmunk calls for natural selection to be effective. Perhaps as Paterson (1982) suggested, signaling systems such as represented by this call system, are subject to stabilizing selection because of co-adaptations which exist between the signaling and receiving systems within species and between sexes.

Incipient Species and Dispersal Barriers

Chipmunks are likely candidates for resource limitation and character displacement studies because of the few number of species competing for seed resources in any given area. Although there are many kinds of chipmunks, there are many combinations of only a few taxa at any one time and place. Many of the differences between species and populations are subtle, perhaps because of the fairly recent nature of isolation and diversification (Brown, 1971; Nadler et al., 1985). Subtle differences in the morphology of *os genitalia* and behavioral repertoire of vocalizations are important diagnostic features (Sutton, 1995).

Are these lines of contact between species a result of competitive exclusion or could they also be a result of environmental or habitat discontinuities presenting physical or physiological barriers to these animals (Chappell, 1978)? The temperate river systems of the Pacific coast of North America present a situation whereby a large number of allopatric or parapatric species and subspecies exist (Ayers and Clutton–Brock, 1992). Although results are equivocal at present for the effectiveness of riverine systems in the Amazon Basin, the differences between tropical and temperate systems may turn out to be the extent and degree of flooding combined with a more homogeneous mosaic of overstory in tropical systems. Even habitat alterations related to logging practices of the Pacific Northwest since the 1800's may have ramifications on chipmunk evolution. Morphological and allozymic shifts have been observed in other sciurids (e.g., *Sciurus niger* and *S. carolinensis* are affected by the Mississippi River; Moncrief, 1993). Due to relatively recent post-glacial conditions, many forms of chipmunks in this temperate region are cryptic and undoubtedly will require patient teasing apart to analyze them. The Townsend-group chipmunks of the Pacific Northwest are a case in point.

CONCLUSIONS

Variation in call characters suggests that chipmunks respond ecologically to the physical nature of a riverine filter, slowing mixing of chipmunk populations. This study concludes that:

1) Species-specific call types change in characters measured from "chip" vocalizations and exhibit significantly reduced variation at parapatric riverine edges.
2) Call variation increases with distance from these parapatric riverine edges.
3) Compression of the variance about call characters has been shown between species pairs *T. senex* and *T. siskiyou*, and *T. siskiyou* and *T. townsendii*, most likely facilitated by limited dispersal of individuals across rivers.
4) In most cases, call characters are less divergent from opposing species at far localities compared to near ones.

Many investigators have claimed that competition drives character displacement. Although this study cannot exclude the role of competition in this phenomenon, other processes such as environmental constraints on the acoustic community, speciation, or simply a reaction of one species maintaining species identity at "noisy" edges, can be used as competing hypotheses to test these patterns further. Chipmunks provide an excellent assemblage of species for systematic studies of behavioral ecology.

ACKNOWLEDGMENTS

I thank Lisa T. Arciniega, Jorge Salazar-Bravo, Richard B. Forbes, William T. Stanley, Gordon L. Kirkland, Jr., M. Scott Burt, and Damien T. Scott for comments that greatly improved the initial manuscript and two anonymous reviewers for critical and helpful comments. Timothy E. Lawlor mentored the initial development of this project. Lisa A. Valle contributed technical advice, commented on parts of the statistical analysis, and provided options I never would have dreamed. Computer access for analysis of sound was provided by the Yates Program, The University of New Mexico.

Jim Findley has taught me things not by what he says, but by what he does. Both the elegance and focus of his writing and the broad view by which he can place an extrapolated fact will continue to greatly inspire my career. Finally, part of this work was completed while working on my Master's degree at Humboldt State University, Arcata, CA. The University of New Mexico Department of Biology staff, in particular Anne Rice, assisted in preparation of figures and camera-ready text.

LITERATURE CITED

Ayres, J.M., and T.H. Clutton–Brock. 1992. River boundaries and species range size in Amazonian primates. The American Naturalist, 140:531-537.

Blankenship, D.J., and L.R. Brand. 1987. Geographic variation in vocalizations of California chipmunks *Tamias obscurus* and *T. merriami*. Bulletin of the Southern California Academy of Science, 86:126-135.

Bolles, K. 1988. Evolution and variation of antipredator vocalizations of antelope squirrels, *Ammospermophilus* (Rodentia: Sciuridae). Zeitschript fur Saugetierkunde, 53:129-147.

Brand, L.R. 1976. The vocal repertoire of chipmunks (genus *Tamias*) in California. Animal Behavior, 24:319-335.

Brown, J.H. 1971. Mechanisms of competitive exclusion between two species of chipmunks. Ecology, 52:305-311.

Brown, W.L., and E.O. Wilson. 1956. Character displacement. Systematic Zoology, 5:49-64.

Chappell, M.A. 1978. Behavioral factors in the altitudinal zonation of chipmunks (*Eutamias*). Ecology, 59:565-579.

Dayan, T., D. Simberloff, E. Tchernov, and Y. Yom–Tov. 1989. Inter- and intraspecific character displacement in mustelids. Ecology, 70:1526-1539.

Dunford, C., and R. Davis. 1975. Cliff chipmunk vocalizations and their relevance to the taxonomy of coastal Sonoran chipmunks. Journal of Mammalogy, 56:207-212.

Findley, J.S. 1993. Bats: A community perspective. Cambridge University Press, Cambridge, 167 pp.

Futuyma, D.J. 1986. Evolutionary biology (second edition). Sinaur Associates, Inc, Sunderland, MA, 600 pp.

Gannon, W.L., and T.E. Lawlor. 1989. Variation in the chip call vocalization of three species of Townsend's chipmunk (genus *Eutamias*). Journal of Mammalogy, 70:740-753.

Gannon, W.L., and M.S. Foster. 1996. Recording mammal calls. Monitoring biodiversity. Pp. 311-326, *in* Measuring and monitoring biological diversity: Standard methods for mammals (D.E. Wilson, F.R. Cole, J.D. Nichols, R. Rudran, and M.S. Foster, eds.), Smithsonian Institution Press, 400 pp.

Gannon, W.L., and W.T. Stanley. 1991. Chip vocalization of Palmer's chipmunk (*Tamias palmeri*). The Southwestern Naturalist, 36:315-317.

Grant, P.R. 1972. Convergent and divergent character displacement. Biological Journal of the Linnean Society, 4:39-68.

Hall, E.R. 1981. The mammals of North America. John Wiley & Sons, New York, 690 pp.

Heller, H.C. 1971. Altitudinal zonation of chipmunks (*Eutamias*): Interspecific aggression. Ecology 52:312-319.

Johnson, D.H. 1943. Systematic review of the chipmunks (genus *Tamias*) of California. University of California, Publications in Zoology, 48:63-148.

Jones, J.K., Jr., R.S. Hoffmann, D.W. Rice, C. Jones, R.J. Baker, and M.D. Engstrom. 1992. Revised checklist of North American mammals north of Mexico, 1991. Occasional Papers, The Museum, Texas Tech University, 146:1-23.

Kain, D.E. 1985. The systematic status of *Eutamias ochrogenys* and *Eutamias senex* (Rodentia: Sciuridae). Unpublished Master of Science thesis, Humboldt State University, Arcata, California, 67 pp.

Kelt, D.A., M.L. Taper, and P.L. Meserve. 1995. Assessing the impact of competition on community assembly: A case study using small mammals. Ecology, 76:1283-1296.

Kirkland, G.L., Jr. 1985. Small mammal communities in temperate North American forests. Australian Mammalogy, 8:137-144.

Meredith, D.H. 1977. Interspecific antagonism in two parapatric species of chipmunks, *Eutamias*. Ecology, 58:423-430.

Moncrief, N.D. 1993. Geographic variation in fox squirrels (*Sciurus niger*) and gray squirrels (*S. carolinensis*) of the lower Mississippi River valley. Journal of Mammalogy, 74:547-576.

Morton, E.S. 1975. Ecological sources of selection on avian sounds. The American Naturalist, 108:17-34.

Nadler, C.F., R.S. Hoffmann, J.H. Honacki, and D. Pozin. 1977. Chromosomal evolution in chipmunks, with special emphasis on A and B karyotypes of the subgenus *Neotamias*. The American Midland Naturalist, 98:343-353.

Nadler, C.F., R.S. Hoffmann, and H. Levenson. 1985. Biochemical and morphological relationships among Holarctic chipmunks. Acta Zoologica Fennica, 170:19-23.

Norris, R.M., and R.W. Webb. 1990. Geology of California (second edition). John Wiley and Sons, Inc., New York, New York, 541 pp.

Paterson, H.E.H. 1982. Perspective on speciation by reinforcement. The South African Journal of Science, 78:53-57.

Patterson, B.D., and L.R. Heaney. 1987. Preliminary analysis of geographic variation in red-tailed chipmunks (*Eutamias ruficaudus*). Journal of Mammalogy, 68:782-791.

SAS Institute, Inc. 1985. SAS user's guide: Statistics. Version 5 ed. SAS Institute, Inc., Cary, North Carolina, 956 pp.

Schluter, D. 1993. Adaptive radiation in sticklebacks: Size, shape, and habitat use efficiency. Ecology, 74:699-709.

Schluter, D., and J.D. McPhail. 1992. Ecological character displacement and speciation in sticklebacks. The American Naturalist, 140:85-108.

Sharples, F.E. 1983. Habitat use by sympatric species of *Eutamias*. Journal of Mammalogy, 64:572-579.

Slatkin, M. 1980. Ecological character displacement. Ecology, 61:163-177.

Smith, S.F. 1978. Alarm calls, their origin, and use in *Eutamias sonomae*. Journal of Mammalogy, 59:888-893.

Strong, D.R. 1984. Natural variability and the manifold mechanisms of ecological communities. Pp. 56-80, *in* Ecology and evolutionary biology: A round table on research (G.W. Salt, ed.). The University of Chicago Press, Chicago, IL. 130 pp.

Strong, D.R., L.A. Szyska, and D.S. Simberloff. 1979. Tests of community-wide character displacement against null hypotheses. Evolution, 33:897-913.

Sullivan, R.M. 1985. Phyletic, biogeographic, and ecologic relationships among montane populations of least chipmunks (*Eutamias minimus*) in the Southwest. Systematic Zoology, 34:419-448.

Sullivan, R.M., and K.E. Petersen. 1988. Systematics of Southwestern populations of least chipmunks (*Eutamias mimimus*) reexamined: A synthetic approach. Occasional Papers of the Museum of Southwestern Biology, 5:1-27.

Sutton, D.A. 1987. Analysis of Pacific coast Townsend chipmunks (Rodentia: Sciuridae). The Southwestern Naturalist, 32:371-376.

Sutton, D.A. 1992. *Tamias amoenus*. Mammalian Species, 390:1-8.

Sutton, D.A. 1995. Problems of taxonomy and distribution in four species of chipmunks. Journal of Mammalogy, 76:843-850.

Sutton, D.A., and C.F. Nadler. 1974. Systematic revision of three Townsend chipmunks (*Eutamias townsendii*). The Southwestern Naturalist, 19:199-212.

Taper, M.L., and T.J. Case. 1992. Models of character displacement and the theoretical robustness of taxon cycles. Evolution, 46:317-333.

Trombulak, S.C. 1985. The influence of interspecific competition on home range in chipmunks (*Eutamias*). Journal of Mammalogy, 66:329-337.

Wilson, D.E., and D.M. Reeder. 1993. Mammal species of the world: A taxonomic and geographic reference. Second ed. Smithsonian Institution Press, Blue Ridge Summit, Pennsylvania, 1206 pp.

Williams, K.M. 1980. The systematic status of three forms of Townsend's chipmunk (*Eutamias townsendii*). Unpublished Master of Science thesis, Humboldt State University, Arcata, California, 44 pp.

William L. Gannon, Museum of Southwestern Biology and Department of Biology, University of New Mexico, Albuquerque, New Mexico 87131-1091

Subnivean Foraging by Abert's Squirrels

Richard B. Forbes

Abstract

This paper describes foraging by Abert's squirrels *(Sciurus aberti)* on an unusually abundant crop of acorns of Gambel's oak (*Quercus gambelii*) buried beneath as much as 30 cm of snow. Possible ancillary benefits of this behavior, such as reduced convective and radiative heat losses and reduced exposure to certain predators, are discussed.

INTRODUCTION

Abert's squirrels (*Sciurus aberti*) are reported to depend primarily on seeds, apical buds, twigs, and cambium of ponderosa pine (*Pinus ponderosa*) for food, especially in winter and spring (e.g., Bailey, 1931; Keith, 1965; Hoffmeister, 1971, 1986; Stephenson, 1974; Hall, 1981; Clark and Stromberg, 1987; Findley, 1987; Snyder, 1993; Snyder and Linhart, 1993). Although acorns of Gambel's oak (*Quercus gambelii*) may make up 40% of the squirrels' diets in years of good mast production (Stephenson, 1974), and fleshy fungi, carrion, bones, antlers, pinyon nuts, and the bark of pinyon twigs also are eaten (Bailey, 1931; Keith, 1965; Reynolds, 1966; Hoffmeister, 1971, 1986; Golightly and Ohmart, 1978), authorities agree that Abert's squirrels depend on ponderosa pine for long-term survival. This paper reports that in the winter of 1992-93 Abert's squirrels in the Sandia Mountains of New Mexico appeared to feed primarily on an unusually abundant crop of acorns of Gambel's oak. The paper also describes a previously unreported means by which the squirrels foraged for acorns buried under as much as 30 cm of snow, and it considers some possible ancillary benefits of this means of foraging.

MATERIALS AND METHODS

I observed Abert's squirrels in the Sulphur Canyon Picnic Area (106° 22' 30" W, 35° 10' 30" N), Sandia Mountains, Bernalillo Co., New Mexico, during several days in September, on 5 October, 15 and 16 December 1992, and on 28 January and 15 March 1993. Observations were conducted for 3-5 hr each day. Squirrels were observed with the unaided eye and with the aid of a 400mm f3.5 Nikkor lens mounted on a Nikon FE2 35mm camera. More than 100 color slides were taken documenting the observations described below.

The Sandia Mountains received an unusually large amount of snow in mid-December 1992. On 15 and 16 December, snow in Sulphur Canyon Picnic Area was approximately 25-30 cm deep. About 20 cm of soft snow overlay a thin layer of crusted snow, which in turn overlay more soft snow extending to the soil surface. Air temperatures on both days were well below freezing. The sky was overcast and snow fell throughout my observations (3 hr) on the afternoon of 15 December. Although 16 December was sunny, wind blew snow off the trees almost constantly throughout my 5-hr observation period. The sky was generally clear and the weather seasonable during my other visits.

In December, January, and March, I observed at least eight Abert's squirrels foraging on approximately 800 sq m of the west-facing 30° slope just beyond a gate across the road into the picnic area. The elevation of this slope is 2200 ± 5 m. Dominant trees where the squirrels were foraging are Gambel's oak, Rocky Mountain juniper (*Juniperus scopulorum*), and one-seeded juniper (*J. monosperma*). Ponderosa pine (*Pinus ponderosa*), white fir (*Abies concolor*), and Douglas-fir (*Pseudotsuga menziesii*) are dominant higher on the slope and on the east-facing slope across the road. Red squirrels

Life Among the Muses: Papers in Honor of James S. Findley. Edited by T.L. Yates, W.L. Gannon & D.E. Wilson.

Fig. 1. An Abert's squirrel (*Sciurus aberti*) consuming acorns of Gambel's oak (*Quercus gambelii*) at Sulfur Springs Campground in the Sandia Mountains of New Mexico on 15 December 1992 at the entrance of a subnivian tunnel it had apparently excavated.

(*Tamiasciurus hudsonicus*) occur at higher elevations in Sulphur Canyon, but I saw none on the slope just described. Rock squirrels (*Spermophilus variegatus*) and Colorado chipmunks (*Tamias quadrivittatus*) also occur in Sulphur Canyon but were not seen in the winter months; presumably, they were hibernating.

RESULTS

In September and October 1992, while photographing Abert's squirrels in and near Sulphur Canyon, I noticed that acorns of Gambel's oak were exceptionally abundant. I made no measurements of their abundance, but acorns nearly covered the soil beneath and near Gambel's oaks. Except for individuals chasing each other, the Abert's squirrels I found were widely dispersed; those I saw foraging were taking cones and twigs of ponderosa pine. On only one occasion in September and October did I see an Abert's squirrel consume an acorn.

In contrast, upon entering Sulphur Canyon on 15 December 1992, I immediately noticed a number of Abert's squirrels on the slope described above. The periodic disappearance of individuals made a precise count of them difficult, but at least eight were present. The squirrels were feeding at the entrances of many tunnels into the snow that covered the slope (Fig. 1). Small amounts of soil and duff around the openings of the tunnels revealed that they extended to the soil surface. Most tunnels appeared to have only one opening, but some may have had at least one other opening several meters from the one most used. Typically, a foraging squirrel would disappear into a tunnel for ca. 15 sec–2 min, then emerge, tail first, bearing an acorn in its mouth. The acorn usually was eaten immediately, either at the tunnel entrance or just outside it. Occasionally a squirrel carried an acorn up a tree and ate it while sitting on a branch. When the acorn was consumed, the squirrel usually re-entered the tunnel, apparently to resume foraging. Judging from the time squirrels occasionally spent beneath the snow, I thought that some tunnels might be several m long, but I have no idea of their form. I never saw one of the squirrels initiate a tunnel, but I observed no animals other than Abert's squirrels using them. Each squirrel defended a tunnel while using it, but if one squirrel left a tunnel, another squirrel often entered it. One squirrel foraged in the same tunnel for sessions exceeding one hr, then climbed a tree and loafed for ca. 10 min before returning to the same tunnel.

On 28 January 1993, snow depth on the slope had decreased to approximately 15 cm. Many tunnels were still present, and the squirrels were using them as described above. On 15 March 1993, although there was no snow on the ground, six squirrels were seen searching for and eating acorns, which remained abundant. On 28 May 1993, I returned to the slope and observed no Abert's squirrels present, although many acorns remained on the ground.

DISCUSSION

Subnivean foraging certainly provided the Abert's squirrels I observed with access to an unusually abundant, energy-rich winter food. Subnivean foraging may also have served to reduce the squirrels' convective and radiative heat losses and their exposure to predators. Golightly and Ohmart (1978), noting that Abert's squirrels are not known to store large amounts of food and carry little or no subcutaneous fat, concluded that these squirrels must forage in all weather conditions in order to survive. These authors observed that although fresh snow restricted the movements of Abert's squirrels, the animals they studied foraged even during severe snowstorms. Golightly and Ohmart (1978) observed Abert's squirrels excavating acorns buried beneath 10-15 cm of snow but speculated that food buried under deep snow would be inaccessible to the squirrels.

Golightly and Ohmart (1978) reported that body temperatures of free-ranging Abert's squirrels in the field varied from 35.2° to 41.1°C (mean = 39.0°) and that of six variables influencing body temperature in these squirrels, the most important was activity. Golightly and Ohmart found that body temperatures of squirrels foraging in trees were lower than those of squirrels foraging on the ground, and body temperatures of ground-foraging squirrels were similar to those of running squirrels. The authors did not report body temperatures of squirrels excavating acorns from beneath snow, but they stated that at low ambient temperatures, windy conditions resulted in declines in body temperatures of Abert's squirrels. The authors also noted that squirrels foraging in trees during windy conditions "would generally locate themselves on the protected side of a tree for shelter and draw themselves into a ball."

Since Abert's squirrels must forage even during cold, windy conditions, and the only way to meet the cost of such activity is through the energy input supplied by successful foraging, the squirrels I studied appeared to have achieved a simultaneous solution to at least two problems: 1) in this instance, tunneling through deep snow to the soil surface provided access to an unusually abundant energy-rich winter food supply; and 2) foraging beneath the snow, and possibly digging through it, presumably generated a significant amount of metabolic heat within a microenvironment that minimized convective and radiative heat losses. Easy access to tunnels for escape as well as subnivean foraging might also reduce the squirrels' exposure to certain predators, although I have no observations on this point.

Given the usual winter weather throughout much of the geographic range of Abert's squirrels, it seems remarkable that significant subnivean foraging by these animals has not previously been reported. Bakken (1959) and Barkalow and Shorten (1973) reported subnivean foraging by gray squirrels (*Sciurus carolinensis*). Hoffmeister (1986) noted that red squirrels use subnivean tunnels to reach their large caches. Of the references I examined for *S. aberti*, however, only Hoffmeister (1971), who noted that in winter Kaibab squirrels (*S. aberti kaibabensis*) could be seen digging up buried seeds at the snow-free bases of trees, and Golightly and Ohmart (1978) discussed ground-foraging by Abert's squirrels during the winter.

My observations raise several questions in need of further study. First, these observations should be quantified and extended in order to confirm them and to ascertain whether the behavior here described is more common than has been reported. Perhaps subnivean foraging is merely an opportunistic response to the combination of an unusual abundance of acorns and deep snow. In order to test the possible thermal benefits of subnivean foraging during cold, windy weather, it is necessary to measure body temperatures of squirrels active beneath as well as above the snow. Finally, it would be interesting to know what nutritional considerations, aside from the irregular abundance of acorns, might cause Abert's squirrels to turn toward or away from acorns as a major food resource.

CONCLUSIONS

Although Abert's squirrels over the long term depend on products of ponderosa pine for food, my observations show that these mammals may feed primarily or even exclusively on acorns of Gambel's oak in winter when those nuts are abundant. My observations confirm that Abert's squirrels forage extensively even during cold, snowy, windy weather and reveal that these squirrels will forage in tunnels through snow at least 30 cm deep in order to reach an abundant supply of acorns. In addition to providing access to an abundant, energy-rich food supply, foraging beneath deep snow presumably generates a significant amount of metabolic heat while reducing convective and radiative heat losses. Reduced exposure to certain predators may be yet another benefit of subnivean foraging.

ACKNOWLEDGMENTS

I thank William L. Gannon, Stanley S. Hillman, Robert O. Tinnin, and two anonymous reviewers for comments and suggestions that significantly improved earlier versions of this manuscript. I remain indebted to James S. Findley who introduced me to the discipline of mammalogy and who, through precept and example, continues to encourage and inspire my interest in mammalian natural history.

LITERATURE CITED

Bailey, V. 1931. Mammals of New Mexico. North American Fauna 53, 412 pp.

Barkalow, F.S., Jr., and M. Shorten. 1973. The world of the gray squirrel. J.B. Lippincott Co., Philadelphia and New York, 160 pp.

Bakken, A. 1959. Behavior of gray squirrels. Pp. 393-407, *in* Symposium on the gray squirrel (V. Flyger, ed.). Contribution Number 162, Maryland Department of Research and Education. Extract from Proceedings of the Thirteenth Annual Conference of the Southeastern Association of Game and Fish Commissioners, pp. 355-407.

Clark, T.W., and M.R. Stromberg. 1987. Mammals in Wyoming. University of Kansas Museum of Natural History, Lawrence, 314 pp.

Findley, J.S. 1987. The natural history of New Mexican mammals. The University of New Mexico Press, Albuquerque, 164 pp.

Golightly, R.T., Jr., and R.D. Ohmart. 1978. Heterothermy in free-ranging Abert's squirrels (*Sciurus aberti*). Ecology, 59:897-909.

Hall, J. 1981. A field study of the Kaibab squirrel in Grand Canyon National Park. Wildlife Monographs, 75, 54 pp.

Hoffmeister, D.F. 1971. Mammals of Grand Canyon. University of Illinois Press, Urbana, 183 pp.

Hoffmeister, D.F. 1986. Mammals of Arizona. University of Arizona Press and Arizona Game Fish Department, Tucson, 602 pp.

Keith, J.O. 1965. The Abert squirrel and its dependence on Ponderosa pine. Ecology, 46:150-163.

Reynolds, H.G. 1966. Abert's squirrels feeding on pinyon pine. Journal of Mammalogy, 47:550-551.

Snyder, M.A. 1993. Interactions between Abert's squirrel and ponderosa pine: The relationship between selective herbivory and host plant fitness. American Naturalist, 141:866-879.

Snyder, M.A., and Y. Linhart. 1993. Barking up the right tree. Natural History, 102(9):44-49.

Stephenson, R.L. 1974. Seasonal food habits of Abert's squirrels, *Sciurus aberti*. Journal of the Arizona Academy of Science, 9. Proceedings Supplement.

Richard B. Forbes, Department of Biology, Portland State University, Portland, Oregon 97207